Atmospheric Dispersion and Chemistry Models: Advances and Applications

Atmospheric Dispersion and Chemistry Models: Advances and Applications

Editor

Daniel Viúdez-Moreiras

MDPI

Basel • Beijing • Wuhan • Barcelona • Belgrade • Novi Sad • Cluj • Manchester

Editor
Daniel Viúdez-Moreiras
Instituto Nacional de Técnica
Aeroespacial–INTA
Torrejón de Ardoz
Spain

Editorial Office
MDPI
St. Alban-Anlage 66
4052 Basel, Switzerland

This is a reprint of articles from the Special Issue published online in the open access journal *Atmosphere* (ISSN 2073-4433) (available at: www.mdpi.com/journal/atmosphere/special_issues/Atmospheric_Dispersion_Models).

For citation purposes, cite each article independently as indicated on the article page online and as indicated below:

Lastname, A.A.; Lastname, B.B. Article Title. *Journal Name* **Year**, *Volume Number*, Page Range.

ISBN 978-3-0365-8755-4 (Hbk)
ISBN 978-3-0365-8754-7 (PDF)
doi.org/10.3390/books978-3-0365-8754-7

Contents

About the Editor

Daniel Viúdez-Moreiras

Dr. Daniel Viúdez-Moreiras is an atmospheric and instrument scientist with broad experience in the field of modeling and simulation, as well as data processing and analysis. He has authored several refereed scientific articles and conference papers on atmospheric and planetary sciences. His research activities are developed through Instituto Nacional de Técnica Aeroespacial (INTA). He is a team member of the three NASA robotic missions currently operating on the Martian surface (Mars Science Laboratory, InSight and Mars 2020). His main research interests are related to the characterization of planetary atmospheres, focusing on studies of atmospheric chemistry and dynamics, as well as radiative transfer in the atmosphere, based on observational data supported by mathematical models.

atmosphere MDPI

Editorial

Editorial for the Special Issue "Atmospheric Dispersion and Chemistry Models: Advances and Applications"

Daniel Viúdez-Moreiras

Instituto Nacional de Técnica Aeroespacial, Torrejón de Ardoz, 28850 Madrid, Spain; viudezmd@inta.es

Atmospheric dispersion and chemical transport models (CTMs) are a key tool in both atmospheric chemistry and environmental sciences. From urban air pollution modeling to ozone depletion, these models give us a picture, at different scales, of the distribution of species concentrations and pollutant deposition rates, among other relevant quantities. These models help us to interpret observational data which, in some cases, are sparse and incomplete.

Many dispersion models and CTMs have been developed to date, with both Eulerian and Lagrangian approaches, each of which is mostly focused on a particular spatial scale and application. A large portion of them do not generate their own meteorological field, which is previously computed by an external meteorological model, i.e., a complete meteorological prediction is not required to be run for each dispersion simulation. Therefore, these models can significantly minimize the computational times associated with an online approach, allowing for studies that would require enormous times in other types of models, and in applications where feedback from the species distribution to the meteorological field is not necessary.

Their usefulness is not limited to only scientific research, but also to supporting environmental decision making. Therefore, the characterization of model uncertainties and model validation play a central role in the development of model applications.

This Special Issue (SI) of the open access journal *Atmosphere* aims to cover papers related to all aspects involved in the development of atmospheric dispersion models and CTMs, such as the implementation of new physical and chemical schemes, coupling with meteorological models, application studies related to atmospheric transport and chemistry, urban air quality assessments, and model evaluations. This volume comprises 11 high-quality papers that were accepted for publication from the submissions received for this SI. These papers are summarized as follows, in order of their publication.

The first paper published in this SI, by Mazzeo et al. [1], used an air quality model, coupled online with a meteorological model, to simulate the impact of emission reductions on PM2.5 in the West Midlands region of the United Kingdom. They showed several results concerning the effectiveness of mitigation policies in reducing anthropogenic emissions under different simulated scenarios. The second paper, by Chen et al. [2], used two dispersion models to analyze the potential detection of methane emissions by a continuous monitoring sensor network in an oil and gas production region in Texas and to assist in the design of the network. They also analyzed the sensitivity of this network under different scenarios, including meteorological conditions, emission fluxes, and intermittency. In the third paper, Kubas et al. [3] studied the impact of an accidental emission of dangerous substances in the Slovak Republic, in the context of population and environment protection in crisis management and emergency planning. Their results could be useful for authorities and rescue system units. Liu et al. [4] analyzed the relationship between sandstorm periods and the transport and dispersion of particulate emissions from coal bases in northwest China, assisted by a backward trajectory analysis performed with an atmospheric dispersion model. Their results could be useful for preventing possible risks to human health. Cogliati et al. [5] studied the distribution of bioaerosols in the vicinity of a cattle feedlot in Argentina. Bioaerosol emissions from intensive livestock breeding, which include

check for updates

Citation: Viúdez-Moreiras, D. Editorial for the Special Issue "Atmospheric Dispersion and Chemistry Models: Advances and Applications". *Atmosphere* **2023**, *14*, 1275. https://doi.org/10.3390/atmos14081275

Received: 1 August 2023
Revised: 8 August 2023
Accepted: 10 August 2023
Published: 11 August 2023

bacteria, viruses, and other parasites, can cause severe human diseases. They used an atmospheric dispersion model to support their study, predicting the distribution and concentration of bioaerosols as a function of wind patterns. Tølløse et al. [6] focused on the estimation of the source term in atmospheric dispersion models in the case of emissions of hazardous radioactive matter to the atmosphere due to a nuclear accident, when there is no reliable estimation of the source term. The Bayesian inversion method for probabilistic estimation of the source term they developed is intended for operational use in the early stages of such events. Kiselev et al. [7] focused their study on assessing uncertainties in the modeling of wet raindrop deposition. They presented the results of a multi-model ensemble construction to determine the below-cloud scavenging coefficient in atmospheric dispersion and chemistry models, which could be used to improve the current approaches associated with modeling the distribution of pollutants in the atmosphere in the case of an emergency. Parra [8] evaluated several land surface schemes in an online-coupled atmospheric chemical model applied to the complex Andean region of Ecuador, where atmospheric modeling is challenging. His findings provide insights into the influence of land surface schemes on meteorology and air quality modeling. Bukosa et al. [9] used a global CTM to perform chemical simulations of carbon dioxide, carbon monoxide, and methane. They improved the model setup from the standard procedure implemented in the CTM to compute some chemical terms more consistently, discussing the differences between the two approaches. Talafha et al. [10] applied an atmospheric dispersion model to evaluate the behavior of radionuclides released to the atmosphere during a hypothetical accident at a nuclear research reactor located in Jordan. This study could provide insights for emergency preparedness and response planning to mitigate the radiological consequences of a nuclear accident at such a reactor. Finally, Lipták et al. [11] presented an atmospheric dispersion model based on a Lagrangian approach integrated into a nuclear decision support system. Their development focused on strict time constraints, as these would exist when running the necessary simulations right after an event. Their results suggest that the computation of atmospheric dispersion and radiological impacts can be performed using a powerful computer on a timescale equal to that of the actual event.

In summary, the 11 papers included in this Special Issue, "Atmospheric Dispersion and Chemistry Models: Advances and Applications", published between 2022 and 2023, cover several developments and applications related to atmospheric dispersion and chemistry models. These studies highlight the potential benefits of using such models for many scientific and technical applications in atmospheric chemistry and environmental sciences, involving, for example, the characterization of aerosols and chemical species in the atmosphere, and risk analyses and decision making for potential pollutant releases, nuclear accidents, and climate change.

Conflicts of Interest: The author declares no conflict of interest.

References

1. Mazzeo, A.; Zhong, J.; Hood, C.; Smith, S.; Stocker, J.; Cai, X.; Bloss, W.J. Modelling the Impact of National vs. Local Emission Reduction on PM2.5 in the West Midlands, UK Using WRF-CMAQ. *Atmosphere* **2022**, *13*, 377. [CrossRef]
2. Chen, Q.; Modi, M.; McGaughey, G.; Kimura, Y.; McDonald-Buller, E.; Allen, D.T. Simulated Methane Emission Detection Capabilities of Continuous Monitoring Networks in an Oil and Gas Production Region. *Atmosphere* **2022**, *13*, 510. [CrossRef]
3. Kubas, J.; Polorecka, M.; Holla, K.; Soltes, V.; Kelisek, A.; Strachota, S.; Maly, S. Use of Toxic Substance Release Modelling as a Tool for Prevention Planning in Border Areas. *Atmosphere* **2022**, *13*, 836. [CrossRef]
4. Liu, Y.; Zhao, T.; Wang, R.; Ai, X.; Wang, M.; Sun, T.; Jiang, Q. Coupling Effects of Sandstorm and Dust from Coal Bases on the Atmospheric Environment of Northwest China. *Atmosphere* **2022**, *13*, 1629. [CrossRef]
5. Cogliati, M.G.; Paez, P.A.; Pianciola, L.A.; Caputo, M.A.; Mut, P.N. Bioaerosol Concentration in a Cattle Feedlot in Neuquén, Argentina. *Atmosphere* **2022**, *13*, 1761. [CrossRef]
6. Tølløse, K.S.; Sørensen, J.H. Bayesian Inverse Modelling for Probabilistic Multi-Nuclide Source Term Estimation Using Observations of Air Concentration and Gamma Dose Rate. *Atmosphere* **2022**, *13*, 1877. [CrossRef]
7. Kiselev, A.; Osadchiy, A.; Shvedov, A.; Semenov, V. Ensemble of Below-Cloud Scavenging Models for Assessing the Uncertainty Characteristics in Wet Raindrop Deposition Modeling. *Atmosphere* **2023**, *14*, 398. [CrossRef]

8. Parra, R. Assessment of Land Surface Schemes from the WRF-Chem for Atmospheric Modeling in the Andean Region of Ecuador. *Atmosphere* **2023**, *14*, 508. [CrossRef]
9. Bukosa, B.; Fisher, J.A.; Deutscher, N.M.; Jones, D.B.A. A Coupled CH4, CO and CO2 Simulation for Improved Chemical Source Modeling. *Atmosphere* **2023**, *14*, 764. [CrossRef]
10. Talafha, M.; Kim, S.; Suh, K.-S. A Study on Radiological Hazard Assessment for Jordan Research and Training Reactor. *Atmosphere* **2023**, *14*, 859. [CrossRef]
11. Lipták, L'.; Čarný, P.; Marčišovský, M.; Marčišovská, M.; Chylý, M.; Fojciková, E. Dispersion and Radiation Modelling in ESTE System Using Urban LPM. *Atmosphere* **2023**, *14*, 1077. [CrossRef]

atmosphere

MDPI

Communication

Modelling the Impact of National vs. Local Emission Reduction on $PM_{2.5}$ in the West Midlands, UK Using WRF-CMAQ

Andrea Mazzeo [1], Jian Zhong [1], Christina Hood [2], Stephen Smith [2], Jenny Stocker [2], Xiaoming Cai [1] and William J. Bloss [1,*]

1 School of Geography Earth and Environmental Sciences, Edgbaston Campus, University of Birmingham, Birmingham B15 2TT, UK; a.mazzeo@bham.ac.uk (A.M.); j.zhong.1@bham.ac.uk (J.Z.); x.cai@bham.ac.uk (X.C.)

2 Cambridge Environmental Research Consultants, 3 King's Parade, Cambridge CB2 1SJ, UK; chood@cerc.co.uk (C.H.); ssmith@cerc.co.uk (S.S.); jstocker@cerc.co.uk (J.S.)

* Correspondence: w.j.bloss@bham.ac.uk

Abstract: Ambient air pollution from $PM_{2.5}$ is a major risk to human and environmental health, with significant impacts on mortality and morbidity. Mitigation policies—which may be regional or national in extent—need to consider both primary and secondary particles to be effective, balancing within-region emissions and longer-range transport phenomena. The modelling system WRF-CMAQ was used to simulate the impact of emissions reductions in the West Midlands region of the UK, evaluating the change in total $PM_{2.5}$ and in its primary and secondary components. Domestic combustion, road transport and agriculture emissions were reduced individually or in combination, at a national or at local level. Combined reduction of road transport and agriculture emissions showed the strongest reduction (29%) in average $PM_{2.5}$ if applied at national level. At the local level, reductions from domestic combustion were shown to be the most effective policy (13.4% on average). Secondary inorganic fractions of $PM_{2.5}$ are the most abundant, with 25% NO_3^- 21% SO_4^{2-} and 13% NH_4^+ on average. Scenario analysis shows that the contribution of secondary components to the fractional change of $PM_{2.5}$ dominates for national policies (up to 0.86 for NO_3^-) when road transport and agriculture activities are reduced, while at the regional level the elemental and organic carbon fractional changes are dominant (up to 0.64 for organic carbon).

Keywords: air pollution; air quality modelling; CMAQ; WRF; particulate matter; $PM_{2.5}$; West Midlands

Citation: Mazzeo, A.; Zhong, J.; Hood, C.; Smith, S.; Stocker, J.; Cai, X.; Bloss, W.J. Modelling the Impact of National vs. Local Emission Reduction on $PM_{2.5}$ in the West Midlands, UK Using WRF-CMAQ. *Atmosphere* **2022**, *13*, 377. https://doi.org/10.3390/atmos13030377

Academic Editor: Daniel Viúdez-Moreiras

Received: 17 January 2022
Accepted: 21 February 2022
Published: 24 February 2022

Publisher's Note: MDPI stays neutral with regard to jurisdictional claims in published maps and institutional affiliations.

1. Introduction

Several steps have been taken in the last decade to improve air quality in the United Kingdom. Mitigation policies at a national level, alongside technological and societal changes, have led to significant reductions in $PM_{2.5}$ concentrations, by 23 and 26% at urban and roadside locations, respectively [1]. Despite this, the recent changes made by the World Health Organisation (WHO) to guideline levels for the protection of human health, lowered to 5 $\mu g/m^3$ for annual mean $PM_{2.5}$ concentrations, call for further efforts aimed at reducing anthropogenic emissions, especially where these impact urban areas [2].

The West Midlands (WM) is the second-most populous region of the UK after Greater London, with more than 2.9 million inhabitants. It includes the UK's second-largest city, Birmingham, with 1.1 million inhabitants. UK government projections predict the WM to have one of the highest population growth rates (+7.5%) in the period 2015–2025 [3]. This rapid population growth and urbanisation will potentially increase total population exposure to air pollution in the region.

Within the UK, the National Atmospheric Emissions Inventory (NAEI, [4]) indicates that 38% of primary $PM_{2.5}$ emissions in the UK are generated by domestic combustion, including biomass, wood and coal burning in closed stoves and open fires [5]. Road transport also makes a significant contribution to primary $PM_{2.5}$ (12%), despite an 85%

decrease in exhaust emissions since 1996 due to stricter emission standards [6]. Another pollutant contributing to secondary formation of $PM_{2.5}$ is ammonia (NH_3). During the short time this gas persists in the atmosphere (e.g., a few hours) it reacts with other gases such as nitrogen oxides and sulphur dioxide to form secondary PM species such as ammonium nitrate and ammonium sulphate, which remain suspended for a few days in the atmosphere and are often transported over large distances [7]. In total, 88% of UK ammonia is emitted by agricultural activities, with minor contributions from waste (2.5%) and transport (1.7%).

Additionally, approximately 30% of the total $PM_{2.5}$ mass concentration in the UK comprises secondary inorganic aerosols (SIA); this percentage reaches 44% of the total concentration in the city of Birmingham [8]. Studies in the WM have shown that NO_3^-, SO_4^{2-} and NH_4^+ secondary inorganic fractions were the main constituents of $PM_{2.5}$ in WM urban areas, followed by carbonaceous fractions of organic and elemental carbon (OC and EC) [9]. To maximise the effects of national and local environmental policies, it is important to analyse the influence that potential emissions reductions have not only on total $PM_{2.5}$ but also on its individual components. Health impacts are likely to be affected by PM mass concentration ($PM_{2.5}$), composition, particle size and morphology—including ultrafine particle number concentrations—however, our focus here is $PM_{2.5}$ mass concentration and bulk composition.

Chemistry-transport models (CTMs) have frequently been used to simulate aerosol formation, composition, dispersion, and transport. Within the UK/European context, some recent studies have focused on the effect of long-range transport of aerosols from north-west Europe to the UK [10,11] while others focused on the sensitivity of final concentrations to primary $PM_{2.5}$ emission reductions for present and future periods [12]. The impact of policy options and in particular of anthropogenic emission reduction has been investigated using different types of models and CTMs (e.g., among the most recent [13–15]). Finally, while some works have focused on high-resolution numerical simulations over the city of London [16,17], none have previously addressed the impact of national vs. regional primary $PM_{2.5}$ emission reductions on total and individual secondary inorganic fractions in the West Midlands.

In this work we use the modelling system WRF-CMAQ to simulate average concentrations of $PM_{2.5}$ and the main fractions of NO_3^-, SO_4^{2-}, NH_4^+, EC and OC for January and July 2016, representing winter and summer conditions. We simulate $PM_{2.5}$ changes for scenarios with reduced anthropogenic emissions from road transport, agriculture, and domestic combustion activities, applying a reduction to (i) the emissions from these sectors across the whole UK, and (ii) emissions limited to the WM area. Finally, we evaluate the changes in monthly average concentrations of $PM_{2.5}$ and its main individual chemical components.

The paper is organised as follows: Section 2 describes the main characteristics of the modelling system and the configuration used for the simulations. Section 3 shows the results of evaluation of the modelling system in comparison with observations, using different metrics, the results of scenarios with reduced anthropogenic emissions and the fractional concentration change of $PM_{2.5}$ components. Finally, Section 4 summarises the conclusions and proposed future developments.

2. Materials and Methods

2.1. Modelling System

Meteorology and chemistry transport processes over the West Midlands have been simulated using the Weather Research and Forecasting (WRF) model, version 3.9.1 [18] and the Community Multi-Scale Air Quality Model (CMAQ), version 5.2.1 [19]. WRF is a next-generation mesoscale numerical model developed to perform operational forecasts and atmospheric research through weather simulations. WRF incorporates multiple options for different physical parametrisations for the simulation of tropospheric weather fields. CMAQ is an open-source numerical model developed by the USEPA for the simulation of chemistry and transport processes in the low troposphere involving a large range of air pollutants. CMAQ is widely used for research and regulatory purposes by academics and policy makers for the simulation of air pollution levels, creations of forecasts, and scenarios with reduced emissions for policy making.

Both models have been configured to run simulations on 4 nested domains at increasing spatial resolutions. A coarse domain at 27 × 27 km covers most of western continental Europe, two intermediate domains at 9 × 9 km and 3 × 3 km are centred on the UK and Southern England, while the finest domain at 1 × 1 km is centred on the West Midlands area. (Figure 1). The WRF-CMAQ grid includes 30 vertical levels with the first at 20 m from the ground and 9 in total below 1 km height.

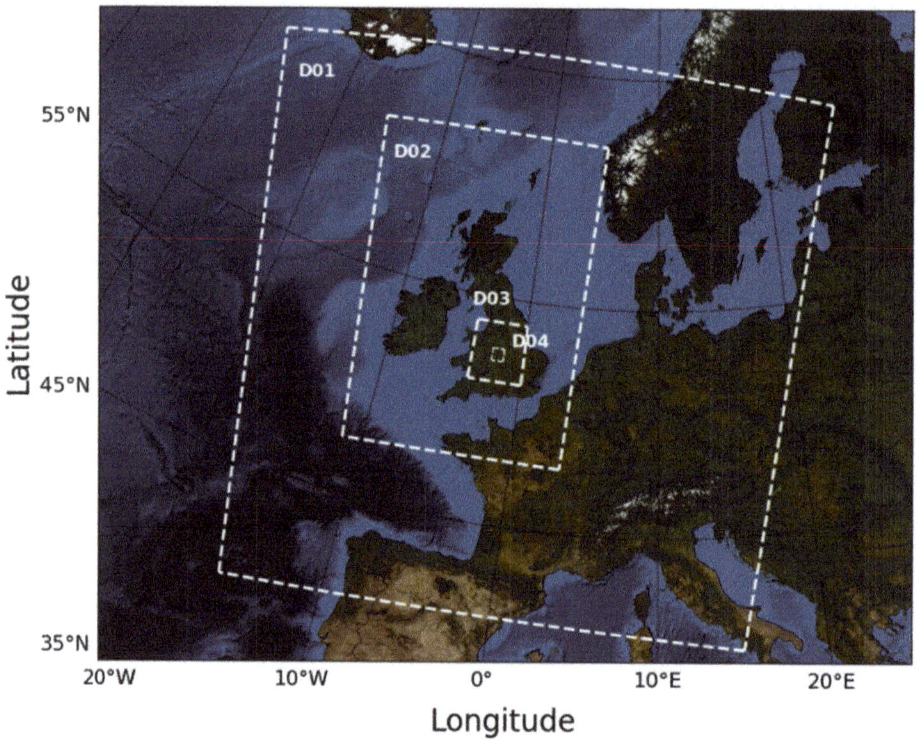

Figure 1. Geographic domains used for CMAQ simulations. The first domain (D01) has spatial resolution of 27 × 27 km; the first nested domain (D02) centred on the UK has 9 × 9 km resolution. The second and third nested domains centred on the WM area have 3 × 3 km (D03) and 1 × 1 km (D04) spatial resolution, respectively.

The adopted WRF configuration follows the parameters recommended by the "CMAQ Development for UK National Modelling Report" (CMAQ4UK) [20,21]. Initial and boundary conditions are derived from the European Centre for Medium-Range Weather Forecast (ECMWF) ERA 5 reanalysis [22]. These IC/BC are created using forecasts at 31 km resolution (one-fourth the spatial resolution of the operational model). They integrate 137 hybrid sigma-pressure levels in the vertical, up to 0.01 hPa. The choice of the ECMWF IC/BC is motivated by the evidence shown in previous works focused on the optimisation of the WRF configuration for the UK, relating to the influence that initial and boundary conditions used in WRF have on both meteorological patterns and on conditions of regional air quality [23]. Grid nudging has been applied every 6 h to constrain WRF outputs to observations, with nudging coefficients defined for U and V wind components, temperature (T) and water-vapour-mixing ratio (Q). The process permits to constrain the values of selected variables (e.g., U, V, T and Q) calculated by WRF to the original re-analysis value from the data used by WRF (e.g., ECMWF) with a certain frequency of time (e.g., 6 h).

The CMAQ configuration uses initial and boundary conditions for the outermost domain created using seasonal average hemispheric CMAQ outputs for the year 2016, distributed through the CMAS data warehouse. These data were generated using CMAQv5.3 with spatial resolution of 108 × 108 km on a polar stereographic grid covering the northern hemisphere. Species concentrations have been mapped to the CB05 mechanism using the 'combine' program from the CMAQ post-processing toolkit before being used to create the initial and boundary conditions for the domain at 27 km resolution. The internal domains at 9 km, 3 km, and 1 km resolution draw initial and boundary conditions from the respective parent domain. These IC/BC have been created using the ICON and BCON modules internal to CMAQ. The Carbon Bond 05 (CB05) chemical mechanism has been adopted for all simulations. It was developed in 2005 and is a condensed mechanism of atmospheric oxidant chemistry for 51 species and 156 reactions, suitable for modelling ozone, particulate matter, visibility, acid deposition and air toxics issues [24]. A summary of the main settings of the WRF-CMAQ configuration is given in Table 1.

The UK NAEI [4] has been merged with the regional emission inventory CAMSv3.1 [25] to provide a comprehensive description of anthropogenic emissions for the UK and northwest Europe. Both emission inventories provide annual totals of anthropogenic sources for the year 2016 and have been disaggregated spatially and temporally over the simulation domains using appropriate pre-processing tools: EMIT [26] for NAEI and HERMES [27] for CAMSv3.1, then merged by pollutant on each grid. EMIT and HERMES were used to disaggregate the emission rates from the original emission inventories (in annual totals) on spatial grids at different resolution (from 27 × 27 km to 1 × 1 km). Moreover, the tools also provide temporal and vertical profiles of the emissions from annual totals to hourly fluxes according to emission coefficients diversified by pollutant and by sector. These profiles for the disaggregation of both emission inventories have been taken from EMEP model inputs [28]. Finally, biogenic emissions used in this work come from MEGAN software, version 3.1 [29]. The leaf area index data for 2016 has been taken from the European Union's Earth observation programme Copernicus [30] and implemented in MEGAN for the calculation of biogenic emissions.

Table 1. WRF and CMAQ configuration used for simulations and scenarios.

WRF Configuration		CMAQ Configuration	
WRF version	3.9.1	CMAQ version	5.2.1
IC/BC	ECMWF ERA5	Sp. Projection	Lambert Conformal Conic
Land use	USGS	IC/BC	CMAQ Hemispheric Outputs
Urban Physics	BEP	Chemical Scheme	CB05e51_ae6_aq
Boundary Layer	BouLac	Anth. Emissions	CAMS3.1/NAEI
Surface Layer	Monin	Temp. Profiles	Simpson et al., 2012 [28]
Land surface	NOAH	Natural Emis.	MEGAN3.1
Vertical Levels	30	Vertical Levels	30

2.2. Simulation Period and Observation Sites

The simulations using WRF and CMAQ were conducted for two monthly periods representative of winter and summer conditions of 2016, namely January and July, applying a spin-up period of 5 days before the formal start of the simulations of both models.

The simulation months were chosen as those showing the highest mean temperature during summer and lowest during winter (around 17 and 5 °C, respectively for the domain in Figure 2) in comparison to the average annual value of 10 °C. Moreover, during these two months no extreme weather events (e.g., rainstorms, heat waves) impacting the common weather condition were recorded. The simulation year of 2016 was defined to make use of the most up-to-date nationally ratified anthropogenic emissions for the UK available at the time of model development.

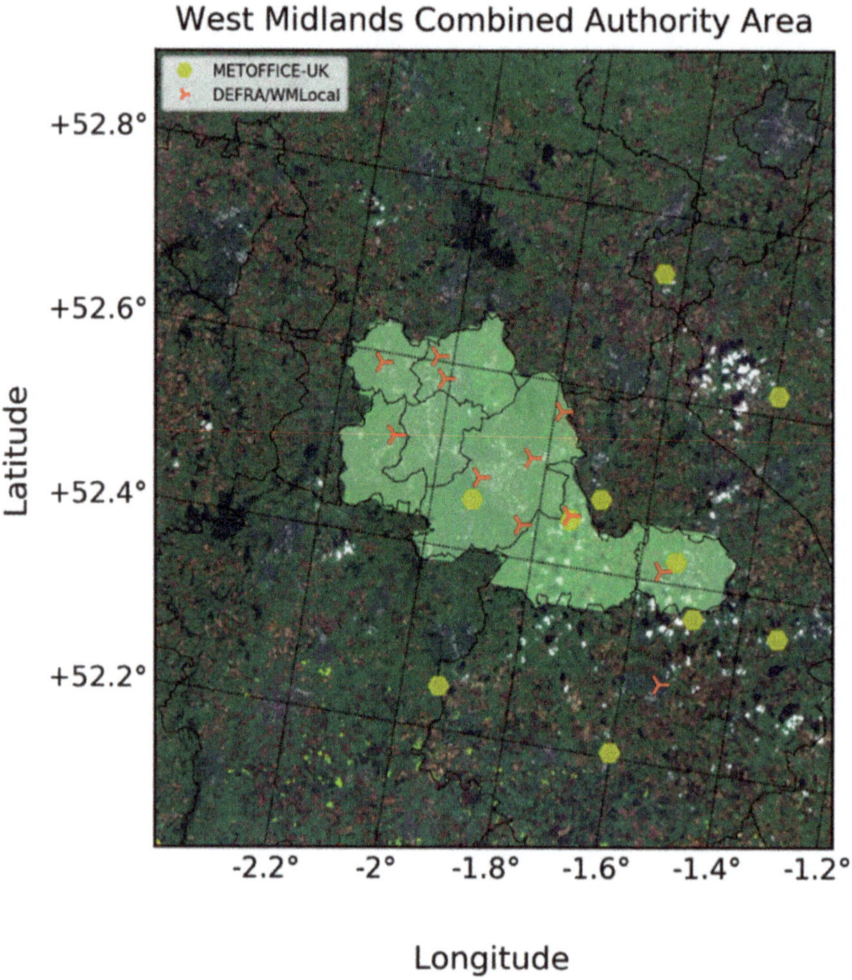

Figure 2. Map showing the modelled area relative to the West Midlands Combined Authority boundaries (in light green). Area used as mask for the reduction of the emissions in WM case scenarios. Yellow spots show the location of weather observation points from the UK Met Office, while red crosses show the position of $PM_{2.5}$ observation points from AURN-DEFRA network.

Ten meteorological measurement stations in the WM have been used for the validation of WRF (Figure 2). Surface temperature, wind speed and direction data used for the validation come from the Met Office UK database for 2016 [31], while relative humidity was calculated using the coefficients proposed by Alduchov and Eskridge [32] based on hourly observed values of surface and dew point temperatures. U and V vector components of wind speed were calculated by combining observed values of wind speed and direction. A total of 11 stations were used for the validation of $PM_{2.5}$ in CMAQ. All stations are representative of urban background: 7 from the West Midlands local authority network and 4 from the AURN-DEFRA national network (Figure 2).

2.3. Scenario Design

According to the UK Clean Air Strategy 2019 [5], 38% of primary $PM_{2.5}$ emissions are generated by domestic wood and coal burning, followed by industrial combustion (16%) and road transport (12%), among others. Besides this, secondary PM is formed in the atmosphere through chemical reactions between gaseous pollutants such as NO_X, SO_2 and NH_3 generated by road transport, industrial and agricultural activities, and following the chemical processing and condensation of organic components.

Three scenarios have been created considering the following emissions changes: 85% reduction of all emissions from the SNAP2 sector (A), corresponding approximately to removal of domestic combustion activities related to coal, coke and wood burning; 30% reduction of ammonia emissions (only) from the SNAP10 (agriculture) sector (B); and 30% reduction of (all) road transport emissions (SNAP7, C). A fourth scenario combining the reductions in SNAP7 and SNAP10 (D) was created to consider the combined effect of possible mitigation policies (Table 2). Scenarios A and C were designed by reducing primary emissions of all pollutants included in the respective sectors, while in scenario B, emissions from NH_3 alone were reduced.

Table 2. Percentage of reduction of sector emissions calculated for each scenario simulated in CMAQ.

Label	Sector	Description	Reduction
A	SNAP2	Domestic Combustion	85%
B	SNAP7	Road Transport	30%
C	SNAP10	NH_3 agriculture	30%
D	SNAP7+10	Road transport + NH_3 agriculture	30 + 30%

Scenario A was designed to explore a near-total removal of solid fuels from domestic combustion activities. Wood, coal and coke burning represent the highest source of emissions connected to domestic combustion in the NAEI and can impact both primary and secondary formation of $PM_{2.5}$. According to the NAEI, wood burning generates approximately 85% of primary emissions of $PM_{2.5}$ from the whole sector, hence this magnitude of emissions reduction was selected. Coal and coke burning are responsible for 22 and 56% of SO_2 and 3 and 2% of NO_X emissions (respectively) from SNAP2 [4] and so represent probable further contributors to PM secondary formation in the atmosphere. Approximately 30% of wood fuel used in UK is sourced from the informal "grey" wood market, and 90% of domestic wood users use logs either solely or in conjunction with other fuels (pellets, briquettes, waste wood, gathered wood, and wood chips) [33]. Due to these complications, the reduction of emissions of individual pollutants by fuel type (e.g., wood vs. coal) is difficult to estimate and therefore reduction in all primary emissions from the SNAP2 sector was chosen.

Scenario B for road transport emissions represents reduction of all emissions across the vehicle fleet, including emissions from exhaust, brakes and tyres, and different fuel types. In this respect, it does not reflect the expected transition to electric vehicles, for which most non-exhaust particulate emissions would remain. The reduction of 30% was selected to align with the anticipated impact of the UK Clean Air strategy to meet National Emission Ceiling Regulation limits in 2030 for the road-transport sector [34]. Similar to scenario A, the reduction in primary emissions (30%) was applied to each pollutant present in the sector. Finally, scenario C was designed to reflect changes in agricultural practices reducing emissions of NH_3 only, while keeping all other sector pollutant emissions unchanged. Ammonia emissions from this sector arise from plant production, fertilisation, and livestock manure and the reduction is an important component of the UK Net Zero Strategy [35]. Similarly to scenario B, the reduction of 30% ammonia emissions in scenario C was set considering the ammonia emission reduction ceiling planned for the year 2030 and included in the Clean Air Strategy [5].

The four scenarios have been applied to represent policy applications on either a local/regional or a national basis. The emissions have been manipulated in two different ways: (1) a comprehensive reduction of UK emissions in all domains (hereafter called UK case), simulating national policy effects. Emissions from CAMSv3.1 inventory for northwest Europe included in the 27 and 9 km domains were unchanged (no reduction). (2) A reduction of emissions only within the masked area of the West Midlands, (hereafter called WM case), simulating the effects of potential regional (only) policies (Figure 2).

Percentage reductions of total NH_3, NO_X, SO_2 and primary $PM_{2.5}$ emissions are shown for scenarios A and B, while for scenarios C and D we show only the reduction in NH_3 (Figure 3). The domestic combustion emission reduction (A) has the strongest effects on SO_2 and $PM_{2.5}$ emissions both in winter and summer, with average emission reductions of 62 and 21% for the UK case and 27 and 16% for the WM case, respectively. The reduction in road transport emissions (B) affects both NO_X and primary $PM_{2.5}$, with average emission reductions of 23 and 18% for the UK case and 10 and 8% for the WM case, respectively.

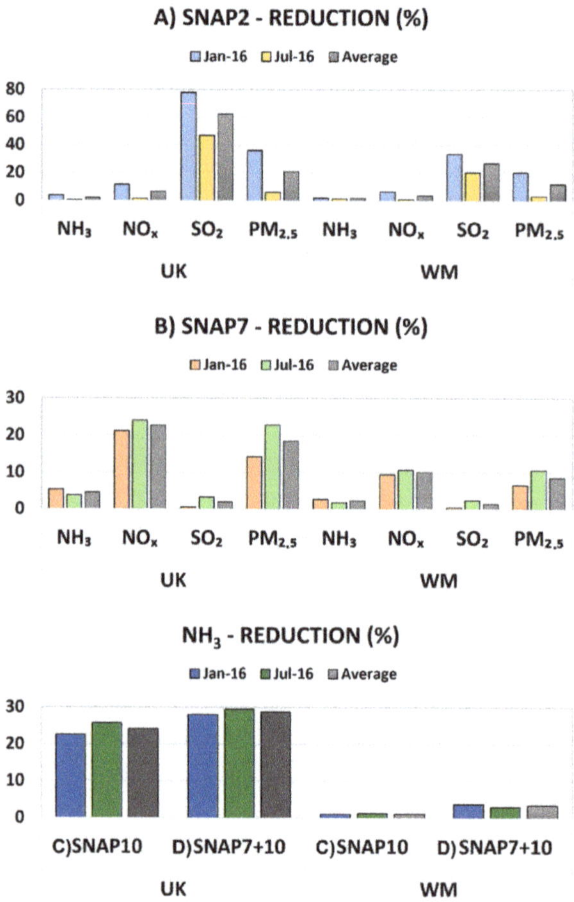

Figure 3. Percentage reduction of the total monthly emissions (January, July and average), across all sectors combined, for scenarios A (SNAP2, (**top**)) and B (SNAP7, (**middle**)), for NH_3, NO_X, SO_2 and primary $PM_{2.5}$. The percentage reduction of NH_3 from scenarios C (SNAP10) and D (SNAP7+10) is shown on the (**bottom**) panel for January, July, and their average. The reductions are shown for the UK-wide reductions (UK) and for local reduction (WM) options.

Finally, the proportional reductions of NH_3 emissions in scenarios C and D for different domains are substantially different: 24 and 29% for the UK case and 1 and 3% for the WM case, respectively. This difference in emission reduction is connected to the limited extent of agricultural activity inside the WM borders. The majority of these (agricultural) emissions are, in fact, included in the 1 × 1 km domain but outside the WM masked area and therefore altered only in the UK scenarios.

The effect of the emissions reductions from the four scenarios on $PM_{2.5}$ concentrations have been analysed in term of the most abundant components: NO_3^-, SO_4^{2-}, NH_4^+, EC and OC for both UK and WM cases using the fractional change in concentrations (FC) (Equation (1)):

$$FC = \frac{1}{N}\sum_{i=1}^{N}\frac{B_i - C_i}{B_i} \quad (1)$$

where N is the total number of ground level computational cells within the domain, B_i is the base case predicted value of the pollutant concentration in cell i and C_i is the predicted value of the pollutant concentration in cell i for the relevant scenario.

3. Results

3.1. Modelling System Validation

The validation of the combined WRF-CMAQ modelling system has been carried out for the domains at 9, 3 and 1 km resolution. In this work we present the results of the validation of the finest resolution domain at 1 × 1 km for both models, limited to surface data due to the absence of sites providing vertical sounding inside the domain area.

Mean normalised bias (MNB), root mean square difference (RMSD), index of agreement (IOA) and Pearson's coefficient (R) have been used to quantify the performance of the models against observations (Table 3).

Table 3. Statistical operations used for the validation of the modelling system WRF-CMAQ for the simulation periods. M_i is the modelled value at the time i, O_i is the observed value at the time i.

Operation	Formula
Mean Normalised Bias (MNB)	$\frac{\sum_{i=1}^{n}(M_i - O_i)}{\sum_{i=1}^{n}(O_i)}$
Root Mean Square Difference (RMSD)	$\sqrt{\frac{\sum_{i=1}^{n}(M_i - O_i)^2}{n}}$
Index of Agreement (IOA)	$1 - \left[\frac{\sum_{i=1}^{n}(O-M)^2}{\sum_{i=1}^{n}\left(\lvert M-\overline{O}\rvert + \lvert O-\overline{O}\rvert\right)^2}\right]$
Pearson's Coefficient (R)	$\frac{n(\sum_{i=1}^{n} M_iO_i) - (\sum_{i=1}^{n} M_i)(\sum_{i=1}^{n} O_i)}{\sqrt{[n\sum_{i=1}^{n} M_i^2 - (\sum_{i=1}^{n} M_i)^2][n\sum_{i=1}^{n} O_i^2 - (\sum_{i=1}^{n} O_i)^2]}}$

The performance of WRF in simulating temperature and relative humidity shows a correlation (R) between 0.95 and 0.90 for the former and 0.57 and 0.69 for the latter. While the surface temperature tends to be underestimated in winter (−0.13) and in summer (−0.15) from the mean normalised bias (MNB), the opposite is found for the relative humidity (0.06 in winter and 0.18 in summer). The index of agreement (IOA) for these two variables is higher for the temperature (93% in winter and 90% in summer) than for relative humidity (52% in winter and 69% in summer) (Table 4).

WRF is able to reproduce the main wind speed and direction with correlation (R) between 0.71 and 0.72 for wind speed and between 0.72 and 0.77 for wind direction that tends to be better reproduced in January than in July. The MNB is found to lie between 0.13 and 0.19 and between 0.003 and 0.005 for wind speed and direction, respectively. Finally, the IOA for both variables is between 52 and 55% (wind speed) and 51 and 60% (wind direction) suggesting that the model better reproduces the wind components during the summer period.

Table 4. Statistical evaluation of WRF calculated for 2016 for surface parameters of Temperature (°C), relative humidity (%), wind speed (ms^{-1}) and direction (degrees) and U and V components of wind (ms^{-1}).

Jan-16	V	U	W Sp.	W Dir.	Temp.	RH
Mean Obs	2.04	0.82	3.74	197.43	5.27	89.54
Mean Model	1.93	1.01	4.47	197.06	4.59	94.98
MNB	−0.05	0.23	0.19	0.003	−0.13	0.06
RMSD	2.33	1.88	2.14	66.7	1.50	8.78
IOA	0.70	0.80	0.55	0.60	0.93	0.52
R	0.80	0.88	0.72	0.77	0.95	0.57
Jul-16	V	U	W Sp.	W Dir.	Temp.	RH
Mean Obs	0.96	1.99	2.96	240.69	16.99	76.64
Mean Model	1.23	2.24	3.34	241.74	14.33	90.60
MNB	0.28	0.12	0.13	0.005	−0.15	0.18
RMSD	1.50	1.45	1.64	55.1	3.27	17.7
IOA	0.76	0.66	0.52	0.51	0.90	0.69
R	0.87	0.81	0.71	0.72	0.82	0.60

The statistics for wind speed and direction are confirmed in the decomposition of the winds into U and V vector components. The correlation (R) is found between 0.80 and 0.88 in January and 0.81 and 0.87 in July. The MNB is found positive in July for both U and V (0.12 and 0.28, respectively) while in January it is positive for U (0.23) and negative for V (−0.05). The IOA is between 70 and 76% for V and between 66 and 88% for U between the two periods.

The statistical evaluation of CMAQ in reproducing $PM_{2.5}$ concentrations in January and July 2016 is shown in Table 5. The model tends to underestimate the average concentration during winter (−0.38) and summer (−0.42), according to the MNB values. Despite this higher correlation (R) and index of agreement values are found during January (0.67, 72%) than July (0.41, 57%). The reason for this difference can be attributed to the higher photochemistry acting in the atmosphere in July that could have an influence on the secondary formation of aerosol components not well captured by the model (Table 5).

Table 5. Statistical evaluation of CMAQ calculated for January and July 2016 for $PM_{2.5}$ from urban background stations in the 1 × 1 km domain shown in Figure 1.

$PM_{2.5}$	Jan-16	Jul-16
Mean Obs	7.95	6.23
Mean Model	4.93	3.60
MNB	−0.38	−0.42
RMSD	2.19	1.55
IOA	0.72	0.57
R	0.67	0.41

3.2. $PM_{2.5}$ Changes for Each Scenario

The effects of the emission reduction scenarios on concentrations of $PM_{2.5}$ have been tested. The percentage reductions of concentrations have been calculated for the WM area as shown in Figure 1, excluding all the cells outside the region.

Scenarios simulating possible national mitigation policies (UK case) show that of the scenarios considered, the combined reduction of road transport and agriculture sectors provide the largest decrease of $PM_{2.5}$ in both simulated periods (Figure 4, top). Scenarios representing mitigation policies applied at the local level only (WM case) show that the scenario with strongest effect on the final $PM_{2.5}$ concentrations within the region was the SNAP2 reduction (Figure 4, bottom). Comparing the difference in $PM_{2.5}$ reduction from the UK to the WM case, we find that scenario A leads to, on average, 4.2% difference between national and regional-only emissions changes, while scenarios B and C show a higher difference between these two approaches of around 18%. Finally, scenario D shows the greatest difference between the UK and WM-only cases, of around 20%.

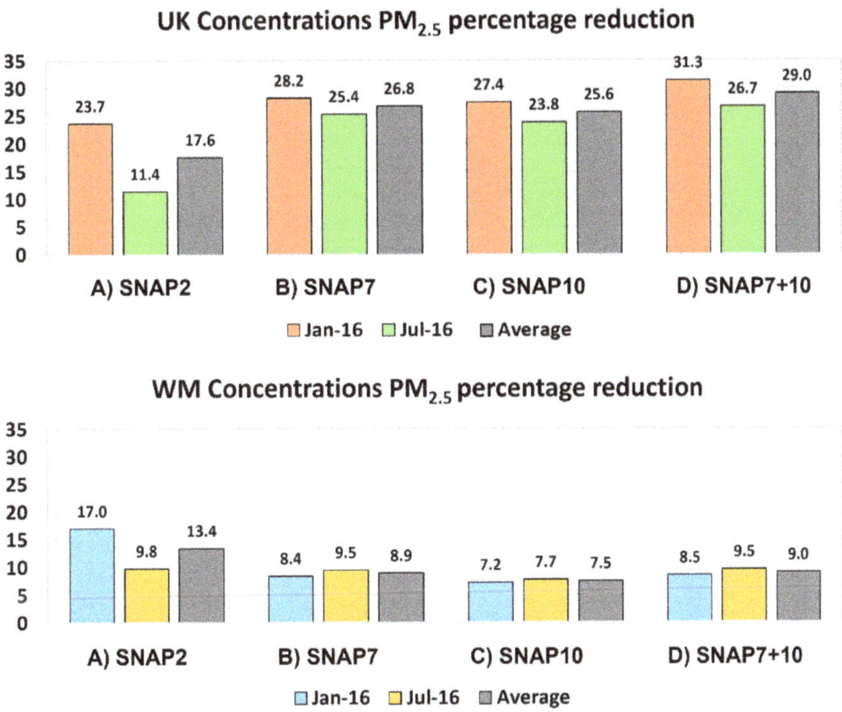

Figure 4. Percentage reductions of $PM_{2.5}$ from all scenarios, calculated from the monthly average inside the masked area for January, July, and their average. (**Upper**) panel: the percentage reductions for all scenarios (A to D) with emissions reduced in all domains (UK case); (**Lower**) panel: the reductions for all scenarios (A to D) with emissions reduced only inside the WM masked area (WM case).

The difference in $PM_{2.5}$ concentration reductions between the UK and WM cases for scenarios B, C and D highlights that agriculture and road transport emissions outside the WM area make a substantial contribution to the final concentrations of $PM_{2.5}$ within the region. For scenario B, this is linked to the main road arteries connecting the West Midlands with the north, east (the M6) and south part of the country (the M40 and M5) extending outside the WM mask and not considered in the WM reduction cases. For scenario C, ammonia emissions are located almost completely outside the WM borders, due to the largely urban character of the WM region. The impact of ammonia reduction on $PM_{2.5}$ was already highlighted by Vieno et al., 2016 [12] as one of the most influential sources in agricultural and natural areas. Hence the reduction of agricultural ammonia alone, or in combination with road transport reductions, would be more effective as a national policy (UK case). In contrast, the domestic combustion scenario (A), despite being a source with high seasonal variability, shows the largest reduction in $PM_{2.5}$ in response to WM region-only mitigation policies (13.4% on average, with substantially larger benefits in winter, when $PM_{2.5}$ concentrations are greatest), of the scenarios considered. The reduction in $PM_{2.5}$ achieved for region-only domestic combustion emission reductions is similar to that found for equivalent national policies (17.6%) suggesting that the main influence comes from sources located inside the WM region, which can effectively be addressed by local and regional mitigation policies.

The effects of reductions in primary NO_X, SO_2 and NH_3 emissions on concentrations of $PM_{2.5}$ in the UK context have previously been highlighted by the Air Quality Expert Group (AQEG, [36]). The greatest impact upon PM for reduction of a single species' emissions corresponded to reduction in primary ammonia, which was followed by the reduction in SO_2 and substantially higher than the reduction in NO_X only. Results obtained by AQEG showed also that lowest concentrations of $PM_{2.5}$ comes from the combined reduction of primary $PM_{2.5}$ and NH_3. This result is in line with the UK scenario with agriculture (NH_3 only from B) and road transport activities (all primary emissions including $PM_{2.5}$ from C) simultaneously reduced by 30%.

3.3. Scenario Effects on $PM_{2.5}$ Components

Model outputs for the reduced emissions scenarios have been analysed to assess the change in individual components of $PM_{2.5}$, calculated from the base case simulations inside the WM masked area (Figure 5). Results show the importance of NO_3^-, SO_4^{2-} and NH_4^+ followed by elemental and organic carbon fractions (EC and OC) contributing to $PM_{2.5}$ mass concentrations. In winter there is a predominance of NO_3^- while SO_4^{2-} has the highest influence in summer.

The percentage of NO_3^-, SO_4^{2-} and NH_4^+ in total $PM_{2.5}$ was modelled as 34, 15 and 14%, respectively in January and 12, 29 and 11%, respectively, in July. Elemental and organic carbon (EC and OC) follows with 9 and 7% in January and 6 and 10% in July, respectively (Figure 5). These fractions are similar to the ambient measurement results obtained by Yin et al. [9] for the observationally derived source apportionment of $PM_{2.5}$ in the West Midlands. The authors highlighted the predominance of sulphates and nitrates in $PM_{2.5}$, followed by high level of carbonaceous species, particularly in urban areas. Secondary $PM_{2.5}$ in the UK can also be influenced by meteorological conditions. The contribution of $PM_{2.5}$ transported to UK from north west Europe has been quantified as between 21 and 30% and about 15% from natural sources [36]. However, these long-range transport events generally occur during March/April so in January and July the production of NO_3^- is considered largely local.

The results for fractional changes in the predicted individual SIA fractions for all scenarios and for the UK and WM cases are shown in Table 6. The highest fractional changes in PM composition in the WM case in January come from the domestic combustion scenario (A): EC and OC show the largest fractional reductions of around 33%, highlighting the strong impact that solid fuel combustion has on this sector in comparison to other fuel types. The SO_4^{2-} is reduced by around 24%, reflecting the 33% reduction in the primary SO_2 emissions. The other three scenarios show similar values of between 8 and 16% for NO_3^-, SO_4^{2-} and NH_4^+ and lower percentages (between 0.3 and 4%) for OC and EC. Shifts in $PM_{2.5}$ composition in July are dominated by the fractional concentration changes of NO_3^- in all scenarios (around 40%) followed by NH_4^+ (16%) and, for scenario A, by EC (11%). All the other components reduce by between 2 and 8%. For the UK case, larger fractional reductions are found for NH_4^+, SO_4^{2-} and NO_3^- in scenario D both in January (between 0.50 and 0.58) and in July (between 0.17 and 0.86). High change comes also from EC (0.52) and OC (0.64) in scenario A but limited to January. Scenarios B and C show similar fractional concentration change for NH_4^+, NO_3^- and SO_4^{2-} in January (between 0.44 and 0.54). In July the strongest reduction for D is visible for NO_3^- (0.84 and 0.83), followed by NH_4^+ (0.41 and 0.43) and SO_4^{2-} (0.16).

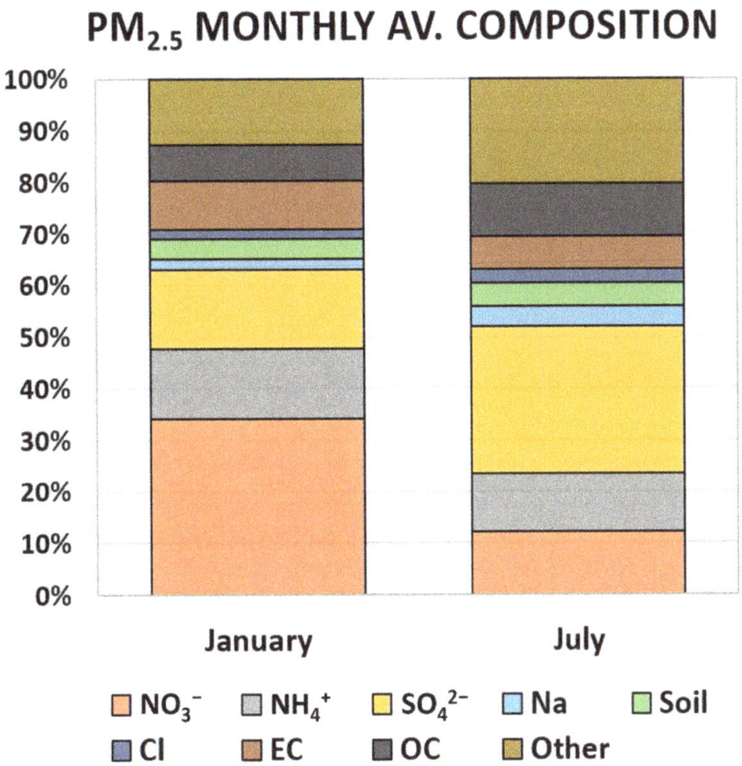

Figure 5. Individual $PM_{2.5}$ fractions calculated from the base case simulations in CMAQ for January and July 2016. The fractions have been calculated for the concentrations inside the WM masked area only.

Table 6. Fractional concentration change (FC) calculated for each scenario (A to D) for the WM (top) and UK (bottom) cases. The FC values are shown for the secondary fractions NO_3^-, NH_4^+, SO_4^{2-}, EC and OC of $PM_{2.5}$ for the months of January and July 2016.

	WM	(A) SNAP2	(B) SNAP7	(C) SNAP10	(D) SNAP7+10
Jan-16	NO_3^-	0.09	0.08	0.08	0.09
	NH_4^+	0.15	0.13	0.13	0.13
	SO_4^{2-}	0.24	0.16	0.15	0.16
	EC	0.33	0.04	0.003	0.04
	OC	0.33	0.01	0.004	0.01
Jul-16	NO_3^-	0.40	0.42	0.40	0.42
	NH_4^+	0.16	0.15	0.15	0.15
	SO_4^{2-}	0.03	0.02	0.02	0.02
	EC	0.11	0.08	0.01	0.08
	OC	0.06	0.02	0.01	0.02

Table 6. *Cont.*

	UK	(A) SNAP2	(B) SNAP7	(C) SNAP10	(D) SNAP7+10
Jan-16	NO_3^-	0.16	0.45	0.44	0.50
	NH_4^+	0.23	0.53	0.54	0.58
	SO_4^{2-}	0.34	0.47	0.49	0.50
	EC	0.52	0.06	0.01	0.06
	OC	0.64	0.02	0.006	0.01
Jul-16	NO_3^-	0.41	0.84	0.83	0.86
	NH_4^+	0.15	0.41	0.43	0.44
	SO_4^{2-}	0.03	0.16	0.16	0.17
	EC	0.14	0.12	0.02	0.12
	OC	0.13	0.07	0.02	0.07

4. Conclusions

A WRF-CMAQ modelling system based on the NAEI has been implemented and validated for simulation of meteorology and air quality over the area of the West Midlands, UK.

Scenarios with reduced emissions from changes in road transport, agricultural activities and domestic combustion have been designed to test the impact of possible mitigation policies at a national or local level on ambient concentrations of $PM_{2.5}$.

Results show that, of the cases considered, combined mitigation policies to reduce both road transport and agricultural emissions would have the strongest effect on the average $PM_{2.5}$ levels both in winter and in summertime if applied at a national level (UK cases). Conversely, mitigation policies to reduce domestic solid fuel combustion inside the WM area would result in the most effective policy if applied on a regional level only (WM case), of the scenarios considered.

The effects of emission reduction scenarios have also been evaluated in terms of the chemical components of $PM_{2.5}$. The main fractions simulated by CMAQ show a similar magnitude to findings obtained by experimental field campaigns in urban background areas of the West Midlands.

The reduction of primary emissions from domestic combustion of solid fuels in scenario A (wood, coal, and coke) shows the largest reduction in modelled EC and OC in the WM case, as these fractions are mostly locally generated (primary), while the secondary inorganic fractions of NH_4^+, NO_3^- and SO_4^{2-} form over larger time and spatial scales and therefore their reductions became more effective for emission reductions applied at a national level. This is particularly evident for scenario C (agriculture) considering the low primary emissions of NH_3 predicted by the NAEI within the WM borders. The results obtained in this work show that the effectiveness of possible mitigation policies reducing anthropogenic emissions to improve air quality in the WM are dependent not only on the targeted emissions sector but also on the spatial extent of the reduction. Combined reduction of transport emissions and ammonia from agriculture (scenario D) can have a greater impact on $PM_{2.5}$ concentrations if applied nationally. In contrast, local/regional reductions in emissions from domestic combustion of solid fuels (scenario A) represents an effective mitigation measure to reduce $PM_{2.5}$ concentrations locally, even if applied only within the WM area.

Future work may enable a more detailed analysis of the photochemical effects contributing to the formation of secondary inorganic and organic aerosols. The analysis will be extended from two monthly periods to the annual level and multiple years to explore the variation of the concentrations of $PM_{2.5}$ and its main inorganic and organic components over different time periods. Finally, CMAQ will also be used to test the impact of national and/or local mitigation policies on additional pollutants such as ozone.

Author Contributions: Conceptualization, X.C.; methodology, A.M. and C.H.; software, J.S., S.S. and C.H.; validation, A.M.; formal analysis, A.M.; investigation, A.M.; resources, X.C., J.Z., S.S. and C.H.; data curation, A.M., J.Z. and C.H.; writing—original draft preparation, A.M. and C.H.; writing—review and editing, J.Z., W.J.B. and C.H.; visualization, A.M.; supervision, W.J.B., X.C. and J.S.; project administration, J.S. and X.C.; funding acquisition, W.J.B. All authors have read and agreed to the published version of the manuscript.

Funding: This research was funded by the UK Natural Environment Research Council (NERC) project WM-Air, grant number NE/S003487/1.

Institutional Review Board Statement: Not applicable.

Informed Consent Statement: Not applicable.

Data Availability Statement: Weather observations are available from the MIDAS UK database of the Met Office: https://catalogue.ceda.ac.uk/uuid/dbd451271eb04662beade68da43546e1 (accessed on 19 November 2019). Anthropogenic emissions used for the CMAQ Simulations can be downloaded from the NAEI website: https://naei.beis.gov.uk/ (accessed on 20 May 2021). EMIT Tool is CERC software available under license at: http://www.cerc.co.uk/environmental-software/EMIT-tool.html (accessed on 25 May 2021). Leaf Area Index data have been downloaded from the Copernicus database: https://land.copernicus.eu/global/products/lai (accessed on 10 April 2021).

Acknowledgments: The authors would like to acknowledge the School of Geography, Earth and Environment Sciences of the University of Birmingham and the BlueBEAR HPC service (http://www.bear.bham.ac.uk, accessed on 30 November 2021) for providing the computational resource. The authors would like to acknowledge national and local authorities within WM for provision of national and local air quality measurement and for their review of anthropogenic emissions data.

Conflicts of Interest: The authors declare no conflict of interest.

References

1. DEFRA. Air Quality Statistics—An Annual Update on Concentrations of Major Air Pollutants in the UK. Available online: https://www.gov.uk/government/statistics/air-quality-statistics (accessed on 5 October 2021).
2. WHO. WHO Global Air Quality Guidelines: Particulate Matter ($PM_{2.5}$ and PM_{10}), Ozone, Nitrogen Dioxide, Sulfur Dioxide and Carbon Monoxide. Available online: https://apps.who.int/iris/handle/10665/345329 (accessed on 10 November 2021).
3. Gov.UK. Trend Deck 2021: Urbanisation. Available online: https://www.gov.uk/government/publications/trend-deck-2021-urbanisation/trend-deck-2021-urbanisation (accessed on 5 October 2021).
4. NAEI. National Atmospheric Emission Inventory for UK. Available online: https://naei.beis.gov.uk/ (accessed on 20 May 2021).
5. DEFRA. Clean Air Strategy. Available online: https://www.gov.uk/government/publications/clean-air-strategy-2019 (accessed on 6 October 2021).
6. Gov.UK. Emissions of Air Pollutants in the UK—Particulate Matter (PM_{10} and $PM_{2.5}$). Available online: https://www.gov.uk/government/statistics/emissions-of-air-pollutants/emissions-of-air-pollutants-in-the-uk-particulate-matter-pm10-and-pm25 (accessed on 12 December 2021).
7. Gov.UK. Emissions of Air Pollutants in the UK—Ammonia (NH_3). Available online: https://www.gov.uk/government/statistics/emissions-of-air-pollutants/emissions-of-air-pollutants-in-the-uk-ammonia-nh3 (accessed on 14 December 2021).
8. AQEG. Fine Particulate Matter ($PM_{2.5}$) in the United Kingdom. Available online: https://uk-air.defra.gov.uk/assets/documents/reports/cat11/1212141150_AQEG_Fine_Particulate_Matter_in_the_UK.pdf (accessed on 23 November 2021).
9. Yin, J.; Harrison, R.M.; Chen, Q.; Rutter, A.; Schauer, J.J. Source apportionment of fine particles at urban background and rural sites in the UK atmosphere. *Atmos. Environ.* **2010**, *44*, 841–851. [CrossRef]
10. Vieno, M.; Heal, M.R.; Twigg, M.M.; MacKenzie, I.A.; Braban, C.F.; Lingard, J.J.N.; Ritchie, S.; Beck, R.C.; Móring, A.; Ots, R.; et al. The UK particulate matter air pollution episode of March–April 2014: More than Saharan dust. *Environ. Res. Lett.* **2016**, *11*. [CrossRef]
11. Vieno, M.; Heal, M.; Hallsworth, S.; Famulari, D.; Doherty, R.; Dore, A.; Tang, Y.; Braban, C.; Leaver, D.; Sutton, M. The role of long-range transport and domestic emissions in determining atmospheric secondary inorganic particle concentrations across the UK. *Atmos. Chem. Phys. Discuss.* **2013**, *13*, 33433–33462. [CrossRef]
12. Vieno, M.; Heal, M.R.; Williams, M.L.; Carnell, E.J.; Nemitz, E.; Stedman, J.R.; Reis, S. The sensitivities of emissions reductions for the mitigation of UK $PM_{2.5}$. *Atmos. Chem. Phys.* **2016**, *16*, 265–276. [CrossRef]
13. Chemel, C.; Fisher, B.E.A.; Kong, X.; Francis, X.V.; Sokhi, R.S.; Good, N.; Collins, W.J.; Folberth, G.A. Application of chemical transport model CMAQ to policy decisions regarding $PM_{2.5}$ in the UK. *Atmos. Environ.* **2014**, *82*, 410–417. [CrossRef]
14. Carnell, E.; Vieno, M.; Vardoulakis, S.; Beck, R.; Heaviside, C.; Tomlinson, S.; Dragosits, U.; Heal, M.R.; Reis, S. Modelling public health improvements as a result of air pollution control policies in the UK over four decades—1970 to 2010. *Environ. Res. Lett.* **2019**, *14*. [CrossRef]

15. Chalabi, Z.; Milojevic, A.; Doherty, R.M.; Stevenson, D.S.; MacKenzie, I.A.; Milner, J.; Vieno, M.; Williams, M.; Wilkinson, P. Applying air pollution modelling within a multi-criteria decision analysis framework to evaluate UK air quality policies. *Atmos. Environ.* **2017**, *167*, 466–475. [CrossRef]
16. Sokhi, R.S.; San José, R.; Kitwiroon, N.; Fragkou, E.; Pérez, J.L.; Middleton, D.R. Prediction of ozone levels in London using the MM5–CMAQ modelling system. *Environ. Model. Softw.* **2006**, *21*, 566–576. [CrossRef]
17. Beevers, S.D.; Kitwiroon, N.; Williams, M.L.; Carslaw, D.C. One way coupling of CMAQ and a road source dispersion model for fine scale air pollution predictions. *Atmos Environ.* **2012**, *59*, 47–58. [CrossRef] [PubMed]
18. Skamarock, W.C.; Klemp, J.B.; Dudhia, J.; Gill, D.O.; Barker, D.M.; Wang, W.; Powers, J.G. *A Description of the Advanced Research WRF Version 3*; NCAR Technical note-475+ STR; National Center for Atmospheric Research: Boulder, CO, USA, 2008.
19. Byun, D.; Schere, K. Review of the Governing Equations, Computational Algorithms and Other Components of the Models-3 Community Multiscale Air Quality (CMAQ) Modeling System. *Appl. Mech. Rev.* **2006**, *59*, 1–77. [CrossRef]
20. Fraser, A.M.T.; Abbot, J.; Venfield, H.; Beevers, S.; Kitwiroon, N.; Beddows, A.; Carslaw, D.; Good, N.; Chemel, C.; Xavier, F.; et al. *CMAQ-UK Development Phase 1 Summary Report Final*; AEA/ENV/R/3334; 23/10/2012; Department for Environment Food & Rural Affairs (DEFRA): London, UK, 2012.
21. Fraser, A.M.; Beevers, S.; Kitwiroon, N.; Xavier, F.; Sokhi, R.; Derwent, R. *CMAQ-UK Development Phase 2 Report Final*; Department for Environment Food & Rural Affairs (DEFRA): London, UK, 2014.
22. Malardel, S.; Wedi, N.; Deconinck, W.; Diamantakis, M.; Kuhnlein, C.; Mozdzynski, G.; Hamrud, M.; Smolarkiewicz, P. ERA5 Reanalysis—European Centre for Medium-Range Weather Forecasts. 2017, updated monthly. ERA5 Reanalysis. Research Data Archive at the National Center for Atmospheric Research, Computational and Information Systems Laboratory. NCAR-UCAR, Ed. 2017. Available online: https://rda.ucar.edu/datasets/ds630.0/ (accessed on 21 November 2020).
23. Francis, X.; Sokhi, R. CMAQ Development for UK National Modelling: Technical Report on the Influence of Boundary Conditions on Regional Air. 2013. Available online: https://uk-air.defra.gov.uk/assets/documents/reports/cat20/1511260911_AQ0701_BoundaryConditions_Mar2013_Final.pdf (accessed on 14 December 2020).
24. Luecken, D.; Sarwar, G.; Yarwood, G.; Whitten, G.Z.; Carter, W.P.L. Impact of an Updated Carbon Bond Mechanism on Predictions from the CMAQ Modeling System: Preliminary Assessment. *J. Appl. Meteorol. Climatol.* **2008**, *47*, 3–14. [CrossRef]
25. Granier, C.; Darras, S.; van der Gon, H.D.; Jana, D.; Elguindi, N.; Bo, G.; Michael, G.; Marc, G.; Jalkanen, J.-P.; Kuenen, J. *The Copernicus Atmosphere Monitoring Service Global and Regional Emissions (April 2019 Version)*; Copernicus Atmosphere Monitoring Service (CAMS) report; European Centre for Medium-Range Weather Forecasts: Reading, UK, 2019. [CrossRef]
26. CERC. EMIT—Emissions Inventory Toolkit. Available online: http://www.cerc.co.uk/environmental-software/EMIT-tool.html (accessed on 25 May 2021).
27. Guevara, M.; Martínez, F.; Arévalo, G.; Gassó, S.; Baldasano, J.M. An improved system for modelling Spanish emissions: HERMESv2. 0. *Atmos. Environ.* **2013**, *81*, 209–221. [CrossRef]
28. Simpson, D.; Benedictow, A.; Berge, H.; Bergström, R.; Emberson, L.D.; Fagerli, H.; Flechard, C.R.; Hayman, G.D.; Gauss, M.; Jonson, J.E. The EMEP MSC-W chemical transport model-technical description. *Atmos. Chem. Phys.* **2012**, *12*, 7825–7865. [CrossRef]
29. Guenther, A.; Jiang, X.; Heald, C.L.; Sakulyanontvittaya, T.; Duhl, T.; Emmons, L.; Wang, X. The Model of Emissions of Gases and Aerosols from Nature version 2.1 (MEGAN2. 1): An extended and updated framework for modeling biogenic emissions. *Geosci. Model. Dev.* **2012**, *5*, 1471–1492. [CrossRef]
30. Copernicus. Copernicus Global Land Service—Leaf Area Index. Available online: https://land.copernicus.eu/global/products/lai (accessed on 10 April 2021).
31. Met Office. *Met Office Integrated Data Archive System (MIDAS) Land and Marine Surface Stations Data (1853–Current)*; NCAS British Atmospheric Data Centre: Met Office: London, UK, 2012.
32. Alduchov, O.E.R. Improved Magnus Form Approximation of Saturation Vapor Pressure. *J. Appl. Meteorol. Climatol.* **1996**, *35*, 601–609. [CrossRef]
33. Gov.UK. Summary Results of the Domestic Wood Use Survey. Available online: https://www.gov.uk/government/publications/summary-results-of-the-domestic-wood-use-survey (accessed on 28 December 2021).
34. Gov.UK. The National Emission Ceilings Regulations. 2018. Available online: https://www.legislation.gov.uk/uksi/2018/129/contents/made (accessed on 4 January 2022).
35. Net Zero Strategy: Build Back Greener. Available online: https://assets.publishing.service.gov.uk/government/uploads/system/uploads/attachment_data/file/1033990/net-zero-strategy-beis.pdf (accessed on 17 May 2021).
36. AQEG. Mitigation of United Kingdom $PM_{2.5}$ Concentrations. Available online: https://uk-air.defra.gov.uk/library/reports.php?report_id=827 (accessed on 28 December 2021).

atmosphere

MDPI

Article

Simulated Methane Emission Detection Capabilities of Continuous Monitoring Networks in an Oil and Gas Production Region

Qining Chen, Mrinali Modi, Gary McGaughey, Yosuke Kimura, Elena McDonald-Buller and David T. Allen *

Center for Energy and Environmental Resources, University of Texas at Austin, Austin, TX 78758, USA; qnchen@utexas.edu (Q.C.); mrinalimodi@utexas.edu (M.M.); garym@mail.utexas.edu (G.M.); yosuke@austin.utexas.edu (Y.K.); ecmb@mail.utexas.edu (E.M.-B.)

* Correspondence: allen@che.utexas.edu

Abstract: Simulations of the atmospheric dispersion of methane emissions were created for a region containing 26 oil and gas production sites in the Permian Basin in Texas. Virtual methane sensors were placed at 24 of the 26 sites, with at most 1 sensor per site. Continuous and intermittent emissions from each of the 26 oil and gas production sites, over 4 week-long meteorological episodes, representative of winter, spring, summer, and fall meteorology, were simulated. The trade-offs between numbers of sensors and precision of sensors required to reliably detect methane emissions of 1 to 10 kg/h were characterized. A total of 15 sensors, able to detect concentration enhancements of 1 ppm, were capable of identifying emissions at all 26 sites in all 4 week-long meteorological episodes, if emissions were continuous at a rate of 10 kg/h. More sensors or sensors with lower detection thresholds were required if emissions were intermittent or if emission rates were lower. The sensitivity of the required number of sensors to site densities in the region, emission dispersion calculation approaches, meteorological conditions, intermittency of the emissions, and emission rates, were examined. The results consistently indicated that, for the conditions in the Permian Basin, a fixed monitoring network with approximately one continuous monitor per site is likely to be capable of consistently detecting site-level methane emissions in the range of 5–10 kg/h.

Keywords: oil and gas; methane; atmospheric dispersion modeling; sensor network

Citation: Chen, Q.; Modi, M.; McGaughey, G.; Kimura, Y.; McDonald-Buller, E.; Allen, D.T. Simulated Methane Emission Detection Capabilities of Continuous Monitoring Networks in an Oil and Gas Production Region. *Atmosphere* **2022**, *13*, 510. https://doi.org/10.3390/atmos13040510

Academic Editor: Daniel Viúdez-Moreiras

Received: 17 February 2022
Accepted: 21 March 2022
Published: 23 March 2022

1. Introduction

Methane emissions from oil and gas production sites are detected and quantified in a variety of ways [1]. Some sensing approaches are short-duration measurements of methane concentrations using equipment deployed on foot, on drones, on ground vehicles, on aircraft, or on satellites. These short-term measurements, which employ a wide variety of detection technologies, capture instantaneous snapshots of methane concentrations, which can be used to estimate emission rates. Since many emission sources in upstream oil and gas operations are intermittent, short-term measurements may not detect all emissions from a site or may observe an intermittent emission that is then interpreted as persistent. In addition, since most short-term measurements are deployed on a monthly, quarterly, semi-annual, or annual basis, emissions that develop between measurements could persist undetected until the next scheduled measurement. If these emission rates are large, total emissions could be dominated by sources that develop between scheduled measurements. These limitations of periodic, short-duration measurements have driven interest, including regulatory initiatives [2,3], in continuous monitoring of emissions, using networks of sensors. The primary advantage of a continuous monitoring network is that it may be able to detect methane emissions much more quickly than detection methods based on short sampling times that are periodically repeated. The disadvantage of such networks is the cost of deploying the large numbers of sensors which would be required to enable a network to reliably and quickly detect unintended emissions.

This work will examine how many sensors, with what precision, would be required in a prototypical continuous monitoring network for methane emissions. The effectiveness of a prototypical network for detecting methane emissions in the Permian Basin oil and gas production region in West Texas will be assessed. The Permian Basin is one of the largest oil and gas production regions in the world, with 2019 gas production, in just the portion of the Permian Basin in the State of Texas, of approximately 11.8 billion cubic feet/day (~4 trillion cubic feet/year, approximately 10% of total US production) and oil production of approximately 3 million barrels per day (approximately a quarter of total US production) [4]. Recent assessments of methane emissions from the Permian Basin, based on TROPOMI satellite data, have estimated the methane emission rate as 2.9 Tg per year, a rate equivalent to approximately 3.7% of the volume of gas produced [5]. In addition to having methane emissions and methane emission intensities (methane emitted/natural gas produced) that are among the largest in the United States, the Permian Basin has simple topography and persistent winds, making the basin an ideal location for continuous monitoring using networks of fixed sensors.

The multiple operators and close proximity of sites in the Permian Basin also enhance the advantages of a shared network of sensors. As will be demonstrated in this work, because of the density and proximity of sites (>3 site/km^2 in the region examined), most unintended emissions could, in principle, be detected without having multiple sensors surrounding every site. Compared to isolated sites requiring sensors located at multiple cardinal directions (e.g., sensors north, south, east, and west of each site), sensors in the Permian Basin could be located at nearby sites (e.g., sites to the north, south, east, and west of a central site) to replace some of the information that would be provided by multiple sensors for isolated sites.

To evaluate the ability of a fixed network of methane sensors to detect emissions in the Permian Basin, a domain consisting of 26 oil and gas production sites and a flare site, near Midland Texas, was chosen. Four meteorology episodes, each one week in duration, were chosen as representative of seasonal variability in wind speed and direction. The dispersion of simulated emissions from each of the 26 oil and gas production sites was modeled for each week-long episode. The concentration enhancements due to emissions, from each source location, that would be detected at monitoring sites located near each source location were estimated. Analyses were performed for individual and collective weeks of meteorology, each representative of a season. Details of the analyses are described in the Methodology Section and in the Supplementary Materials. The overall goal of the analysis, however, is to characterize trade-offs between numbers of sensors and the precision of sensors required in order to reliably detect emissions, of various magnitudes, using a fixed monitoring network in an oil and gas production region.

This analysis represents a best-case scenario for fixed methane monitoring networks. The region chosen for modeling has site densities of >3 per square kilometer. Winds are generally strong and persistent. Emissions, some continuous and some intermittent, are assumed to persist for the week-long periods modeled. Dispersion models are assumed to accurately characterize the relationship between emission rates and methane concentrations at sensor sites. Sensors are assumed to accurately track simulated concentrations, and data analytics are assumed to be available such that concentration enhancements above the sensor's precision would reliably be attributed to an emission source. The response of the number of sensors required in the network to the conditions used in the simulations is considered in sensitivity analyses, however even the most idealized analysis characterizes important and basic trade-offs in the design of continuous monitoring networks for methane. This work demonstrates an approach to the design of low-cost methane sensing networks that minimizes the number of sensors required while meeting objectives of emission detection. The results show that, in principle, a relatively small number of sensors is capable of detecting site-level methane emissions of 5–10 kg/h in the Permian

Basin. In the few production regions that have had large, multi-scale campaigns to measure methane emissions, this level of emissions has been shown to represent a significant fraction of total emissions. For example, in the Barnett Shale of North Central Texas, the 10% of sites with emissions above 5 kg/h were estimated to account for approximately a third of all emissions, and a much larger fraction of emissions attributed to abnormal operations [6]. In addition, the framework described here for the design of continuous monitoring of methane emissions could be applied to continuous monitoring of other air pollutants, using low-cost sensors. The analyses in this manuscript represent a proof of concept for the design of fixed monitoring networks for methane or other air pollutants. Sensor deployment, testing, and emission detection studies are ongoing, but are beyond the scope of this paper.

2. Materials and Methods

2.1. Modeled Domain

A 1.9 by 3.7 km rectangular region in Midland County in Texas was chosen as the modeling domain. The region contains well sites, centralized facilities that collect production from multiple wells (tank battery sites), and a flare site (assumed idle). A satellite image of the region is shown in Figure 1, with wells and tank battery sites identified as emission sites. Three operators have a total of twenty-four wells and four tank battery sites in the domain. Wells appearing on the same well pad are clustered as a well site. The 24 wells in the domain are clustered into 22 well sites. This specific region was chosen since the clustering and relative positioning of the wells and tank batteries are representative of the asset designs used by multiple operators in the Permian Basin.

Figure 1. Satellite image of the modeling domain in the Midland Basin. Emissions were simulated at the 22 well sites and 4 tank battery sites shown on the map.

2.2. Emission Detection

Available methane sensors that might be deployed in a fixed monitoring network vary in their precision [7]. The precision of the sensor influences the ability of the sensor to detect emissions. Therefore, a variety of emission detection thresholds were considered. In this work, an emission was counted as detectable if during a one-week period, any sensor location in the network is predicted to observe a concentration enhancement greater than either 200, 500, or 1000 ppb (0.2, 0.5, or 1 ppm) for at least 1 minute. These thresholds for detection were based on the variability in background concentrations of methane observed in the region. A recent multi-month field study [8], evaluating low-cost methane sensors, found that daily minimum concentrations of methane (representative of background concentrations) in the region were 1.96 ppm, with a standard deviation of 0.05 ppm. An emission was assumed to be undetectable if no sensors in the network observe a concentration enhancement greater than these thresholds over a one-week period. This work examines the number of sensors and detection thresholds required to detect emissions from all sources in the region if methane is emitted at a rate of 10 kg/h. The analysis was repeated for emission rates of 5 and 1 kg/h.

2.3. Emissions and Sensor Placement

Emissions from each of the 22 well sites and the 4 tank battery sites in the domain were modeled. Emissions were not modeled for a flare site in the region. The flare was assumed to be idle at most times, and if it were to be combusting gas, unintended emissions would be due to unlit operation or very low combustion efficiencies. Unlit or low combustion efficiency emissions would lead to ambient concentrations of methane that would be readily detectable by the sensor network. The emissions from each of the well and tank battery sites were modeled separately (different simulations for each location). Emissions in each simulation were continuous at flow rates of 10, 5, and 1 kg/h, or were intermittent emissions at the same instantaneous flow rate. The intermittent emissions were assumed to occur for 1 min of every hour (e.g., minute 30 of hour 1, minute 30 of hour 2, and minute 30 of all subsequent hours), and when the emissions were occurring, they were assumed to emit at an instantaneous rate equal to the continuous emissions (10, 5, and 1 kg/h). Emissions were assumed to be released from the center of each well site, or tank battery site. Continuous releases from both well sites and tank battery sites and intermittent releases from well sites were assumed to occur at a height of 0.2 m, consistent with emissions associated with leaks from ground-level equipment and piping. Intermittent releases from tank battery sites were assumed to occur at a height of 5.5 m, representative of the height of tanks which have intermittent emissions due to the volatilization and release of methane discharged periodically to the tanks from high-pressure separators. The domain was gridded into 100 by 100 m cells and virtual sensors were placed in the center of the grid cell immediately to the north of each site to take advantage of prevailing southerly wind directions. If another emission site was located in the grid cell immediately to the north of a site, no sensor was placed in the cell. This led to a total of 24 sensor locations, as shown in Figure 2a. A modeling domain with reduced site density (~1 site/km^2) was also investigated to assess the importance of the site density. Seven emission sites out of the twenty-six sites in the base case were selected, with all sites separated by ~1 km, as shown in Figure 2b. Sensors were placed in the gird cell immediately to the north of the selected sites.

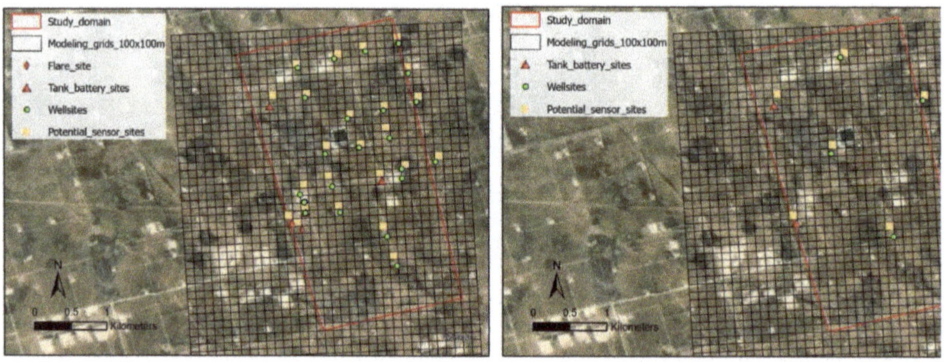

(a) (b)

Figure 2. (**a**) Dispersion modeling base case domain and all sites and sensor locations in the domain. Emissions were sourced from well sites (dots) and tank batteries (triangles). (**b**) Dispersion modeling domain with reduced site density.

2.4. Meteorological Episode Selection

Four time periods of one-week duration, from four calendar quarters, were identified to be representative of meteorological conditions during 2019. The year was first divided into four quarters, followed by an evaluation of wind speeds and wind directions observed in each seven-day period within the quarter against the variability seen in the entire quarter. The weeks that captured a reasonable representation of the range and frequency of wind speeds and wind directions observed during each of those quarters were selected as representative weeks (Figure S1). The dataset employed for this selection was obtained from one of the ground-based monitoring stations in the Midland-Odessa area (Continuous Ambient Monitoring Station, CAMS 47, Chennai, India), operated by the Texas Commission on Environmental Quality [9]. Temperature, surface pressure, dew point temperature, and precipitation were compared to observations at the National Weather Service (NWS) Station at Midland International Airport. As shown in Supplementary Figures S1 and S2, the week-long periods were found to be representative of the annual variability in meteorology.

2.5. Dispersion Modeling

To predict concentrations of methane in the modeling domain from each of the emission sources, atmospheric dispersion modeling was utilized. The atmospheric dispersion models use the emissions, together with a representation of 3D meteorological conditions and geophysical and land surface characteristics, to predict downwind concentrations throughout the modeling domain. While the geophysical and surface characteristics are assumed to remain constant, the meteorological conditions can vary considerably depending on the time of day and the time of year. Two dispersion models were used, HYSPLIT (Hybrid Single-Particle Lagrangian Integrated Trajectory, v5.0.0) and CALPUFF (v7.2.1_L150618). The HYSPLIT Model, developed by the National Oceanic and Atmospheric Administration, is one of the most extensively used atmospheric dispersion models [10]. The model calculation method is a hybrid between Lagrangian and Eulerian approaches. It calculates the dispersion of a pollutant by assuming either puff mode or particle mode, or both. For this work, HYSPLIT was used in puff mode. CALPUFF, on the other hand, is a non-steady state, Lagrangian puff modeling system (Exponent, 2014) [11]. The three-dimensional meteorological fields used to drive these models are derived from the output of the North American Mesoscale (NAM) Model Analysis. NAM is a weather model run by the National Centers for Environmental Prediction (NCEP) for producing weather forecasts (out to 84 h) every 6 hours at 0, 6, 12, and 18 UTC (Coordinated Universal Time) [12]. The NAM dataset used in this work is a blend of 4-time daily analyses of observations (at 0, 6, 12, and 18 h) with 3-hour forecasts at 3, 9, 18, and 21 h. It has a spatial resolution of 12 km and a temporal resolution of 3 h. For evaluating this dataset, the observations of wind speed and wind directions at the nearest grid cell were extracted and compared to the observations at the monitoring site maintained by the Texas Commission on Environmental Quality (TCEQ) for the four representative meteorological weeks [9]. The NAM Analysis represented observations with high fidelity. The primary analyses presented in this work will use the CALPUFF dispersion modeling driven by NAM meteorological data. Sensitivity analyses will examine the effect of the choice of dispersion model. The spatial scale of air dispersion modeling for the simulated domain is approximately 7 km^2. At this spatial scale and typical near-surface wind speeds in Midland-Odessa (5–30 km/h), emission plumes from oil and gas sources would be transported across the entire grid domain within a few hours, even during periods of very light wind speeds. Therefore, a time resolution of one minute was chosen for the meteorological modeling.

3. Results

3.1. Base Case

Table 1 reports the number of sensors required to detect emissions from each of the 26 emissions sites in each of the 4 weeks of meteorology (104 total detections). Alternative representation of the results are shown in the Supplementary Figure S3. Simulations were performed for each of the individual sites to determine which sensors, at which times, would detect emissions. Due to the unique orientation of each sensor location to each potential source, the detections of emissions from different sites by a single sensor generally do not interfere with each other. A sensor will detect a site to its south only when the winds are from the south, and this detection will not interfere with the ability of the same sensor, at a different time, to detect an emission from a source that is to the east.

Table 1. Sensors required to detect emissions from each of the 26 emissions sites, in each of the 4 weeks of meteorology evaluated in this work (104 total detections).

Sensor Precision (ppb)	Minimum Number of the 24 Available Sensors Required to Detect Continuous Emissions from 26 Sources in Each of the 4 Week-Long Periods (104 Detections)	Minimum Number of the 24 Available Sensors Required to Detect Intermittent Emissions from 26 Sources in Each of the 4 Week-Long Periods (104 Detections)
	Emission rate of 10 kg/h	
1000	15	24 sensors made 98 detections *
500	7	15
200	4	8
	Emission rate of 5 kg/h	
1000	24 sensors made 103 detections *	24 sensors made 93 detections *
500	15	24 sensors made 98 detections *
200	7	13
	Emission rate of 1 kg/h	
1000	24 sensors made 71 detections *	21 sensors made 25 detections *
500	24 sensors made 96 detections *	24 sensors made 53 detections *
200	24 sensors made 103 detections *	24 sensors made 93 detections *

* Out of possible detection of 26 sources in each of the 4 week-long meteorological episodes (all sources detected is counted as 104 detections). In this case, the number of sensors represents the number of sensors that have detections, not necessarily the minimum number of sensors to achieve the maximum counts of detected sources.

Using this approach, at an emission rate of 10 kg/h per site, a total of 15 sensors are required to detect all continuous emission sources if a detection is defined as at least one concentration enhancement of 1000 ppb, due to continuous emissions from a single site, during each of the 4 week-long meteorological episodes. If the detection threshold is lowered to 500 or 200 ppb, the number of required sensors to detect all continuous emissions is reduced to 7 and 4, respectively. A larger number of sensors or more precise sensors are required if emissions are intermittent. For emissions that occur at a rate of 10 kg/h, but only persist for one minute during each hour, the number of sensors required to detect all emission sources with sensor detection thresholds of 500 and 200 ppb is 15 and 8, respectively. If the detection threshold is 1000 ppb, even if sensors are placed at all 24 possible locations, not all emission sources can be detected within a 1-week period. In this scenario with intermittent emissions and a detection threshold of 1000 ppb, 98 of the 104 possible detections (26 sources in each of 4 weeks) are made.

Similar analyses are reported in Table 1 for emission rates of 5 and 1 kg/h. As emission rates are reduced, either more sensors are required, or better sensor sensitivity is required. Nevertheless, a network of approximately 1 sensor per site, with a sensor detection threshold as high as 1000 ppb, is able to detect a very high fraction of emissions from sites with 5–10 kg/h emission rates. Similar results are obtained using the HYSPLIT dispersion model with the NAM meteorological dataset, as documented in Supplementary Table S1.

3.2. Sensitivity Analyses

Analyses were conducted to assess the importance of site density and meteorology on the required number of sensors. Site density was reduced from 26 sites to 7 sites in the same modeling domain, with all sites approximately 1 km apart from each other, as shown in Figure 2b. The same analyses were performed, and results are shown in Table 2 (alternative representation of the results are shown in the Supplementary Figure S4). The results again indicate that a network of approximately 1 sensor per site is able to detect a very high fraction of emissions from sites with 5–10 kg/h emission rates.

Table 2. Sensors required to detect emissions from each of the 7 emissions sites, in each of the 4 weeks of meteorology evaluated in this work (28 total detections).

Sensor Precision (ppb)	Minimum Number of the 7 Available Sensors Required to Detect Continuous Emissions from 7 Sources in Each of the 4 Week-Long Periods (28 Sources)	Minimum Number of the 7 Available Sensors Required to Detect Intermittent Emissions from 7 Sources in Each of 4 Week-Long Periods (28 Sources)
	Emission rate of 10 kg/h	
1000	7	7 sensors made 25 detections *
500	5	7
200	3	6
	Emission rate of 5 kg/h	
1000	7	7 sensors made 22 detections *
500	7	7 sensors made 25 detections *
200	5	7
	Emission rate of 1 kg/h	
1000	7 sensors made 16 detections *	5 sensors made 5 detections *
500	7 sensors made 24 detections *	6 sensors made 10 detections *
200	7	7 sensors made 22 detections *

* Out of possible detection of 26 source in each of the 4 week-long meteorological episodes (all sources detected is counted as 104 detections). In this case, the number of sensors represents the number of sensors that have detections, not necessarily the minimum number of sensors to achieve the maximum counts of detected sources.

The importance of meteorology was assessed by examining the number of sensors required in each of the one-week periods that were chosen to represent seasonal meteorology. Wind speeds and wind directions vary significantly from season to season in the region, as shown in the Supplementary Figure S2. During summer months, wind directions are consistently from the southwest to the southeast. In contrast, winter wind directions are highly variable. This makes detection during summer months more challenging than detection during winter months. Numbers of sensors required for each season, for both the base case analyses and the reduced site density analyses, are presented in Tables 3 and 4 (alternative representation of the results are shown in the Supplementary Figures S5 and S6). While the number of sensors required in winter months is lower than in summer months, even with the more challenging meteorology of the summer, networks of approximately 1 sensor per site are able to detect a very high fraction of emissions from sites with 5–10 kg/h emission rates.

Table 3. Sensors required to detect continuous and intermittent emissions from each of the 26 emissions sites, for the weeks representing winter, spring, summer, and fall meteorology (26 detections in each week).

Sensor Precision (ppb)	Minimum Number of the 24 Available Sensors Required to Detect Continuous Emissions from 26 Sources in Each Week-Long Period	Minimum Number of the 24 Available Sensors Required to Detect Intermittent Emissions from 26 Sources in Each Week-Long Period
Winter (January/February/March meteorology)		
Emission rate of 10 kg/h		
1000	4	11
500	2	7
200	1	4
Emission rate of 5 kg/h		
1000	9	23 sensors made 25 detections *
500	4	11
200	2	5
Emission rate of 1 kg/h		
1000	20 sensors made 22 detections *	3 sensors made 3 detections *
500	15	14 sensors made 15 detections *
200	9	23 sensors made 25 detections *
Spring (April/May/June meteorology)		
Emission rate of 10 kg/h		
1000	4	16
500	2	10
200	1	6
Emission rate of 5 kg/h		
1000	6	22 sensors made 23 detections *
500	4	16
200	2	8
Emission rate of 1 kg/h		
1000	12 sensors made 13 detections *	No detections
500	22 sensors made 24 detections *	7 sensors made 7 detections *
200	6	22 sensors made 23 detections *
Summer (July/August/September meteorology)		
Emission rate of 10 kg/h		
1000	15	20 sensors made 20 detections *
500	7	13
200	4	8
Emission rate of 5 kg/h		
1000	23 sensors made 25 detections *	18 sensors made 19 detections *
500	15	20 sensors made 20 detections *
200	7	10
Emission rate of 1 kg/h		
1000	9 sensors made 10 detections *	1 sensor made 1 detection *
500	20 sensors made 20 detections *	6 sensors made 6 detections *
200	23 sensors made 25 detections *	18 sensors made 19 detections *

Table 3. *Cont.*

Sensor Precision (ppb)	Minimum Number of the 24 Available Sensors Required to Detect Continuous Emissions from 26 Sources in Each Week-Long Period	Minimum Number of the 24 Available Sensors Required to Detect Intermittent Emissions from 26 Sources in Each Week-Long Period
	Fall (October/November/December meteorology)	
	Emission rate of 10 kg/h	
1000	2	7
500	2	5
200	1	3
	Emission rate of 5 kg/h	
1000	5	11
500	2	7
200	2	4
	Emission rate of 1 kg/h	
1000	15	21 sensors made 21 detections *
500	7	24 sensors made 25 detections *
200	5	11

* Out of possible detection of all 26 sources in each of the week-long meteorological episodes. In this case, the number of sensors represents the number of sensors that have detections, not necessarily the minimum number of sensors to achieve the maximum counts of detected sources.

Table 4. Sensors required to detect continuous and intermittent emissions from each of the 7 emissions sites in the reduced density network, for the weeks representing winter, spring, summer, and fall meteorology (7 detections in each week).

Sensor Precision (ppb)	Minimum Number of the 7 Available Sensors Required to Detect Continuous Emissions from 7 Sources in Each Week-Long Period	Minimum Number of the 7 Available Sensors Required to Detect Intermittent Emissions from 7 Sources in Each Week-Long Period
	Winter (January/February/March meteorology)	
	Emission rate of 10 kg/h	
1000	4	6
500	2	5
200	1	4
	Emission rate of 5 kg/h	
1000	6	5 sensors made 5 detections *
500	4	6
200	2	5
	Emission rate of 1 kg/h	
1000	5 sensors made 5 detections *	No detection
500	6	2 sensors made 2 detections *
200	6	5 sensors made 5 detections *
	Spring (April/May/June meteorology)	
	Emission rate of 10 kg/h	
1000	3	7
500	2	6
200	1	5

Table 4. *Cont.*

Sensor Precision (ppb)	Minimum Number of the 7 Available Sensors Required to Detect Continuous Emissions from 7 Sources in Each Week-Long Period	Minimum Number of the 7 Available Sensors Required to Detect Intermittent Emissions from 7 Sources in Each Week-Long Period
	Emission rate of 5 kg/h	
1000	4	6 sensors detect 6 detections *
500	3	7
200	2	6
	Emission rate of 1 kg/h	
1000	2 sensors made 2 detections *	No detection
500	6 sensors made 6 detections *	1 sensor made 1 source *
200	4	6 sensors made 6 detections *
	Summer (July/August/September meteorology)	
	Emission rate of 10 kg/h	
1000	7	4 sensors made 4 detections *
500	5	6
200	3	6
	Emission rate of 5 kg/h	
1000	7	4 sensors made 4 detections *
500	7	4 sensors made 4 detections *
200	5	6
	Emission rate of 1 kg/h	
1000	2 sensors made 2 detections *	No detection
500	4 sensors made 4 detections *	1 sensor made 1 detection *
200	7	4 sensors made 4 detections *
	Fall (October/November/December meteorology)	
	Emission rate of 10 kg/h	
1000	3	6
500	2	6
200	1	2
	Emission rate of 5 kg/h	
1000	5	7
500	3	6
200	1	4
	Emission rate of 1 kg/h	
1000	7	5 sensors made 5 detections *
500	7	6 sensors made 6 detections *
200	5	7

* Out of possible detection of all 7 sources in each of the week-long meteorological episodes. In this case, the number of sensors represents the number of sensors that have detections, not necessarily the minimum number of sensors to achieve the maximum counts of detected sources.

4. Discussion

As shown in Tables 1 and 2, for both high (>3 site/km^2) and low site density (~1 site/km^2) scenarios, a methane monitoring network with approximately 1 sensor per site is able to detect a large fraction of emissions from sites with 5–10 kg/h emission rates. Site densities in this range are common in the Permian Basin. Figure 3 shows production site densities in the Permian Basin in November 2021 [13]. Production sites in the Permian Basin were aggregated into 4 × 4 km^2 grid cells, and the site density in each grid cell was calculated and mapped. For grid cells containing at least one production site, 58% had site densities less than 1 site/km^2, 28% had site densities between 1 and 3 site/km^2, and 8% had site densities between 3 and 5 site/km^2. Only 1% of the grid cells had site densities over 10 site/km^2. While grid cells with <1 site/km^2 represent more than half of the total grid cells with production reported, cells with >1 site/km^2 account for 85% of the production in the Permian Basin. Therefore, the analyses represent a proof of concept applicable to a large fraction of the production in the Permian Basin.

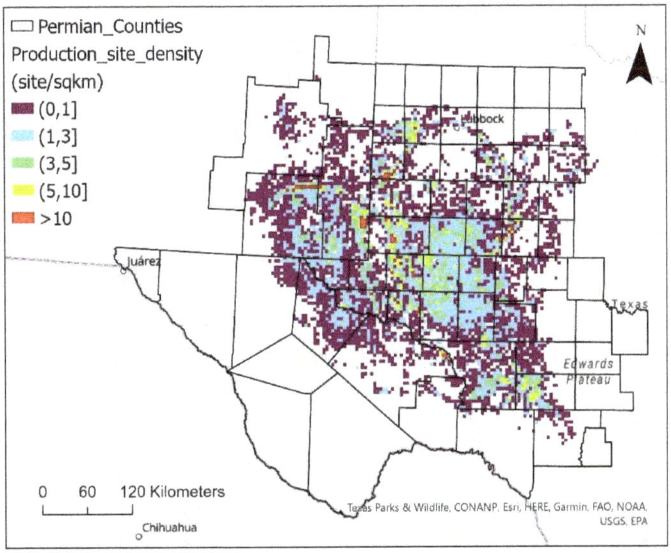

Figure 3. Production site densities in the Permian Basin, with production sites aggregated into 4×4 km^2 grid cells.

As shown in Tables 3 and 4, the effectiveness of detections was affected by meteorology, which varies significantly from season to season in the region. As shown in the Supplementary Figure S2, during spring and summer months (April to September), winds are primarily from the southwest to the southeast. During these seasons, individual sensors are only able to detect emissions from southerly directions. In contrast, during winter months (October to March), winds are variable in direction, allowing sensors to detect emissions from sources in multiple directions. Therefore, compared to summer months, winter months require less sensors in the network to detect all the emissions in the region. However, even during the summer months, a network with approximately one sensor per site is still able to detect most of the sources with emission rates between 5 and 10 kg/h.

Two different dispersion models were used in this work to evaluate sensor placements. Although the counts of sensors required to detect all emission sources vary with the use of different dispersion models, the conclusion that approximately one continuous monitor per site is able to detect a large fraction of emissions in the range of 5–10 kg/h is valid with either dispersion model applied. Figure 4 shows a comparison of the plumes predicted by the two models at a minute instance during the simulation week in winter, due to a continuous emission source with an emission rate of 25 kg/h at the tank battery site in the middle of the study domain, with the same meteorological dataset. Predictions of vertical and horizontal plume width, and other parameters, vary between the dispersion models. The HYSPLIT model led to more grid cells and sensor locations with concentration enhancements above the detection thresholds. For a single emission source, with HYSPLIT simulations, more sensor locations would be able to consistently detect the emissions. Therefore, analyses conducted with the HYSPLIT dispersion model indicated that slightly fewer sensors are required to be able to consistently detect emissions in the range of 5–10 kg/h, compared to the analyses with the CALPUFF Model (results are shown in Supplementary Table S1). Many different combinations of parameters, used as input to dispersion models, could lead to the slightly different numbers of sensors predicted by the two dispersion models. Overall, however, the analyses consistently indicate that, for the conditions in the Permian Basin, a fixed monitoring network with approximately one continuous monitor per site is likely to be capable of consistently detecting site-level methane emissions in the range of 5–10 kg/h.

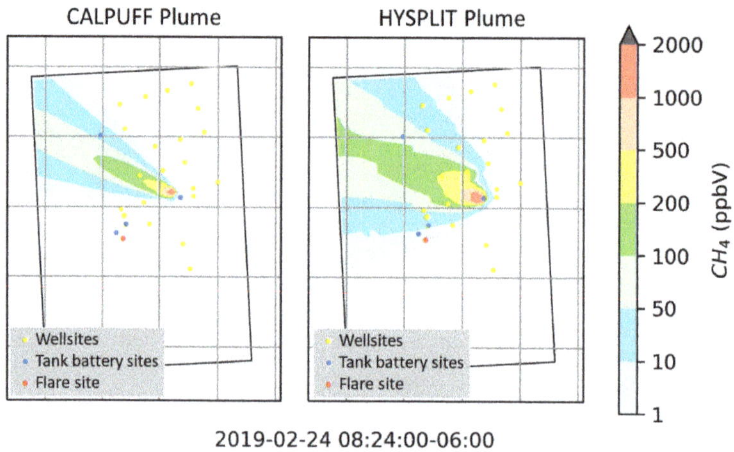

Figure 4. Comparison of methane plumes predicted by CALPUFF and HYSPLIT models at a minute instance during the simulation week in winter, due to a continuous emission source with an emission rate of 25 kg/h at the tank battery site in the middle of the study domain.

5. Conclusions

This paper characterized trade-offs between numbers of sensors and the precision of sensors required to reliably detect methane emissions, ranging from 1 to 10 kg/h, using a fixed continuous methane monitoring network, based in a representative oil and gas production region in the Permian Basin. It demonstrates an approach to the design of methane sensing networks that minimizes the number of sensors required while meeting objectives of emission detection. The number of sensors required to detect continuous and intermittent emissions with various emission rates was examined with two representative site densities, under four meteorological conditions representing four seasons, and with two dispersion models. The results show that although the number of sensors required to detect the emissions varies from case to case, networks with approximately one continuous monitoring sensor per site are capable of detecting site-level methane emissions of 5–10 kg/h in the Permian Basin.

Supplementary Materials: The following Supplementary Materials can be downloaded at: https://www.mdpi.com/article/10.3390/atmos13040510/s1: Figure S1: Wind rose diagrams, showing wind speed frequency data, by wind direction for the four representative weeks and for observational annual average data. Figure S2: Wind roses for each quarter (left) and representative week (right) for quarters 1 (a), 2 (b), 3 (c), and 4 (d). Figure S3: Alternative representation of the results in Table 1. Figure S4: Alternative representation of the results in Table 2. Figure S5: Alternative representation of the results in Table 3. Figure S6: Alternative representation of the results in Table 4. Table S1: Sensors required to detect continuous emissions from each of the 26 emissions sites, in each of the four weeks of meteorology evaluated in this work (104 total detections), using multiple dispersion models.

Author Contributions: Conceptualization, D.T.A.; methodology, all authors; software, Q.C., M.M., G.M. and Y.K.; validation, Q.C., M.M. and Y.K.; formal analysis, Q.C. and M.M.; investigation, Q.C., M.M., G.M. and Y.K.; resources, D.T.A.; data curation, Q.C., M.M. and Y.K.; writing—original draft preparation, Q.C., M.M., G.M. and D.T.A.; writing—review and editing, Q.C. and D.T.A.; visualization, Q.C.; supervision, D.T.A. and E.M.-B.; project administration, D.T.A. and E.M.-B.; funding acquisition, D.T.A. All authors have read and agreed to the published version of the manuscript.

Funding: This research was funded by the University of Texas through its Energy Institute.

Institutional Review Board Statement: Not applicable.

Informed Consent Statement: Not applicable.

Data Availability Statement: Not applicable.

Acknowledgments: The authors thank Ling Huang at Shanghai University for providing simulations using the HYSPLIT Model and collaborators at ExxonMobil Upstream Research Company and Pioneer Natural Resources for guidance on typical site configurations.

Conflicts of Interest: The authors declare the following competing financial interest(s): One of the authors (D.T.A.) has served as chair and is currently a member of the Environmental Protection Agency, Science Advisory Board; in this role, he is a Special Governmental Employee. D.T.A. has current research support from the National Science Foundation, the Department of Energy, the Texas Commission on Environmental Quality, the Gas Technology Institute, the Collaboratory to Advance Methane Science, the National Institute of Clean and Low Carbon Energy (NICE), the ExxonMobil Upstream Research Company, Pioneer Natural Resources, and the Environmental Defense Fund. He has also worked on methane emission measurement projects that have been supported by multiple natural gas producers and the Environmental Defense Fund. D.T.A. has done work as a consultant for multiple companies, including British Petroleum, Cheniere, Eastern Research Group, ExxonMobil, KeyLogic, and SLR International.

References

1. Methane Guiding Principles, Reducing Methane Emissions Best Practice Guide: Identification, Detection, Measurement and Quantification. September 2020. Available online: https://methaneguidingprinciples.org/wp-content/uploads/2020/09/Reducing-Methane-Emissions_Identification-Detection-Measurement-and-Quantification_Guide.pdf (accessed on 10 August 2021).
2. State of Colorado. Senate Bill 19-181. Available online: https://leg.colorado.gov/sites/default/files/2019a_181_signed.pdf (accessed on 10 August 2021).
3. U.S. Environmental Protection Agency (EPA). Standards of Performance for New, Reconstructed, and Modified Sources and Emissions Guidelines for Existing Sources: Oil and Natural Gas Sector Climate Review. 2021. Available online: https://www.federalregister.gov/documents/2021/11/15/2021-24202/standards-of-performance-for-new-reconstructed-and-modified-sources-and-emissions-guidelines-for (accessed on 15 November 2021).
4. Texas Railroad Commission (TRRC). Permian Basin Information (Website). 2020. Available online: https://www.rrc.state.tx.us/oil-gas/major-oil-and-gas-formations/permian-basin-information (accessed on 10 August 2021).
5. Zhang, Y.; Gautam, R.; Pandey, S.; Omara, M.; Maasakkers, J.D.; Sadavarte, P.; Lyon, D.; Nesser, H.; Sulprizio, M.P.; Varon, D.J.; et al. Quantifying methane emissions from the largest oil-producing basin in the United States from space. *Sci. Adv.* **2020**, *6*, eaaz5120. [CrossRef] [PubMed]
6. Zavala-Araiza, D.; Alvarez, R.A.; Lyon, D.R.; Allen, D.T.; Marchese, A.J.; Zimmerle, D.J.; Hamburg, S.P. Abnormal process conditions required to explain emissions from natural gas production sites. *Nat. Commun.* **2017**, *8*, 14012. [CrossRef] [PubMed]
7. Fox, T.A.; Barchyn, T.E.; Risk, D.; Ravikumar, A.P.; Hugenholtz, C.H. A review of close-range and screening technologies for mitigating fugitive methane emissions in upstream oil and gas. *Environ. Res. Lett.* **2019**, *14*, 053002. [CrossRef]
8. Methane Sensor Intercomparison. Available online: http://dept.ceer.utexas.edu/ceer/astra/showdown/showdown.cfm (accessed on 10 August 2021).
9. Texas Commission on Environmental Quality (TCEQ). CAMS 47 Site. 2021. Available online: https://www.tceq.texas.gov/cgi-bin/compliance/monops/site_photo.pl?cams=47 (accessed on 28 November 2020).
10. Stein, A.F.; Draxler, R.R.; Rolph, G.D.; Stunder, B.J.B.; MCohen, D.; Ngan, F. NOAA's hysplit atmospheric transport and dispersion modeling system. *Bull. Am. Meteorol. Soc.* **2015**, *96*, 2059–2077. [CrossRef]
11. Exponent. Official CALPUFF Modeling System. 2014. Available online: http://www.src.com/ (accessed on 1 December 2020).
12. North American Mesoscale Forecast System (NAM). National Centers for Environmental Information (NCEI). 2021. Available online: https://www.ncdc.noaa.gov/data-access/model-data/model-datasets/north-american-mesoscale-forecast-system-nam (accessed on 1 December 2020).
13. IHS Markit Enerdeq Browser (Licensed). Production Allocated, Permian Basin. Monthly Production Data in November 2021. Available online: https://my.ihs.com/energy (accessed on 17 February 2022).

atmosphere MDPI

Article

Use of Toxic Substance Release Modelling as a Tool for Prevention Planning in Border Areas

Jozef Kubas [1,*], Maria Polorecka [2], Katarina Holla [1], Viktor Soltes [3], Alexander Kelisek [1], Simeon Strachota [2] and Stanislav Maly [4]

1 Department of Crisis Management, Faculty of Security Engineering, University of Žilina, 010 26 Žilina, Slovakia; katarina.holla@uniza.sk (K.H.); alexander.kelisek@uniza.sk (A.K.)
2 Department of Fire Engineering, Faculty of Security Engineering, University of Žilina, 010 26 Žilina, Slovakia; maria.polorecka@uniza.sk (M.P.); strachota.simeon@gmail.com (S.S.)
3 Department of Security and Safety Research, Faculty of Security Engineering, University of Žilina, 010 26 Žilina, Slovakia; viktor.soltes@uniza.sk
4 Occupational Safety Research Institute, Jeruzalemska 1283/9, Praha 1, 110 00 Nove Mesto, Czech Republic; malys@vubp-praha.cz
* Correspondence: jozef.kubas@uniza.sk; Tel.: +42-141-513-6705

Abstract: The paper deals with the protection of the population and the environment in crisis management and emergency planning. It includes a proposal for an auxiliary tool for crisis managers and commanders to increase the safety of the population and the environment in the evaluated area. The proposal was developed thanks to a detailed analysis of the border area in selected regions of Slovakia, where extraordinary events may occur during the cross-border transport of hazardous substances. The actual outputs are maps of area-border crossings, including the places of transport of hazardous substances specifying a range of possible adverse effects on the endangered area. The modelling process was based on real conditions in the given area. Various scenarios of the possible occurrence of the release of hazardous substances were developed. The scenarios were applied in the ALOHA CAMEO software. Using the software output, it was possible to draw the most probable emergency scenarios with a cross-border effect. Cross-border impacts are crucial challenges in dealing with an emergency, as there is a need to ensure cooperation and coordination of emergency services in two different countries. The outputs proposed by the authors are a tool suitable not only for taking preventive measures but also as an aid in repressive activities. It is, therefore, suitable both for reducing the probability of the occurrence of given emergencies and minimizing its consequences.

Keywords: crisis management; safety; emergency event; civil protection; ALOHA software; scenario; case study; simulation; risk

Citation: Kubas, J.; Polorecka, M.; Holla, K.; Soltes, V.; Kelisek, A.; Strachota, S.; Maly, S. Use of Toxic Substance Release Modelling as a Tool for Prevention Planning in Border Areas. *Atmosphere* **2022**, *13*, 836. https://doi.org/10.3390/atmos13050836

Academic Editors: Daniel Viúdez-Moreiras and Alexandra Monteiro

Received: 23 February 2022
Accepted: 18 May 2022
Published: 20 May 2022

Publisher's Note: MDPI stays neutral with regard to jurisdictional claims in published maps and institutional affiliations.

1. Introduction

Emergencies do not affect a secure society. A minimum system of tasks and measures is necessary to protect life, health and property. The system is reduced to the analysis of possible threats and, consequently, reduces the probability of occurrence and elimination of consequences. Act 42/1994 Coll. on Civil Protection is valid in the Slovak Republic. According to this act, an emergency event means a natural disaster, accident, catastrophe, threat to public health of II. degree, the mass arrival of foreigners on the territory of the Slovak Republic, or a terrorist attack. Due to extensive problems, the article will focus on the crash accident. According to this law, an accident is an emergency event that causes a deviation from the steady operating state. As a result, hazardous substances are released or are affected by other destructive factors that affect life, health or property [1].

Concerning dangerous substances that can occur in an accident, it is necessary to focus on prevention, especially with those that are dangerous to life and human health or toxic to the environment. It is also significant to pay attention to substances that are difficult to dispose of or remove from the environment and can worsen living conditions in the long-term. The last group consists of hazardous substances, which, although not extremely dangerous, are transported or stored in large quantities. The effects of these substances may thus be significant due to their quantity [2].

An accident involving a dangerous substance can occur during the storage, use or transport of dangerous substances. Storage and production facilities or equipment are among the stationary sources. In addition to stationary sources, however, crisis managers often deal with leaks of hazardous substances during their transport [3,4]. This type of leakage is the subject of our paper, where we focus on road transport by tankers between several countries.

Crisis management deals with such events. It is an inseparable part of modern society and the concept of a smart city, which wants to create a safe environment for its inhabitants [5]. A new trend increasingly used in smart cities and crisis management is digitization. It allows us to predict and manage various events at the same time [6,7]. Crisis management planning is mainly concerned with the readiness to react in the event of an emergency so that the system remembers it as quickly as possible. Risk management and crisis management are not separate functions but rather two complementary phases of the same function. They consist of learning aimed at developing the organization's ability to deal with emergencies and disruption without consequences or with minimal consequences. Subsequently, it is possible to recover quickly from various crises [8]. Therefore, it is necessary to have proper knowledge of the environment and the ability to anticipate all possible negative impacts that could upset the balance and escalate into a disaster [9,10].

Dangerous goods (DG) are transported in various modalities from one or more places of origin to a destination where they are needed. As people need DG to live and work, they are widely distributed in different environments. Dangerous goods transportation (DGT) concerns all kinds of goods such as liquids, gasses, and solids, including radioactive, flammable, explosive, corrosive, oxidizing, asphyxiating, biohazardous, toxic, pathogenic, or allergenic materials. All the mentioned substances have latent or non-latent serious health risks that can harm humans and the environment and destroy surrounding properties or other living organisms [11]. Accidents involving DG often require a response from various emergency services. Therefore, procedures for the mutual exchange of information and coordination should be adopted [12]. The severity of the consequences related to road accidents in DGT and the large number of lorries that transport DG daily require the development and use of tools for risk assessment and treatment that can define the risk as a product of event frequency/probability and its consequences [13]. The case studies in this area illustrate the factors leading to crisis events and their impacts on facilities, the environment and nearby communities. The studies also specify steps that can be taken by prevention decision makers and local and state emergency managers to reduce the risks posed by emergency events through changes in procedures and investment in equipment [14].

Risks related to DGT can be characterised by various aspects, such as the type and quantity of dangerous goods, vehicle and road characteristics, traffic and weather conditions and population density. Some of these issues are time-dependent (traffic and weather conditions, population density). This aspect implies the complexity of such analysis and supports the use of modelling, and, thus, simulation models. Many approaches to simulation modelling have recently been adopted, such as:

- System Dynamics (SD), which is characterised by a high level of abstraction, low details and a strategic level [15].
- Discrete Event (DE), a mostly discrete model characterised by a medium abstraction level, medium details and a tactical level [16].
- Agent-Based (AB), also referred to as ABM (agent-based modelling), ABS (agent-based systems or simulation), and IBM (individual-based modelling), which are mostly decentralized and preferred for complex systems [17].

According to McKinsey, 70% of senior managers have begun planning to adopt digital risk management. The use of risk management software is becoming increasingly important in maintaining resilience to crisis events [18].

The main advantage of simulation is that it provides more information on the behaviour of the DGT system in different situations, considering involved decision makers, various what-if hypotheses and numerous approaches. In terms of emergency management or planning activities, e.g., operative training, models and simulations could be applicable for technicians, drivers, controllers, firefighters, emergency operators and other DGT subjects.

In the context of developing a decision-making support tool for final users working in the field of emergency response planning, we propose a comprehensive system.

The system consists of three available basic tools. One of them we apply in our case study:

- The Transport Integration Platform (TIP);
- The Areal Location of Hazardous Atmospheres (ALOHA);
- a GIS (or WEB GIS) platform for visualization of all collected data and display of results [19].

We decided to demonstrate by modelling with the ALOHA software used in this area. In our case study, we also consider the impacts on neighbouring states.

The article is focused on the use of possibilities of modelling the dispersion of toxic substances in the ALOHA software in border areas, with an emphasis on the preparedness of crisis managers and improving the prevention and response of the population to crisis events.

2. Materials and Methods

Each country should analyse the potential risks to its environment. Significant attention needs to be paid to the atmosphere, as leakages of dangerous substances in the border area can have a negative impact on several countries. In our paper, we focused on the territory of Slovakia. Slovakia has eight regions (NUTS 3), one of which was selected as a model: the Žilina Region. There were several reasons for this decision, including the fact that the workplace of several authors is located there. The authors also regularly communicate with the regional management in this locality. This region is bordered by two other states. Its location indicates the possibility of transporting dangerous substances from the surrounding states. Due to the fact that dangerous goods are transported through the Žilina Region only by road and rail, it is possible to deal only with the ADR agreement for road transport and the RID order for rail transport. Due to the wide range of issues, the paper will focus on the road transport of dangerous goods.

The simulation software tool ALOHA CAMEO was used to create the map. This programme is available for free download in the current version for Windows at www.epa.gov (accessed on 18 May 2022). It is regularly updated and supplemented with new chemicals. As it features an extensive library, it can evaluate many chemicals. If a substance is not included in the library, it can be added to the list based on knowledge of the important physicochemical and toxicological data provided for each dangerous substance. The programme is used to evaluate the leakage of a dangerous substance and allows the selection of location and atmospheric conditions. Based on the tank dimensions, it determines the filling height of the tank and is able to evaluate the cloud trace. The obtained results are then projected into Google Maps, so it is possible to obtain an up-to-date map showing the impact of the emergency in the selected area.

The elaboration of a map by simulating various scenarios of the leakage of hazardous substances at border crossings points to the real threat of the leakage of hazardous substances in the territory of neighbouring states, which would also affect the environment of the Slovak Republic. In the case of an accident, the states should inform the possibly affected surrounding countries. Significant information for further procedures can be obtained by determining the distance of the escaping substance.

The initial phase of the problem solution includes the identification and analysis of relevant documents. Subsequently, Slovak border crossings located in the territory of the Žilina Region were characterised, focusing on the quality and load-bearing capacity of roads and the possibilities of transporting dangerous goods. The analysis specified their surroundings, terrain fragmentation, the presence of highly frequented roads, watercourses and protected landscape areas. Considered data allow us to identify the factors influencing the effects of the leakage of a dangerous substance. The information gathered on the factors influencing the dispersion of hazardous substances was the basis needed to create a model and simulate the release of a hazardous substance. Using the ALOHA CAMEO evaluation programme, we created danger zones for escaped hazardous substances. By comparing the impacts of the leakage of a dangerous substance at individual border crossings, it was possible to point out the negative impacts on the population and the environment.

The transport of dangerous goods is accompanied by numerous risks. The release of the dangerous substance, its dispersion and its effects on the population and the environment are influenced by the factors specific to the leakage area. The effects of the leakage of a dangerous substance will be less severe in an uninhabited unprotected area without watercourses than in a densely populated area or the vicinity of protected natural areas. Therefore, it is very important to identify possible risks arising from the activities related to the transport of dangerous goods and to evaluate the impacts on the population and environment at specific border crossings using simulations in the ALOHA CAMEO programme. Based on these results, it will be possible to assess the suitability of individual border crossings for the transport of dangerous goods. Software-simulated infestation areas could be the basis for collaborative tactical exercises, while rescue services could deepen their practical skills and knowledge in real-world situations.

The selection of the dangerous substance was based on a document of the Ministry of the Interior of the Slovak Republic on the most frequently transported dangerous substances [20] and a publication by Ján Buzalka [21]. These sources state that ammonia is one of the most frequently transported substances in the Slovak Republic. Meteorological conditions play a key role in the dispersion of dangerous substances. The simulation aims to create the most favourable conditions for the dispersion of a dangerous substance and to determine the direction of the wind directing the substance fumes toward human dwellings. We entered the following input parameters into the programme for modelling the evaluation of the area of danger in case of the leakage of hazardous substances [22]:

- Wind speed in m/s or km/h;
- Wind direction in angular degrees;
- Vertical air stability (inversion, isothermia, convection);
- Air temperature in degrees Celsius;
- Altitude above ground at which meteorological parameters are measured;
- Nature of weather (clouds, rain, snow).

The dispersion of a hazardous substance of these parameters is most affected by wind speed, wind direction and inversion. With low wind speeds and constant stability, the range of vapours of hazardous substances is extended. On the contrary, strong and unstable winds shorten their range [23]. These findings were also confirmed in preliminary attempts to simulate the release of a hazardous substance in the ALOHA CAMEO evaluation programme. We determined that at a wind speed of 2 m/s, a dangerous substance dispersed over the longest distance and the largest area. At lower wind speeds, the substance hardly dispersed to the surroundings, only in a circle around the point of leakage. A speed higher than 2 m/s caused a narrowing of the vapour area of the hazardous substance.

The vertical stability of the atmosphere significantly influences the dispersion of substances in space. In the given area, the conditions for the occurrence of an emergency are highly probable. The inversion occurs about an hour before sunset and disappears within one hour after sunrise. Convection occurs about two hours after sunrise and disappears about 2 to 2.5 h before sunset. Hazardous chemicals disperse over longer distances in inverse conditions. On the contrary, in convection, they reach smaller distances [23]. Due to this, the presence of inversion will be taken into account when simulating the leakage of a dangerous substance. From modelling in the ALOHA CAMEO evaluation programme, we determined that the lower the inversion above the surface is, the further the hazardous substance disperses. However, the height of the inversion was adjusted so that it was located above the tank from which the dangerous substance would escape, i.e., at the height of 50 m. The last factor that played a significant role in the modelling is the ambient temperature. Hazardous substances evaporate faster at high temperatures, leading to the formation of a mass of air with a high concentration of the hazardous substance. On the contrary, low temperatures prolong the stability of the hazardous substance in the field, but due to low temperatures, the evaporation of hazardous substances is lower, so the contaminated space will be smaller than it would be at high temperatures [23].

The scenarios in this work are based on a simple syllabus containing:

- The cause of the leakage of the dangerous substance;
- The direction of dispersion of the leaked dangerous substance;
- Extent of the infestation.

First of all, it was necessary to determine how the leakage of the dangerous substance would take place. In our work, we tried to simulate the largest possible leakage of hazardous substances. We proceeded from the knowledge that the amount of leaked hazardous substance per unit time is directly proportional to the size of the hole from which the hazardous substance escapes. In our work, we considered a hole with a diameter of 25 cm. However, such a large opening can only be caused by the action of destructive forces on the tank body. Therefore, in every single scenario, we used a traffic accident as the initiator of the leakage of a hazardous substance. Here, we assumed that it could result in a crack in the tank of the required size. The cause of the accident at the individual border crossings was determined on the basis of an assessment of their surroundings. At the border crossings with connecting lanes, we considered the collision of cars. At border crossings with sharp turns, we considered the skidding of the vehicle and the same procedure was followed at other border crossings. However, the cause of the traffic accident does not affect the results of individual simulations of leaks of hazardous substances. Their goal was to bring our scenarios as close as possible to situations that may occur.

In modelling, we use the toxic limits AEGL (Acute Exposure Guideline Levels), where AEGL-1 means the concentration of a dangerous substance in the air at which the general population, including receptive individuals, is expected to experience obvious discomfort, irritation, or certain sensory symptoms. The effects are not debilitating, they are transient and reversible upon discontinuation of exposure. It is marked yellow in the final drawings.

AEGL-2 means the concentration of a dangerous substance in the air at which the general population, including receptive individuals, is expected to experience irreversible or other serious, long-lasting adverse health effects or impaired escape potential. It is marked orange in the final drawings.

AEGL-3 means the concentration of a dangerous substance in the air at which the general population, including receptive individuals, is expected to experience life-threatening health effects or death. It is marked red in the final drawings.

In the individual scenarios, we chose the direction of the dispersion of the escaped dangerous substance so that the toxic fumes would hit as many inhabited objects, watercourses and protected landscape areas as possible, or the road on which the vehicles could be located. The extent of the infestation and the individual danger zones were determined on the basis of the results of the CAMEO ALOHA evaluation software exported to Google Earth. The processing algorithm is demonstrated in Figure 1.

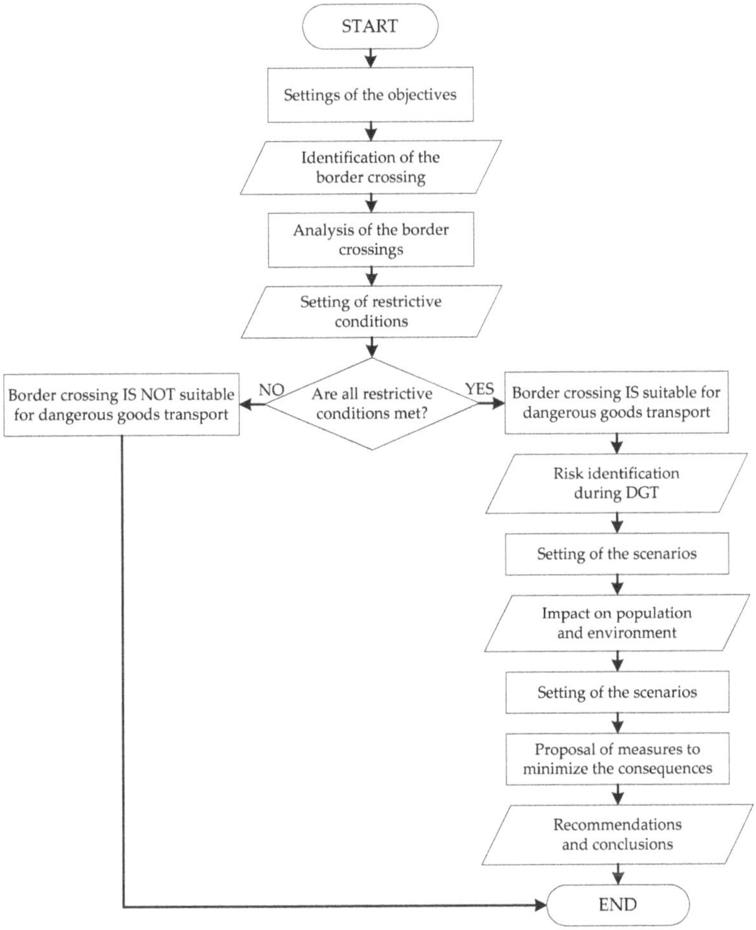

Figure 1. Scheme of the research processing procedure.

The procedure for processing the issue is demonstrated in Figure 1. The first step is setting the objectives of the process, primarily to minimize the likelihood and consequences of an emergency with a leak of a hazardous substance at border crossings points during transport. In the diagram, we see the first step, where we determined the activity, which is the development of a realistic overview of the possibilities of transporting hazardous substances through border crossings in a given area. After obtaining background information such as the passage for trucks, load capacity, width and overall condition of the road, we proceeded to the analysis and determination of restrictive conditions, on the basis of which we selected passages suitable for the transport of hazardous substances. Subsequently, at selected border crossings, risks are identified with the subsequent creation of scenarios according to the described principles. After evaluating the scenarios and modelling them in the ALOHA programme, we identified the expected consequences for the population and the environment and proposed appropriate measures.

3. Results

Based on the procedures and methodology described above, we applied the obtained inputs to modelling in the ALOHA CAMEO software, while ensuring that the scenarios were realistic but sufficiently pessimistic, which ensured a necessary level of vigilance.

3.1. Simulation of a Leak of a Dangerous Substance and Development of a Scenario for Individual Border Crossings

For the simulations, it was necessary to create many scenarios of the possibility of the leakage of dangerous substances for all border crossings, which are suitable for the transport of dangerous goods in terms of surface quality and road width. The simulation of the leakage of a dangerous substance was performed in the ALOHA CAMEO evaluation programme. This step enabled us to assess the impact on the population and environment in the endangered area. At each border crossing point were considered the same amount and type of dangerous substance, the size of the hole from which the substance will leak and the same meteorological conditions. We expected that the initial atmospheric conditions were similar in the given localities. The considered chemical was ammonia (molecular weight 17.03 g/mol) as it is the most frequently transported substance through the mentioned border crossings. In order to provide real data on the atmosphere for the given environment, we communicated with experts from the Slovak Hydro-meteorological Institute. Among the atmospheric data, we specified a wind speed of 2 m/s, air temperature of 5 °C, inversion at the height of 50 m, cloud cover large, vertical stability class F-inversion, and relative humidity of 90%. The leak came from a 25 cm diameter hole in a horizontal cylindrical tank. The flammable substance escaped from the tank with a diameter of 2.66 m and a length of 11 m. The volume of the tank was 61 cubic meters. The weight of the chemicals in the tank was 242 kg, the duration of the leak was 1 min and the rate of leakage was 3.2 kg per second. The only variable condition for individual simulations was the wind direction that we adapted to the direction of the impact on the side of the Slovak Republic, as we addressed prevention in the Žilina Region. In the final version, we removed the conditions for preventive use, circular shape, as the direction of the wind before the accident may be different. Certain inaccuracies must be considered when determining the effects of a leak of a dangerous substance on humans and the environment. There are imperfections when examining vulnerable areas in the 3D and 2D modes in Google Maps. Due to the low quality of the photo, it was sometimes not possible to reliably identify the object. It was also not possible to determine the number of persons in buildings affected by fumes and gases of a dangerous substance. Therefore, we will not count the number of affected persons, but the number of affected objects (buildings) with assumed inhabitants or occupants. Due to the size of the area contaminated by a leak, it was not possible to manually count the number of all objects affected. Objects will be manually counted based on Google Earth Maps only in the red and orange area. For the yellow zone, this number will be determined based on available statistical data on the number of inhabitants of the municipality affected by the leakage of a dangerous substance.

3.1.1. Leakage of a Dangerous Substance at the Makov Border Crossing

The driver of the tank carrying ammonia across the Makov border crossing veered off the road due to a micro-sleep and crashed into a marble monument that was located by the road. A crack with a diameter of 25 cm formed on the tank, from which ammonia began to leak. The wind flowing from the northwest shifted the toxic cloud to the southeast, as depicted in Figure 2.

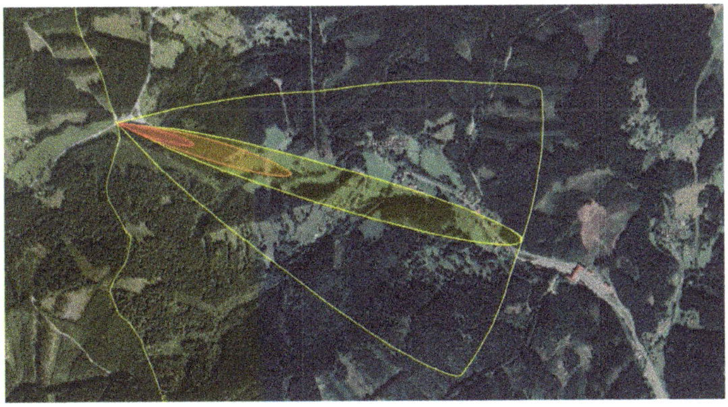

Figure 2. Graphical representation of the dispersion of a dangerous substance at the Makov border crossing [24,25].

There are no buildings in the endangered red zone with a concentration of 1100 ppm reaching a distance of 311 m. In this zone, there is only the driver of the tank from which the dangerous substance leaks and persons on the road near the place of the accident. As the road leading to the border crossing is in the shape of the letter "Y", vehicles can bypass the crash site and continue driving to safety. There is only 1 object in the orange zone with a concentration of 160 ppm reaching a distance of 775 m. This is the customs building itself. The village of Makov with 1710 inhabitants is located in a yellow zone, extending to a distance of 1.87 km from the site of leak with a concentration of 30 ppm. Ammonia gas is dispersed in the Kysuce Protected Landscape Area. No watercourse in the vicinity would be endangered by the leakage of a hazardous substance [26–28].

3.1.2. Leakage of a Dangerous Substance at the Border Crossing Makov-Velké Karlovice-Bíla–Bumbálka

The driver of a tanker carrying ammonia across the border crossing lost control of the vehicle in a sharp turn due to speeding. The vehicle lost stability and overturned due to centrifugal force and the action of the load. At the same time, there was a collision with a passing vehicle. A crack with a diameter of 25 cm formed in the tank, from which ammonia began to leak in the wind direction to the southeast, as demonstrated in Figure 3.

Figure 3. Graphical representation of the dispersion of a dangerous substance at the border crossing Makov-Velké Karlovice-Bíla-Bumbálka [24,25].

In the red zone with a concentration of 1100 ppm reaching a distance of 389 m, some people were involved in the accident. Toxic ammonia fumes with a concentration of more than 1100 ppm also hit a parked car located in the vicinity and two buildings. There is a high probability that there will be people in the parked car, especially truck drivers, who will be resting there. In the orange zone with a concentration of 160 ppm, there are no more affected objects. In the yellow zone with a concentration of 30 ppm extending up to a distance of 1866 m, there is a recreation centre, Bumbálka, with no permanent residents. Ammonia gas disperses in the Kysuce Protected Landscape Area. There is no watercourse in the area affected by the vapours of the dangerous substance [26–28].

3.1.3. Leakage of a Dangerous Substance at the Klokočov-Bíla Border Crossing

A driver leading a tanker carrying dangerous goods crashed into a car coming out of a side road while crossing a populated area of the border crossing. As a result of the collision, a crack with a diameter of 25 cm formed on the tank, from which ammonia gases began to disperse to the surroundings in a northeast direction, as depicted in Figure 4.

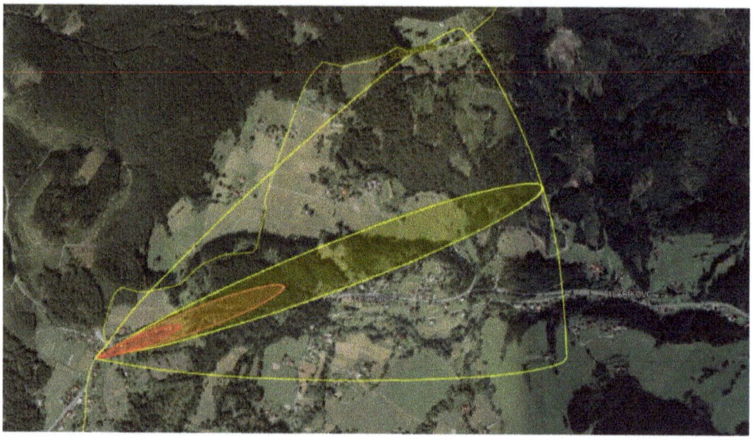

Figure 4. Graphical representation of the dispersion of a dangerous substance at the Klokočov-Bíla border crossing [24,25].

In the red zone with an ammonia concentration of 1100 ppm, which extends up to a distance of 389 m, there are participants in a traffic accident and 10 residential houses, where there is a high presumption that there are people in them. There are 12 objects in the orange zone with a concentration of 160 ppm, reaching a distance of 775 m. The settlement Kornica, which extends near the point of leakage of the dangerous substance, will be hit by ammonia fumes with a concentration of 30 ppm. Kornica is a part of the village, Klokočov, with 2287 inhabitants. The area is located in the Kysuce Protected Landscape Area [26,28].

3.1.4. Leakage of a Dangerous Substance at the Čadca-Milošová-Šance Border Crossing

A car and vehicle carrying ammonia collided at the border crossing. A crack with a diameter of 25 cm formed in the tank, from which vapours and gases escaped into the air, as demonstrated in Figure 5.

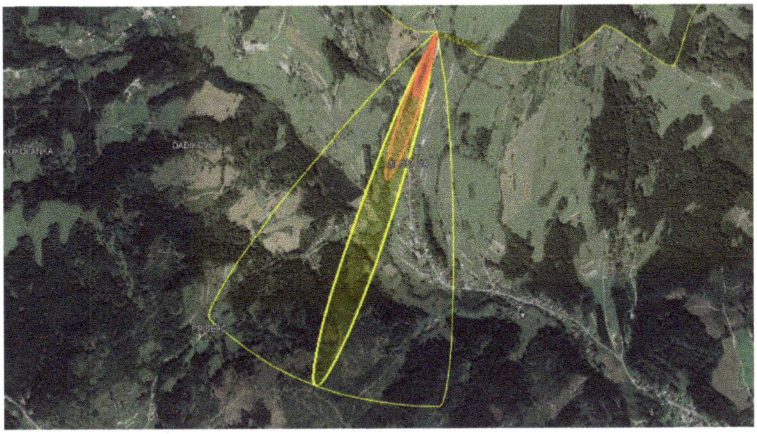

Figure 5. Graphical representation of the dispersion of a dangerous substance at the border crossing Čadca-Milošová-Šance [24,25].

In the red zone with a concentration of 1100 ppm, which extends at the farthest point up to a distance of 389 m, there is a tank driver and a car driver. Ammonia with a concentration of 1100 ppm progressed due to the wind towards the southwest, where no buildings were detected. The orange zone, in which the ammonia concentration is 160 ppm, extends to a distance of 778 m. There are 31 objects in it. These buildings are residential houses and, therefore, it is assumed that there will be people in them. In the yellow zone with a concentration of 30 ppm is located the village of Milošová, part of the town of Čadca, with 23,328 inhabitants. The area affected by the leakage of the dangerous substance is not located in the protected landscape area [26–28].

3.1.5. Leakage of a Dangerous Substance at the Svrčinovec-Mosty u Jablunkova Border Crossing

At the border of the Czech Republic and the Slovak Republic, a car and a tanker collided as a result of the car driver's intention to change direction at the crossroads while entering the oncoming lane. To prevent the collision, the tank driver lost control of the steering and the tank overturned onto its side. A crack with a diameter of 25 cm formed on the side of the tank, from which ammonia began to leak into the surroundings, as demonstrated in Figure 6.

The dangerous substance disperses to the southeast to human settlements and directly across the road crossing the border. Vapours with a concentration of 1100 ppm reached a distance of 389 m from the crash site. In the endangered red zone reach, there is a tank driver trapped in the vehicle, the participants in the traffic accident and two objects, one of them a restaurant. In the orange zone with a concentration of 160 ppm, there are 15 residential houses and buildings used for agricultural purposes. In the yellow zone with a concentration of 30 ppm reaching a distance of 2.02 km, there is the Purašovci settlement, part of the village of Svrčinovec, with 3370 inhabitants. There are also some industrial areas affected. The area is not located in a protected landscape area [26–28].

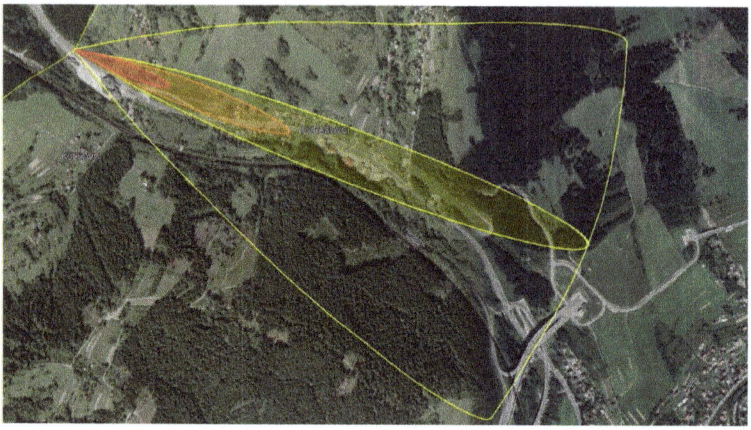

Figure 6. Graphical representation of the dispersion of a dangerous substance at the border crossing Svrčinovec-Mosty u Jablunkova [24,25].

3.1.6. Leakage of a Dangerous Substance at the Skalité-Zwardoň-Myto Border Crossing 2

A traffic accident occurred at the border crossing with the Republic of Poland called Skalité-Zwardoň-Myto 2 due to the carelessness of the tank driver. From the damaged tank, ammonia fumes dispersed to the southwest, as demonstrated in Figure 7.

Figure 7. Graphical representation of the dispersion of a dangerous substance at the Skalité-Zwardoň-Myto 2 border crossing [24,25].

In the red zone, there are ammonia vapours with a concentration of 1100 ppm. In the mentioned area, there is the driver of a tanker carrying a dangerous substance and vehicles passing on the road towards the road leading to the border crossing within a distance of 389 m. There are no objects in this zone. In the orange zone with a concentration of 160 ppm extending to a distance of 856 m, there are no objects that can be affected. In the yellow zone with a concentration of 30 ppm is the village of Skalité with 5247 inhabitants. There is no watercourse in the zone endangered by the leakage of dangerous substances. The area affected by the leakage of the hazardous substance is not located in the protected landscape area [26–28].

3.1.7. Leakage of a Dangerous Substance at the Skalité-Zwadon-Myto Border Crossing

At the border crossing located in the Horná Orava called Skalité, a car and a tanker carrying dangerous substances were involved in an accident. Here, a crack with a diameter of 25 cm formed and ammonia gases began to escape into the air and disperse to the southwest, as depicted in Figure 8.

Figure 8. Graphical representation of the dispersion of a dangerous substance at the Skalité-Zwardoň-Myto border crossing [24,25].

In the red zone, with a concentration of 1000 ppm reaching a distance of 389 m, there is a tank driver, a car driver, vehicles travelling on the road and two objects. In the orange zone with a concentration of 160 ppm, there are three residential houses belonging to the village of Skalité. In the yellow zone, reaching a distance of 2.02 km, there is the village of Skalité with 5247 inhabitants. There is no watercourse in the zone endangered by the leakage of dangerous substance. The area affected by the leakage of the dangerous substance is not located in the protected landscape area [26–28].

3.1.8. Leakage of a Dangerous Substance at the Novoť-Ujsoly Border Crossing

Due to the driver's inattention, the tank carrying the dangerous substance went in the opposite direction, where it collided with a car. This accident caused damage to the tank, from which ammonia began to leak southeast into the surroundings through a crack with a diameter of 25 cm, as depicted in Figure 9.

Vapours with a concentration of 1100 ppm hit the driver of the tank and the car. They dispersed further southeast to a distance of 311 m, where they hit two buildings, one of them a customs building. In the orange zone with a concentration of 160 ppm reaching a distance of 856 m, there are three other objects. The yellow zone with a concentration of 30 ppm affected a nearby wood of the village of Novoť with 3713 inhabitants. Ammonia vapours and gases escaped in the area belonging to the Horná Orava Protected Landscape Area. There is no water stream affected by the leaked hazardous substance. [26–28].

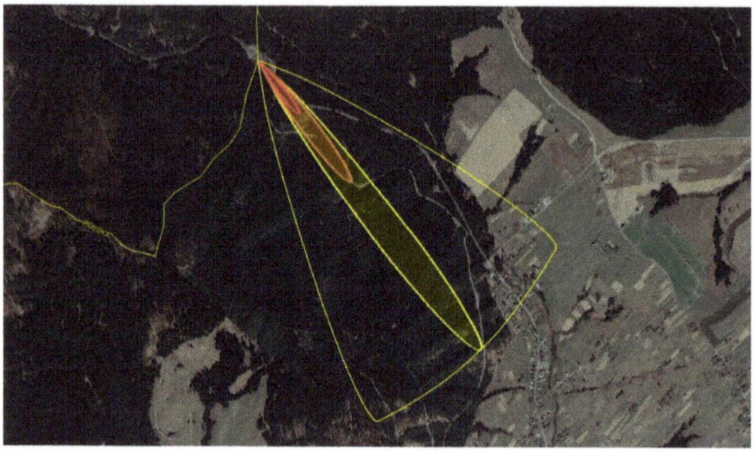

Figure 9. Graphical representation of the dispersion of a dangerous substance at the Novoť-Ujsoly border crossing [24,25].

3.1.9. Leakage of a Dangerous Substance at the Border Crossing Oravská Polhora-Korbielow

An accident involving a tanker carrying ammonia occurred at the Oravská Polhora border crossing. The cause was the driver's micro-sleep. A crack with a diameter of 25 cm formed on the vehicle's cistern, from which a dangerous substance began to evaporate into the air, as demonstrated in Figure 10. Ammonia gas dispersed in a southeast direction.

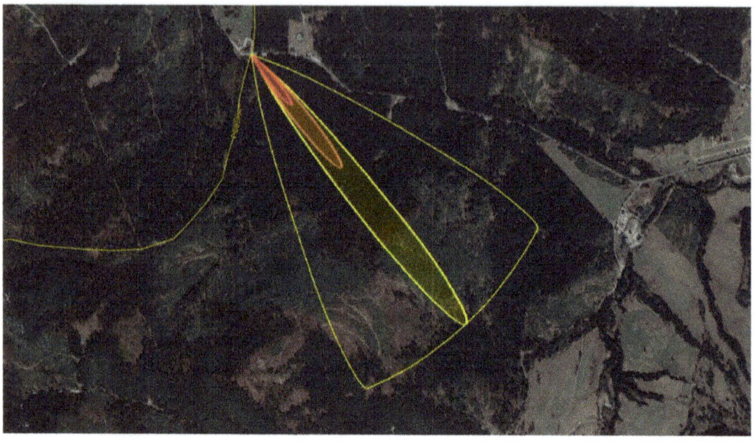

Figure 10. Graphical representation of the dispersion of a dangerous substance at the border crossing Oravská Polhora-Korbielow [24,25].

In the red zone with a concentration of 1100 ppm extending to a distance of 311 m, there is a tank driver trapped in the vehicle. There are no objects in the orange zone, with a concentration of 160 ppm reaching a distance of 856 km. Additionally, in the yellow zone with a concentration of 30 ppm, the area is inhabited. The zone endangered by the leakage of a hazardous substance is located in the Orava Protected Landscape Area. There is no water stream affected by the vapours of the leaked hazardous substance [26–28].

3.1.10. Leakage of a Dangerous Substance at the Bobrov-Wincerówka Border Crossing

The driver of a tanker transporting a dangerous substance through the Bobrov border crossing located in the Orava Protected Landscape Area lost control of the overturned vehicle due to a wet and slippery road. A sharp object pierced the body of the tank, in which a crack with a diameter of 25 cm formed and a dangerous substance began to escape from it into the surroundings in a southwest direction, as depicted in Figure 11.

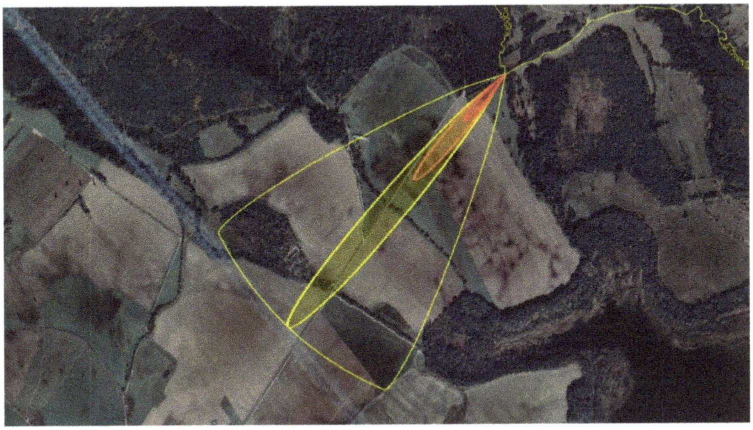

Figure 11. Graphical representation of the dispersion of a dangerous substance at the border crossing Bobrov-Wincerówka [24,25].

In the red zone with a concentration of 1100 ppm extending to a distance of 311 m, there is a tank driver and persons on the road leading through the border crossing. There are no objects in the orange zone with a concentration of 160 ppm reaching a distance of 856 m. This area is formed by an array. However, in this zone, the presence of persons in motor vehicles transported by road or towards the border crossing point is presumed. There are no buildings affected in the yellow zone with a concentration of 30 ppm. An ammonia concentration of 30 ppm will not affect any part of the Orava reservoir [26–28].

3.1.11. Leakage of a Dangerous Substance at the Trstená-Chyžné Border Crossing

The driver of a tanker transporting compressed ammonia across the border of the Slovak Republic and Poland incorrectly entered a lane. Subsequently, there was a collision with a car, causing damage to the tank. A crack with a diameter of 25 cm was formed on the body of the tank, from which ammonia began to leak into the surroundings in a southwest direction, as depicted in Figure 12.

In the zone with an ammonia concentration of 1100 ppm reaching a distance of 311 m, there is a tank driver and a vehicle on the road. The restaurant, which is expected to house a large number of people nearby, is also affected by fumes with a concentration of 1100 ppm, along with six objects. Ammonia fumes are blown to the southwest with a concentration of 160 ppm to a distance of 856 m, where there are no other buildings. The ammonia concentration of 30 ppm in the yellow area also affected the gas station on the road. The leakage of a dangerous substance occurred on the territory of the Horná Orava Protected Landscape Area [26–28].

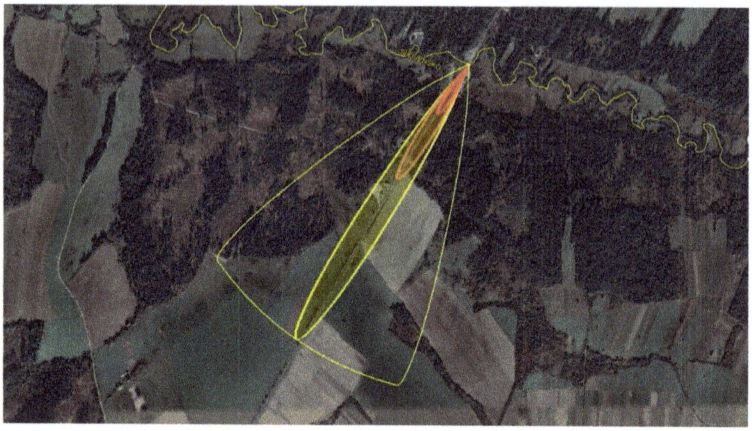

Figure 12. Graphical representation of the dispersion of a dangerous substance at the Trstená-Chyžné border crossing [24,25].

3.1.12. Leakage of a Dangerous Substance at the Suchá Hora Border Crossing

At the Suchá Hora border crossing, a traffic accident occurred due to a violation of road traffic rules. A vehicle carrying a dangerous substance overturned and struck a sharp object. A crack with a diameter of 25 cm formed in the tank, from which ammonia fumes began to disperse to the west, as demonstrated in Figure 13.

Figure 13. Graphical representation of the dispersion of a dangerous substance at the Suchá Hora border crossing [24,25].

In the red zone with a concentration of 1100 ppm reaching a distance of 311 m, there is a tank driver, a car driver and vehicles travelling on the road leading to the border crossing by the Slovak Republic. The concentration of 1100 ppm ammonia will affect the uninhabited area between the borders and the village Suchá Hora. In the orange zone with a concentration of 160 ppm reaching a distance of 778 m, there are 24 buildings with the character of residential houses and agricultural buildings. In the yellow zone, almost the whole centre of the village of Suchá Hora with 1455 inhabitants will be affected. If the wind directions were directed to the southeast in the simulation, the fumes of the dangerous substance would hit TANAP [26–28].

3.2. Map of the Endangered Area Due to the Leakage of a Dangerous Substance in the Border Areas of the Slovak Republic of the Žilina Region

Humans are the worst target in terms of the consequences of exposure to hazardous substances. If the action of dangerous substances causes damage to human health or causes death, it is an irreversible event [29]. During an examination of the possibility of leakage at individual border crossings, it was found that in almost all investigated situations, the most endangered are those who are on the road near the leakage of a dangerous substance in the wind direction and cannot leave it on their own; for example, if they are trapped in a vehicle as a result of an accident or unconsciousness. Few human dwellings located in the red zone at a concentration of 1100 ppm were endangered in a significant way at the Suchá Hora, Čadca-Milošová-Šance border crossings. For an overview, the results are summarised in Table 1. Since the area affected by the fumes of the escaped dangerous substance extends, the numbers of inhabitants in the yellow zone with a concentration of 30 ppm will be given only as an indication, according to the population of the affected municipality.

Table 1. Evaluation of the impacts of the leakage of a dangerous substance on the population and environment at individual border crossings [26–28,30].

Border Crossing	Number of Affected Objects		The Population	Presence of Water Flow	Affected PLA
	AEGL-1	AEGL-2	AEGL -3		
Makov (E442)	0	1	1710	No	Yes
Makov-Velké Karlovice–Bílá-Bumbálka	2	2	N/A	No	Yes
Klokočov-Bílá	10	12	2287	No	Yes
Čadca-Milošová-Šance	0	31	23,328	No	No
Svrčinovec-Mosty u Jablunkova	2	15	3370	No	No
Skalité-Zwardoň–Myto 2	0	0	5247	No	No
Skalité-Zwardoň-Myto	2	3	5247	No	No
Novoť–Ujsoly	2	3	3713	No	Yes
Oravská Polhora-Korbielów	0	0	N/A	No	Yes
Bobrov-Wincerówka	0	0	N/A	No	Yes
Trstená-Chyžné	6	6	N/A	No	Yes
Suchá hora	0	24	1455	No	No

Based on the calculations in the ALOHA programme, the concentration of ammonia in the red zone of AEGL-3 reached the value of 1100 ppm, which is 766.180 mg/m^3. This concentration can lead to death or life-threatening exposure to the human body for 60 min [28]. Even at concentrations above 200 mg/m^3, a person feels intolerable irritation, chest pain, stomach pain, headache, convulsions, respiratory and circulatory disorders, loss of orientation and renal arrest [31]. In the orange zone of AEGL-2 with a concentration of 160 ppm, calculated at 114,444 mg/m^3, exposure for 60 min, can cause severe long-term damage to health and put the driver in a state of inability to move independently without the help of another person [30]. This concentration causes respiratory irritation, tearing and sneezing in humans [32]. The yellow zone of AEGL-1 with a concentration of 30 ppm, which is 20.896 mg/m^3, is a zone in which a person experiences considerable discomfort and irritation during exposure, but it is only temporary [30].

People in enclosed spaces are more protected than people who are outdoors. However, the ALOHA CAMEO evaluation programme does not consider it. By closing the windows, switching off the ventilation and using improvised respiratory protective equipment, it is possible to reduce the effects of ammonia fumes on the human body [31].

In addition to adverse effects on the human body, ammonia vapours can pose a serious threat to nature. Ammonia can change the pH value of aquatic ecological systems and is very toxic to aquatic organisms [33]. The degree of toxicity also depends on the pH and temperature of the water. Aquatic plants are more resistant to ammonia than aquatic animals. The tolerance of individual ammonia concentrations depends on whether they are invertebrates or fish that are less resistant to its effects. Experiments have demonstrated that a lethal concentration of ammonia for fish is from 0.2 to 2.0 ppm. Even at ammonia concentrations above 0.1 ppm, permanent damage to the body organs of fish occurs even with short-term exposure [34]. Ammonia is not harmful to vegetation at low concentrations but is damaging at higher concentrations [35].

The leakage of a dangerous substance will also prevent safe crossing of the border crossing, often without the possibility of bypassing. There is a presumption that the vehicles will be standing in a convoy, and it will be almost impossible to enter the oncoming lane from this convoy and reach safety. It is also assumed that vehicle drivers and their crew will leave the vehicles in a panic and walk on the road, exposing them to the further danger of a collision with moving vehicles. Significant danger is posed by situations where the leakage of a dangerous substance triggers many other adverse events, such as an explosion or fire, as ammonia is a flammable gas which may explode when heated. The scenarios described in this paper would, therefore, be different if the explosion of ammonia vapours and gases were considered together with the occurrence of a fire. However, due to the scale of this work, only the effects of the leakage of the substance itself have been addressed. The results indicate that even a single tank containing ammonia NH3, under favourable meteorological conditions for the dispersion of hazardous substances, poses a significant threat to human, animal and environmental health.

To present the findings concerning the use of border crossings and their usability in the future for the transport of dangerous goods, it is necessary to create a map that will contain all border crossings that are suitable for ADR transport in terms of surface quality and road width. Subsequently, in Figure 14, the representation of the extent of the endangered area by the leakage of a hazardous substance in the direction of wind flow from 0° to 360° in the territory of the Slovak Republic was completed. The yellow semicircles on the map are marked as the zones endangered by the leakage of a dangerous substance in Slovakia, when the wind flows in all directions. Border crossings are marked with yellow pins.

Figure 14. Map of the hazardous area of leakage dangerous substances in the border areas of the Žilina region-representation of the extent of the endangered area by the leakage of a dangerous substance in the direction of wind flow from 0° to 360° [24,25].

Names of border crossing points to the left: Makov, Makov–Velké Karlovice–Bílá–Bumbálka, Klokočov–Bílá, Čadca–Milošová–Šance, Svrčinovec–Mosty u Jablunkova, Skalité–Zwardoň–Myto 2, Skalité–Zwardoň–Myto, Novoť–Ujsoly, Oravská Polhora–Korbielów, Bobrov–Wincerówka and Trstená–Chyžné, Suchá Hora.

4. Discussion

Crisis managers use their skills and abilities to create a safe environment. They pay close attention to the preparedness of the population to deal with various extraordinary events, disasters or emergencies [36]. When focusing on accidents involving the release of hazardous substances, crisis managers use available software that can simulate different accident scenarios. There are many unforeseen events and risks associated with the transport of dangerous goods by road, which can be applied in numerous potential accident scenarios [37]. There are several software programmes available, including a freely distributable and frequently used Aloha programme. Several authors used it to simulate the leakage of hazardous substances in road transport [38–42]. Therefore, it is suitable for crisis managers in public administration who can simulate various leaks with a possible impact on the environment. Such simulations enable taking appropriate measures and planning improvements to the crisis management system.

One of the primary measures is to change the dispersion of the dangerous substance, if possible. Consequently, we must consider that the danger of the dispersion of escaping substances is not limited to their physical-chemical properties but is also affected by various weather conditions [43–45]. Modelling with different input data in the ALOHA CAMEO software determined that the inversion has the most significant effect on the dispersion of the hazardous substance under meteorological conditions.

For the same input data, the presence of inversion could increase the range of the vapours and gases escaped up to three times than in the case of the simulation without the presence of inversion. Various smart technologies should also be used for weather prediction [46,47]. These technologies should be used by hydrometeorology centres and be provided to crisis managers in public administration. Crisis managers would review this information and inform carriers of dangerous substances in their territory, who could consider a plan for the transport of dangerous substances, if the situation allows. Assuming weather conditions in border areas, crisis managers could contact crisis managers in the neighbouring countries.

The expected number of people endangered by the leakage of hazardous substances in the Slovak Republic is approximately one million. In terms of the Žilina Region, the number is approx. 130,000 people. From the above calculations, it is economically unmanageable to provide special means of individual protection for all these persons. Therefore, every vulnerable person must have the correct habits to respond to emergencies caused by leaks of hazardous substances from vehicles [5]. Concentrations are relatively fast (for example, in the red zone, the increase to AEGL3 values ranges from 5 to 15 min), but due to the conditions these values will remain there for a short time, a maximum of 20 min, and then disperse. Thus, we can assume that the respective rescue response will not be sufficiently fast to warn the population. Therefore, informing the public about possible threats concerning the transport of dangerous substances in their surroundings is one of the most significant proposals in prevention, to prepare the population for self-help and self-protection.

However, the readiness of the population to deal with emergencies is still at an insufficient level, despite the guaranteed free access to information by Act no. 211/2000 Coll. on free access to information. The need to prepare the population for emergencies is underestimated by the general public and, in some cases, even facilitated [3]. The use of disaster management scenarios is part of crisis management. Some authors point to urban disaster resilience scenarios [48]. Other authors using the created scenario solution suggest appropriate assistance in dealing with the consequences of disasters [49–51]. Riddele et al. provide several scenarios to assist with more appropriate fund management [52]. Gibsson et al., in turn, examined eight case studies published in a single issue to find a method to combine and use them appropriately [53]. Case studies can be used as examples or as a research tool. We can use them to develop the prevention, preparedness and management of disasters and emergencies. It is necessary that the teaching process at universities addresses current issues and prepares experts [54–56]. It is possible to reduce their negative impact on the environment concerning the life and health of the population located in vulnerable zones [2].

There are many approaches to using scenarios and software in disaster management. However, we must adapt the scenarios to the actual environment. In an emergency or a disaster, several different areas can be affected at once. Therefore, the paper focused on specific events in border areas that could affect the Slovak environment. Crisis management has a specific hierarchy in Slovakia. The district office manages the emergency in its territory. If an extraordinary event affects several districts, it is managed by the district office in the seat of the region. To manage disasters properly, you should also prepare scenarios of potential emergencies and disasters that you share with other district authorities in your region. In the paper, we created the maps of possible extraordinary events in the border areas of the entire Žilina region that had never been created before, using scenarios and Aloha software.

The maps can correctly identify the procedures in the event of an accident with the leakage of hazardous substances at borders. Scenarios further develop the issue and discuss it. Similar maps should be created by other Slovak regions and, thus, a unified map for the whole territory would be created. It is also possible to supplement the impact of safe substances on the territory of neighbouring countries, which would reflect all wind directions. This step will allow for a better response and cooperation between countries in dealing with this type of crisis. The results can also form the basis for individual agreements between them, as well as some important documents such as the agreement on cross-border cooperation between neighbours or Regulation (EU) 2021/836 of the European Parliament and of the Council of 20 May 2021 amending Decision no. 1313/2013/EU on a Union Civil Protection Mechanism.

5. Conclusions

In crisis and adverse events, it is more advantageous to take preventive measures than to eliminate their consequences. However, in case of an emergency, it is necessary to take the most appropriate measures to reduce the negative impact on the environment. To support decision making, crisis managers can use simulation tools and create various scenarios to understand the nature of possible risks. In the paper, we identified the border crossings in the territory of the Žilina Region. In the analysis of border crossings at which ADR transport is or might be performed in the future and simulations of the leakage of dangerous substances on their territory, we pointed out that the differences in impacts on population and nature among individual border crossings are fundamentally different. The most significant impacts are the distance of human dwellings from the place of the leakage of hazardous substances, the presence of watercourses and the location of the border crossing in the PLA, or the vicinity of the national park. A significant factor influencing the extent of the threat is the presence of a high-traffic road near the border crossing. In the event of a leak of a dangerous substance, this road would have to be closed, thus paralyzing traffic. There are also people who may be at risk of escaping a dangerous substance on high-traffic roads.

We developed a map of the Žilina Region, in which the zones of the danger of the leakage of hazardous substances at individual border crossings are marked. A graphical representation of all border crossing points through which dangerous goods are or can be transported in the future, with an indication of the extent of the hazardous areas of the dangerous substance vapours and gases escaping, is a tool that can raise the awareness of the general public on the protection of the population against emergencies. Moreover, this map could also be the basis for tactical exercises of the integrated rescue system units. Additionally, such data can reveal what impact an accident near the border would have on the environment of the Slovak Republic. The solved problem creates space for further research with a focus on other regions, the impact abroad and also on the transport of hazardous substances by rail.

Author Contributions: Conceptualization, J.K. and M.P.; methodology, J.K., M.P. and S.S.; software, J.K., M.P. and S.S.; validation, K.H., J.K. and A.K.; formal analysis, J.K. and S.S.; investigation, J.K., M.P., K.H., V.S., S.M. and A.K.; resources, J.K., M.P., K.H., S.S., V.S., S.M. and A.K.; data curation, J.K., M.P. and S.S.; writing—original draft preparation, J.K., M.P. and S.S.; writing—review and editing, J.K., M.P., K.H., S.S., V.S., S.M. and A.K.; visualization, J.K. and S.S.; supervision, J.K., A.K.; project administration, J.K., M.P. and K.H.; funding acquisition, J.K. and K.H. All authors have read and agreed to the published version of the manuscript.

Funding: This article was created as a one of research project Grant Agency of the Ministry of Education, Science, Research and port of the Slovak Republic VEGA 1/0581/19 Stating the Social and Individual Risk of Employees and Public Resulting from Impacts of the Domino Effects Caused by Industrial Accidents, VEGA No. 1/0628/22 The Security Research in Municipalities with Emphasis on the Citizens' Quality of Life. Grant System of University of Zilina No. 12755/2021 Testing of selected means of individual protection and Grant System of University of Zilina-The project to support young researchers: Methods of Reducing their Possible Escalation and Research into the readiness of municipalities to deal with emergency events with an emphasis on the safety of the inhabitants.

Institutional Review Board Statement: Not applicable.

Informed Consent Statement: Not applicable.

Conflicts of Interest: The authors declare no conflict of interest.

References

1. Act of the National Council of the Slovak Republic No. 42/1994 Coll. on Civil Protection. Available online: https://www.slov-lex.sk/pravne-predpisy/SK/ZZ/1994/42/20220226 (accessed on 22 October 2021).
2. Makovicka Osvaldova, L.; Markova, I.; Jochim, S.; Bares, J. Experimental Study of Straw-Based Eco-Panel Using a Small Ignition Initiator. *Polymers* **2021**, *13*, 1344. [CrossRef] [PubMed]
3. Polorecka, M.; Kubas, J.; Danihelka, P.; Petrlova, K.; Repkova Stofkova, K.; Buganova, K. Use of Software on Modeling Hazardous Substance Release as a Support Tool for Crisis Management. *Sustainability* **2021**, *13*, 438. [CrossRef]
4. Ballay, M.; Sventeková, E.; Prievozník, P. System approach to assessing the criticality of key elements in road transport. In *Transport Means 2020: Proceedings of the 24th International Scientific Conference*; Kauno Technologijos Universitetas: Kaunas, Lithuania, 2020; pp. 702–706. ISSN 1822-296X.
5. Hassankhani, M.; Alidadi, M.; Sharifi, A.; Azhdari, A. Smart City and Crisis Management: Lessons for the COVID-19 Pandemic. *Int. J. Environ. Res. Public Health* **2021**, *18*, 7736. [CrossRef] [PubMed]
6. Paraskevas, A.; Quek, M. When Castro seized the Hilton: Risk and crisis management lessons from the past. *Tour. Manag.* **2019**, *70*, 419–429. [CrossRef]
7. Titko, M.; Ristvej, J. Assessing Importance of Disaster Preparedness Factors for Sustainable Disaster Risk Management: The Case of the Slovak Republic. *Sustainability* **2020**, *12*, 9121. [CrossRef]
8. Wilkinson, E.; Mysiak, J.; Oliveira, C.S.; Peters, K.; Surminski, S. Managing disaster risk. In *Science for Disaster Risk Management 2017: Knowing Better and Losing Less*; Poljanšek, K., Ferrer, M.F., De Groeve, T., Clark, I., Eds.; European Commission: Brussels, Belgium, 2017; pp. 442–515.
9. Buganova, K.; Luskova, M.; Hudakova, M. Early Warning Systems in Crisis Management. In *PT I Book SeriesLecture Notes in Management Science, Proceedings of the International Conference on Management Innovation and Business Innovation (ICMIBI 2013), Singapore, 21–22 April 2013*; ICMIBI: Singapore, 2013; Volume 15, pp. 218–223.
10. Sovcikova, L.; Coneva, I.; Otakar, J.M.; Sabo, J. Serious Industrial Accidents and Their Consequences. Žilina 2005. FŠI ŽU. Available online: http://fsi.uniza.sk/kpi/dokumenty/zph.pdf (accessed on 28 August 2021).
11. Zografos, K.G.; Vasilakis, G.M.; Giannouli, I.M. Methodological framework for developing decision support systems (DSS) for hazardous materials emergency response operations. *J. Hazard. Mater.* **2000**, *71*, 503–521. [CrossRef]
12. Dangerous Goods Transport Problems in the European Union and Poland. Available online: https://www.researchgate.net/publication/301718625_Dangerous_Goods_Transport_Problems_in_the_European_Union_and_Poland (accessed on 1 November 2021).
13. Kanj, H.; Flaus, J.M. An Agent-based framework for mitigating hazardous materials transport risk. In Proceedings of the IEEE International Conference on Evolving and Adaptive Intelligent Systems, Douai, France, 1–3 December 2015. [CrossRef]
14. Santella, N.; Steinberg, L.J. Accidental Releases of Hazardous Materials and Relevance to Terrorist Threats at Industrial Facilities. *J. Homel. Secur. Emerg. Manag.* **2011**, *8*, 53. [CrossRef]
15. Forrester, J.W. Industrial dynamics: A major breakthrough for decision makers. *Harv. Bus. Rev.* **1958**, *36*, 37–66.
16. Gordon, G. A general purpose systems simulation program. In Proceedings of the December, Eastern Joint Computer Conference: Computers-Key to Total Systems Control, Washington, DC, USA, 12–14 December 1961; ACM: New York, NY, USA, 1961; pp. 87–104.

17. Borshchev, A.; Filippov, A. From system dynamics and discrete event to practical agent based modeling: Reasons, techniques, tools. In Proceedings of the 22nd International Conference of the System Dynamics Society, Oxford, UK, 25–29 July 2004; Volume 22.
18. McKinsey & Company. The Future of Risk Management in the Digital Era. Available online: https://www.mckinsey.com/business-functions/risk-and-resilience/our-insights/the-future-of-risk-management-in-the-digital-era (accessed on 1 November 2021).
19. Tomasoni, A.M. Models and methods of risk assessment and control in dangerous goods transportation (DGT) systems, using innovative information and communication technologies. In *Chemical Sciences*; NNT: 2010ENMP1703; Pastel-00006223; École Nationale Supérieure des Mines de Paris: Paris, France; Università degli studi di Genova—Italie: Liguria, Italy, 2010. (In English)
20. Ministry of Interior of the Slovak Republic. Overview of the Most Frequently Transported Chemical Hazardous Substances [online]. MINV SR. Available online: https://www.minv.sk/swift_data/source/miestna_statna_sprava/ou_trnava/okr/informacie/Chemicke_nebezpecne_latky.pdf (accessed on 10 April 2021).
21. Buzalka, J.; Roth, R. *Analysis and Monitoring of Sources of Threat in Civil Protection*; Police Academy in Bratislava: Bratislava, Slovakia, 2006; ISBN 80-8054-372-0.
22. *Decree of the Ministry of the Interior of the Slovak Republic No. 533/2006 Coll*; On Details of the Protection of the General Public Against the Effects of Dangerous Substances; Ministry of the Interior of the Slovak Republic: Bratislava, Slovakia, 2006.
23. Daloš, A. *Hazardous Substances and Ecological Accidents*; FŠI ŽU: Žilina, Slovakia, 2003; ISBN 80-8070-056-7.
24. ALOHA Software Version 5.4.7. Available online: http://www.epa.gov/oem/content/cameo/aloha.htm (accessed on 20 April 2021).
25. Google Earth. Available online: https://earth.google.com/web/@0,0,0a,22251752.77375655d,35y,0h,0t,0r (accessed on 7 May 2021).
26. Google Maps. Available online: https://www.google.com/maps (accessed on 7 May 2021).
27. *Population by Sex—Municipalities (Per Year)*; Štatistický úrad SR: Bratislava, Slovakua; Available online: http://datacube.statistics.sk/#!/view/sk/VBD_DEM/om7101rr/v_om7101rr_00_00_00_sk (accessed on 20 April 2021).
28. Žilina Region—Protected Natural Areas. Electronic portal Guide to Slovakia. Available online: https://www.sprievodcaposlovensku.com/zilinsky-kraj/prirodne-chranene-uzemia-v-kraji/ (accessed on 10 April 2021).
29. Tureková, I.; Kuracina, R. *Žilina Region—Protected Natural Areas*; Slovenská technická univerzita v Bratislave: Bratislava, Slovakia, 2013; ISBN 978-80-227-4055-5.
30. Acute Exposure Guideline Levels (AEGLs) [Online]. Office of Response and Restoration. Available online: https://response.restoration.noaa.gov/oil-and-chemical-spills/chemical-spills/resources/acute-exposure-guideline-levels-aegls.html (accessed on 15 June 2021).
31. *Information for the Public on the Plan for the Protection of the Population in the Event of an Emergency on the Territory of the Capital City of the Slovak Republic, Bratislava*; Magistrát hlavného mesta SR Bratislavy: Bratislava, Slovakia. Available online: https://www.google.com/url?sa=t&rct=j&q=&esrc=s&source=web&cd=&ved=2ahUKEwjMrOeBhrPwAhWlDGMBHTFdAqAQFjAAegQIBRAD&url=https%3A%2F%2Fbratislava.blob.core.windows.net%2Fmedia%2FDefault%2FDokumenty%2FStr%25C3%25A1nky%2FChcem%2520vybavit%2FPlan%2520ochrany.pdf&usg=AOvVaw0MEOgtjl2KmdNg4Ar8ADHl (accessed on 18 July 2021).
32. Ministry of Interior of the Slovak Republic. Information to the Public. Available online: https://www.minv.sk/swift_data/source/miestna_statna_sprava/ou_partizanske/okr/pril%202%20Karta%20bezpecnostnych%20udajov_amoniak_OUPE.pdf (accessed on 6 May 2021).
33. Ministry of Interior of the Slovak Republic [Online] General Instructions for Ammonia Leakage. Available online: https://www.google.com/url?sa=t&rct=j&q=&esrc=s&source=web&cd=&ved=2ahUKEwj_mqaziLPwAhUjDGMBHSWPB4wQFjAAegQIBRAD&url=https%3A%2F%2Fwww.minv.sk%2Fswift_data%2Fsource%2Fverejna_sprava%2Fobu_bardejov%2Fslobodny_pristup_k_inf%2Focoakr%2FVseobecne_pokyny_obyvatelom_pre_pripad_uniku_AMONIAKU.pdf&usg=AOvVaw0c6dDyy1Q1pkONhjzDMBaW (accessed on 9 November 2021).
34. Oram, B. Ammonia in Groundwater, Runoff, Surface Water, Lakes and Streams. Water Research Center Dost. Available online: https://water-research.net/index.php/ammonia-in-groundwater-runoff-and-streams (accessed on 11 August 2021).
35. Safety Data Sheet: AMMONIA. Unipetrol. 2014. Available online: http://www.unipetrol.sk/SK/PonukaProduktov/Documents/Amoniak_SDS_SK_AnnII2010_8ES.pdf (accessed on 15 August 2021).
36. Ristvej, J.; Lacinak, M.; Ondrejka, R. On Smart City and Safe City Concepts. *Mob. Netw. Appl.* **2020**, *25*, 836–845. [CrossRef]
37. Ylipulli, J.; Luusua, A. Smart cities with a Nordic twist? Public sector digitalization in Finnish data-rich cities. *Telemat. Inform.* **2020**, *55*, 101457. [CrossRef]
38. Center for Chemical Process Safety of the American Institute of Chemical Engineers. *Guidelines for Use of Vapour Cloud Dispersion Models*; Wiley-AIChE: New York, NY, USA, 1996.
39. Wesierski, T.; Gałazkowski, R.; Zboina, J. Determining the danger zone during the chemical rescue action. *Przem. Chem.* **2012**, *91*, 582–584.
40. Węsierski, T.; Majder-Łopatka, M.; Walczak, A. Restrictions on the Use of Specimens Based on Magnesium and Titanium Nanocrystalline Oxides in Incidents Involving Uncontrolled Release of Hazardous Substances. *Appl. Sci.* **2020**, *10*, 267. [CrossRef]
41. Woodall, G.M.; Hoover, M.D.; Williams, R.; Benedict, K.; Harper, M.; Soo, J.-C.; Jarabek, A.M.; Stewart, M.J.; Brown, J.S.; Hulla, J.E.; et al. Interpreting Mobile and Handheld Air Sensor Readings in Relation to Air Quality Standards and Health Effect Reference Values: Tackling the Challenges. *Atmosphere* **2017**, *8*, 182. [CrossRef] [PubMed]
42. Bondzic, J.; Sremacki, M.; Popov, S.; Mihajlovic, I.; Vujic, B. Exposure to hazmat road accidents—Toxic release simulation and GIS-based assessment method. *J. Environ. Manag.* **2021**, *293*, 112941. [CrossRef]

43. Makovicka Osvaldova, L.; Kadlicova, P.; Rychly, J. Fire Characteristics of Selected Tropical Woods without and with Fire Retardant. *Coatings* **2020**, *10*, 527. [CrossRef]
44. Jabbari, M.; Atabi, F.; Ghorbani, R. Key airborne concentrations of chemicals for emergency response planning in HAZMAT road transportation- margin of safety or survival. *J. Loss Prev. Process Ind.* **2020**, *65*, 104139. [CrossRef]
45. Ristvej, J.; Sokolova, L.; Starackova, J.; Ondrejka, R.; Lacinak, M. Experiences with Implementation of Information Systems within Preparation to Deal with Crisis Situations in Terms of Crisis Management and Building Resilience in the Slovak Republic. In Proceedings of the 2017 International Carnahan Conference on Security Technology (ICCST), Madrid, Spain, 23–26 October 2017. [CrossRef]
46. Makovicka Osvaldova, L.; Petho, M. Occupational Safety and Health During Rescue Activities. *Procedia Manuf.* **2015**, *3*, 4287–4293. [CrossRef]
47. Essenfelder, A.H.; Larosa, F.; Mazzoli, P.; Bagli, S.; Broccoli, D.; Luzzi, V.; Mysiak, J.; Mercogliano, P.; dalla Valle, F. Smart Climate Hydropower Tool: A Machine-Learning Seasonal Forecasting Climate Service to Support Cost–Benefit Analysis of Reservoir Management. *Atmosphere* **2020**, *11*, 1305. [CrossRef]
48. Tsai, C.-C.; Hong, J.-S.; Chang, P.-L.; Chen, Y.-R.; Su, Y.-J.; Li, C.-H. Application of Bias Correction to Improve WRF Ensemble Wind Speed Forecast. *Atmosphere* **2021**, *12*, 1688. [CrossRef]
49. Yu, F.; Fan, B.; Li, X.Y. Improving emergency preparedness to cascading disasters: A case-driven risk ontology modelling. *J. Contingencies Crisis Manag.* **2020**, *28*, 194–214. [CrossRef]
50. Emadi, P.; Pasek, Z.J. Scenario Planning for Disaster Relief Networks. In Proceedings of the 9th Annual IEEE Global Humanitarian Technology Conference, Seattle, WA, USA, 17–20 October 2019; pp. 657–660.
51. Akhlaghi, V.E.; Campbell, A.M.; de Matta, R.E. Fuel distribution planning for di-sasters: Models and case study for Puerto Rico. *Transp. Res. Part E-Logist. Transp. Rev.* **2021**, *152*, 102403. [CrossRef]
52. Riddell, G.A.; van Delden, H.; Maier, H.R. Exploratory scenario analysis for di-saster risk reduction: Considering alternative pathways in disaster risk assessment. *Int. J. Disaster Risk Reduct.* **2019**, *39*, 101230. [CrossRef]
53. Gibson, T.D.; Tongwa, A.F.; Bari, S.; Chantry, G.; Gupta, M.; Molina, J.G.; Shresha, N.; Norton, J.; Parajuli, B.; Rahmawati, H.; et al. Drawing the case studies together: Synthesis of case studies and group discussions. *Disaster Prev. Manag.* **2019**, *28*, 93–105. [CrossRef]
54. Stofkova, J.; Stofkova, Z.; Stofko, S. Increasing the quality of the educational process through implementation of practice requirements in the university study program. In Proceedings of the 8th International Conference on Education and New Learning Technologies, Barcelona, Spain, 4–6 July 2016; pp. 1548–1554.
55. Stofko, S.; Stofkova, J. Survey of utilization of the e-learning in university education. In Proceedings of the 6th International Conference on Education and New Learning Technologies, Barcelona, Spain, 7–9 July 2014; pp. 1993–1999.
56. Grynszpan, D.; Murray, V.; Llosa, S. Value of Case Studies in Disaster As-sessment? *Prehospital Disaster Med.* **2011**, *26*, 202–205. [CrossRef] [PubMed]

atmosphere

Article

Coupling Effects of Sandstorm and Dust from Coal Bases on the Atmospheric Environment of Northwest China

Yun Liu [1], Tingning Zhao [1,*], Ruoshui Wang [1,*], Xianfeng Ai [1], Mengwei Wang [1], Tao Sun [2] and Qunou Jiang [1]

1 School of Soil and Water Conservation, Beijing Forestry University, Beijing 100083, China
2 Forest and Grass Investigation and Planning Institute, National Forestry and Grassland Administration, Beijing 100083, China
* Correspondence: zhtning@bjfu.edu.cn (T.Z.); wrsily_2002@163.com (R.W.)

Abstract: The coupling effects of sandstorm and dust from coal bases themselves can have a major impact on the atmospheric environment as well as on human health. The typical coal resource city of Wuhai in Inner Mongolia was selected in order to study these impacts during a severe sandstorm event in March 2021. Particulate matter (PM_1, $PM_{2.5}$ and PM_{10}) and total suspended particulate matter (TSP) samples were collected during the sandstorm event of 15–19 March 2021 and non-sandstorm weather (11–13 March 2021) and analyzed for their chemical composition. The concentrations of PM_1, $PM_{2.5}$, PM_{10} and TSP in Wuhai city during the sandstorm were 2.2, 2.6, 4.8 and 6.0 times higher than during non-sandstorm days, respectively. Trace metals concentrations in particles of different sizes generally increased during the sandstorm, while water-soluble ions decreased. Positive matrix fraction (PMF) results showed that the main sources of particles during both sandstorm and non-sandstorm days were industrial emissions, traffic emissions, combustion sources and dust. The proportion of industrial emissions and combustion sources increased compared with non-sandstorm days, while traffic emissions and dust decreased. The backward trajectory analysis results showed that airflows were mainly transported over short distances during non-sandstorm days, and high concentration contribution source areas were from southern Ningxia, southeast Gansu and western Shaanxi. The airflow was mainly transported over long distances during the sandstorm event, and high concentration contribution source areas were from northwestern Inner Mongolia, southern Russia, northern and southwestern Mongolia, and northern Xinjiang. A health risk analysis showed that the risk to human health during sandstorm days related to the chemical composition of particles was generally 1.2–13.1 times higher than during non-sandstorm days. Children were more susceptible to health risks, about 2–6.3 times more vulnerable than adults to the risks from heavy metals in the particles under both weather conditions.

Keywords: sandstorm; coupling effect; chemical components; pollutants source; health risk

Citation: Liu, Y.; Zhao, T.; Wang, R.; Ai, X.; Wang, M.; Sun, T.; Jiang, Q. Coupling Effects of Sandstorm and Dust from Coal Bases on the Atmospheric Environment of Northwest China. *Atmosphere* **2022**, *13*, 1629. https://doi.org/10.3390/atmos13101629

Academic Editor: Daniel Viúdez-Moreiras

Received: 20 September 2022
Accepted: 30 September 2022
Published: 6 October 2022

1. Introduction

Sandstorms usually occur when strong winds draw large amounts of sand and dust from exposed dry soil into the atmosphere. This can cause wind erosion, sand burial and land degradation. At the same time, the accumulation of large amounts of dust particles can reduce atmospheric visibility [1], damage crop growth and pollute the natural environment. In addition, dust containing various toxic chemicals, germs, etc. can penetrate the human mouth, nose, eyes and ears through protective layers of clothing, and cause respiratory diseases [2–4]. Sandstorms often occur in arid and semi-arid areas due to the sparse vegetation and loose soil on the underlying surface [5–8]. Although some studies have found an overall decreasing trend in the number and intensity of spring sandstorm events in East Asia [9], the study of several dust events in Asia in recent years has shown that the pollution caused by dust aerosols may still have significant impacts on the regional atmosphere, health conditions and socio-economic activities [10–14].

Northwest China has a dry climate and frequent high winds, making it a sandstorm-prone area. Sandstorms mainly occur in spring [15]. With the rapid socio-economic development and urbanization of the area, excessive land development and cultivation have resulted in a large amount of bare and loose soil, which provides a large source of dust to feed sandstorms. In addition, the region is rich in coal resources, with coal production accounting for 30% of the total coal production in China. The region is dotted with numerous open-pit and underground coal mines. Predatory and extensive mining of coal resources has resulted in surface subsidence, reduced vegetation cover and loose soil, the combined effects of which release dust into the atmosphere when sandstorms occur. Coupling effects can occur when sandstorms mix with dust from coal bases themselves, degrading the atmospheric environment in affected regions and even globally [16]. The coupling effects are mainly reflected in the chemical composition of local atmospheric particulate matter and its concentration and pollutant sources, together with the human health response. For instance, the occurrence of sandstorms can lead to an increase in the concentration of some dust tracers (i.e., Al, Ca, Fe, Ti) [17], while the amount of water-soluble ions varies between particle sizes [17,18]. However, the magnitude of the effects arising from this coupling is unknown. Previous studies mainly focused on sandstorms, including the detection of desert dust [5,19,20], analysis of dust composition and properties [7,21–23], and observation and model simulation of long-range dust transport [24–27]. Few studies have been conducted on the source apportionment of pollutants, the local atmospheric environment and the impact on human health under the coupling effects of sandstorm and local dust.

On 15 March 2021, affected by strong wind behind the frontal cyclone cloud system, a severe sandstorm occurred in Inner Mongolia, northern North China, western Northeast China and eastern Northwest China [28]. This was the most intense and widespread dusty weather process that China had experienced in the past 10 years. The optical properties of aerosols and the changes of weather conditions have been studied in detail during this sandstorm occurrence [16]. However, the influence of this sandstorm coupled with the dust of coal mining cities has not yet been studied. The understanding of this coupling effect can provide a reference basis for the prevention of air pollution in coal resource cities, and for the collaborative management of air pollution across regions. In this study, we compared and analyzed the effects of particulate matter on environment and human health during sandstorm and non-sandstorm weather from the perspective of particle composition. Particulate matter (PM_1, $PM_{2.5}$ and PM_{10}) and total suspended particulate matter (TSP) samples were collected and analyzed for their chemical composition during sandstorm and non-sandstorm days in Wuhai, Inner Mongolia, a typical coal resource city in the northwest arid desert region. The main aims of the study were (1) to compare the changes of particulate matter concentrations and elemental concentrations during sandstorm and non-sandstorm days; and (2) to analyze the difference of pollutant sources and risks to human health during sandstorm and non-sandstorm days using a Positive Matrix Fraction (PMF) model and backward trajectory analysis.

2. Data Sources and Methods

2.1. Study Area

Wuhai City (106.82° E, 39.67° N) is an emerging industrial city in the western part of the Inner Mongolia Autonomous Region, with coal reserves of more than 3 billion tons, mainly high-quality coking coal, accounting for about 60% of the coking coal reserves in the whole autonomous region. It is located in a warm temperate zone, which is a continental monsoon climate area. The temperature range is −28.9 to 40.2 °C; the annual average temperature is 10.1 °C. The average annual precipitation is 159.8 mm, and annual average evaporation is 3289 mm. Westerly and northwesterly winds are prevalent in the region, and high wind activity is mainly concentrated in spring (March to May). The wind direction was NNW during this study. Frequent wind and sandy weather there tends to cause more serious wind and sand disasters. The annual number of sandy days is greater than 80 d.

2.2. Satellite Data and Air Pollution Monitoring Data

Since a previous study has found similar spatial and temporal distribution patterns of PM_1, $PM_{2.5}$, PM_{10} and TSP in the region in spring [29], we presume that their transport and source trajectories are also comparable, so this study focuses on the spatio-temporal distribution and backward trajectory analysis of $PM_{2.5}$ and PM_{10}. Sandstorm transport data via satellite came from the Desertification Monitoring Center of the National Forestry and Grassland Administration of China. Hourly average $PM_{2.5}$ and PM_{10} data were obtained from China Air Quality Online Monitoring and Analysis Platform (https://www.aqistudy.cn/). Four mid-volume atmospheric integrated samplers (YR-6120 for PM_1 and $PM_{2.5}$, KC-6120 for PM_{10}, TH-150D II for TSP) were mounted 6.5 m above the ground surface in the urban area (106.83° E, 39.69° N) of Wuhai at the same time to collect PM_1, $PM_{2.5}$, PM_{10} and TSP, respectively, at a flow rate of 100 $L \cdot min^{-1}$. The sampling membrane used 90 mm diameter quartz filters. Air samples were collected during days of sandstorm (15–19 March 2021) and non-sandstorm (11–13 March 2021). Samples were collected twice per day in the daytime (08: 00 to 19:00 local time) and nighttime (19:00 to 08:00 local time), respectively. A total of 64 samples were collected. Before and after sampling, the filter was weighed after being equilibrated in a desiccator for 24 h. The weighed filter was placed in a ziplock bag and stored at 4 °C until the sample was analyzed.

2.3. Chemical Analysis

2.3.1. Metal Elements

One-quarter of the filter membrane sample was selected and placed in a PTFE ablation tube, and 5 mL hydrofluoric acid, 5 mL nitric acid, 5 mL perchloric acid were added. The sample was digested at 200 °C for 4 h until the sample was clarified without precipitation. The cap of the digestion tube was removed, and the process continued to drive the acid at 200 °C until about 1 mL of sample remained, and the volume was fixed to 47 mL. Al, Ba, Fe, Mg, Sr, Ti, Zn, Cr, Cu, Mn, Pb and Sn contents were measured by inductively coupled plasma mass spectrometry (ICP-MS, NexION 350X, PerkinElmer, Germany).

2.3.2. Water-Soluble Inorganic Ions

One-quarter of the filter membrane sample was carefully cut, weighed on an analytical balance, and placed in a sample bottle. Then 100.0 mL of deionized water was added to immerse the filter membrane, which was then covered and soaked for 30 min. It was then placed in an ultrasonic cleaner for ultrasonic extraction for 20 min and the extraction liquid was filtered off by a suction filter and poured into a sample tube. A suite of water-soluble ions (Ca^{2+}, Mg^{2+}, Na^+, K^+, F^-, Cl^-, $NO_3{}^-$ and $SO_4{}^{2-}$) was measured by ion chromatography (ICS-900, Thermo, USA).

2.4. Source Identification Methods

2.4.1. PMF

PMF version 3.0 model software as recommended by the U.S. Environmental Protection Agency (USEPA) requires no source list and does not limit the number of sources [30]. With these strengths, it has been widely applied in source apportionment analyses. In this experiment, samples of the various elements and inorganic water-soluble ions in PM_1, $PM_{2.5}$, PM_{10} and TSP were selected for analysis to determine the main sources of airborne particulate matter components from Wuhai city. The detailed principles of PMF model are described in Supplementary S1.

2.4.2. Backward Clustering Trajectory

MeteoInfo software, based on the HYSPLIT4 model developed by the Chinese Academy of Meteorological Sciences [31] was used to simulate the backward trajectory of air masses. In this study, Wuhai City (106.57° E, 38.19° N) was selected as the receptor point of the backward trajectory. The trajectory was extended backward for 72 h for the sandstorm periods (15–19 March 2021) and non-sandstorm periods (11–13 March 2021). The simulated altitude was 500 m, which represents the effect of pollutant transport above the inversion layer, and optimizes the simulated effect [32,33]. GDAS meteorological data with a spatial resolution of $1^\circ \times 1^\circ$ was provided by the U.S. National Center for Environmental Prediction. Weighted potential source contribution function (PSCF) and Weighted concentration weight trajectories (CWT) can further analyze the source of pollutants, and principles are shown in Supplementary S2 (including Table S1 and Table S2.).

2.5. Health Risk Assessment

Since atmospheric pollutants enter the human body mainly by respiration, this study mainly considers the health risks of metals via the respiratory route for adult males, adult females and children. The calculation of carcinogenic risk (CR) and non-carcinogenic risk (HQ) are shown in Supplementary S3.

3. Results

3.1. Sandstorm Transport Process Based on Remote Sensing Observations

A series of images observed by the Sunflower 8 satellite clearly depicted the transport process of this sandstorm event and the main impact areas in northwest China (Figure 1). The sandstorm event began during the night of 14–15 March 2021 (Figure 1a,b). The dust moved eastward and southward from Mongolia and affected most of northern China. The sandstorm continued from March 15 to 19 (Figure 1c–n), and the most severe dust impact was from March 15 to 16. The sandstorm had the widest impact on 15 March at 6 pm to 16 March at 6 pm (Figure 1c–f), involving eastern Gansu, southern Inner Mongolia, northern Shaanxi, north central Ningxia, northern Shanxi, Hebei and other areas. Wuhai was in the center of the sandstorm at this time and was strongly affected by the sandstorm. From 12:00 to 18:00 on 16 March (Figure 1g,h), the dust continued to affect the northwest including Wuhai and surrounding area, while the impact on the central region decreased. During the night of 17 March (Figure 1i), the dust mainly affected eastern Gansu, southern Inner Mongolia and other areas, while during the daytime (Figure 1j), the dust was mainly concentrated in eastern Gansu, western Inner Mongolia and other areas. Dust in Wuhai also dispersed in the daytime on 17 March. The impact of the dust on Inner Mongolia, Gansu, Ningxia and other areas weakened on 18 March (Figure 1k,l), and Wuhai was near the sandstorm but not affected. The impact of the sandstorm on the northwest region diminished significantly on 19 March (Figure 1m). However, dust seriously impacted Inner Mongolia and Ningxia in the evening (18:00) of the 19th (Figure 1n), while Wuhai was also in the center of the sandstorm. The sandstorm then moved eastward (Figure 1o) and dissipated significantly on the 21st (Figure 1p).

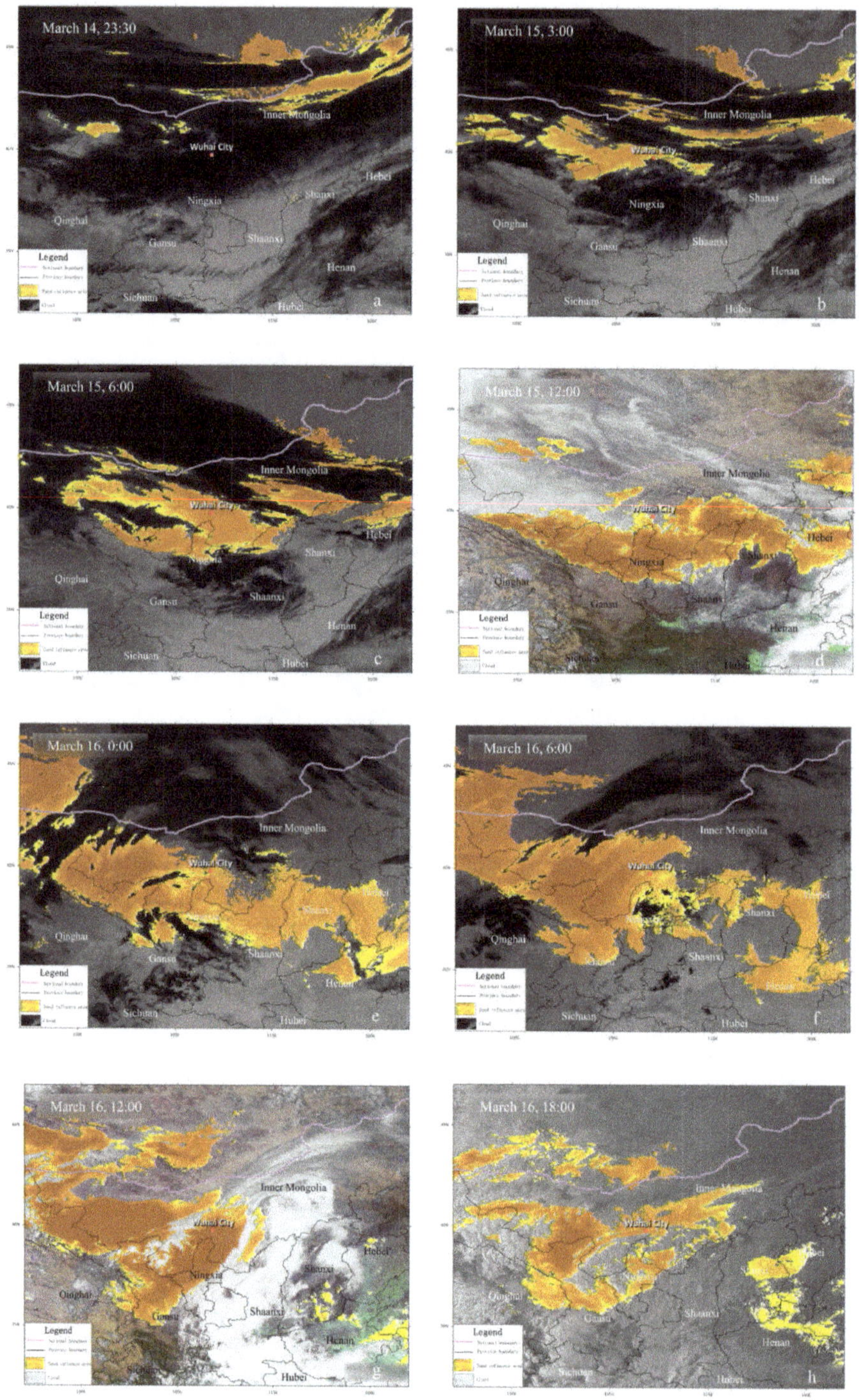

Figure 1. *Cont.*

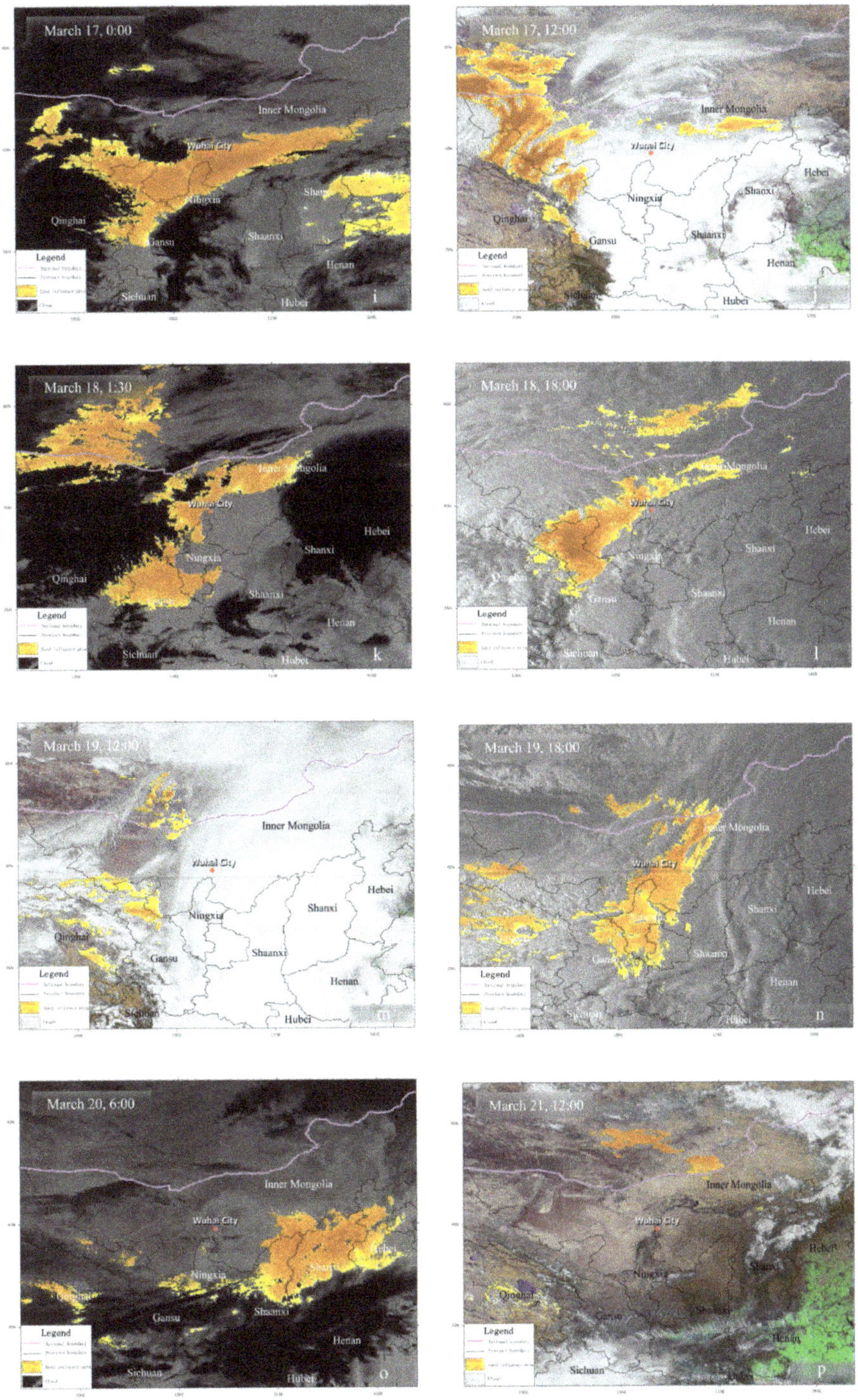

Figure 1. Sandstorm transport process based on remote sensing observations: (**a**,**b**) sandstorm occurred; (**c**–**n**) sandstorm continued; (**o**,**p**) sandstorm dissipation.

3.2. Particulate Matter Concentration Observation

3.2.1. Spatial Distribution

The spatial distribution of particulate matter concentration during the sandstorm (15–19 March) was mapped to study the intensity and extent of this sandstorm (Figure 2). The results showed that the spatial distribution patterns of $PM_{2.5}$ and PM_{10} concentrations were similar. The high concentration values were mainly distributed in the Inner Mongolia, Gansu and Ningxia areas. The PM concentrations in most cities in these areas were well above the Level II limit values set by the Standard for Daily Average Concentration Limit (DACL) of China (GB3095–2012) (75, 150 $\mu g\ m^{-3}$; 24-h average) [34]. The high concentration values on March 15 were distributed in north-central Gansu, northern Ningxia and west-central Inner Mongolia, where $PM_{2.5}$ and PM_{10} concentrations were above 400 $\mu g/m^3$ and 1000 $\mu g/m^3$, respectively. Correspondingly, $PM_{2.5}$ and PM_{10} concentrations in Wuhai also reached 451 $\mu g/m^3$ and 1450 $\mu g/m^3$, 6 and 10 times higher than DACL (China), respectively. The high concentration values region moved westward on March 16, mainly in east-central Gansu. Correspondingly, $PM_{2.5}$ and PM_{10} concentrations in Wuhai were 112 $\mu g/m^3$ and 790 $\mu g/m^3$, respectively, which also far exceeded DACL (China). PM concentration gradually decreased from 17–19 March. Areas of high particulate matter concentration were mainly distributed in east-central Gansu and western Inner Mongolia, where $PM_{2.5}$ and PM_{10} concentrations reached more than 200 $\mu g/m^3$ and 1000 $\mu g/m^3$, respectively. $PM_{2.5}$ and PM_{10} concentrations in Wuhai on 17–18 March remained at around 100 $\mu g/m^3$ and 700 $\mu g/m^3$, respectively. However, the pollution worsened on March 19, with $PM_{2.5}$ and PM_{10} concentrations in Wuhai reaching 160 $\mu g/m^3$ and 1,074 $\mu g/m^3$, respectively, two and seven times higher than DACL (China).

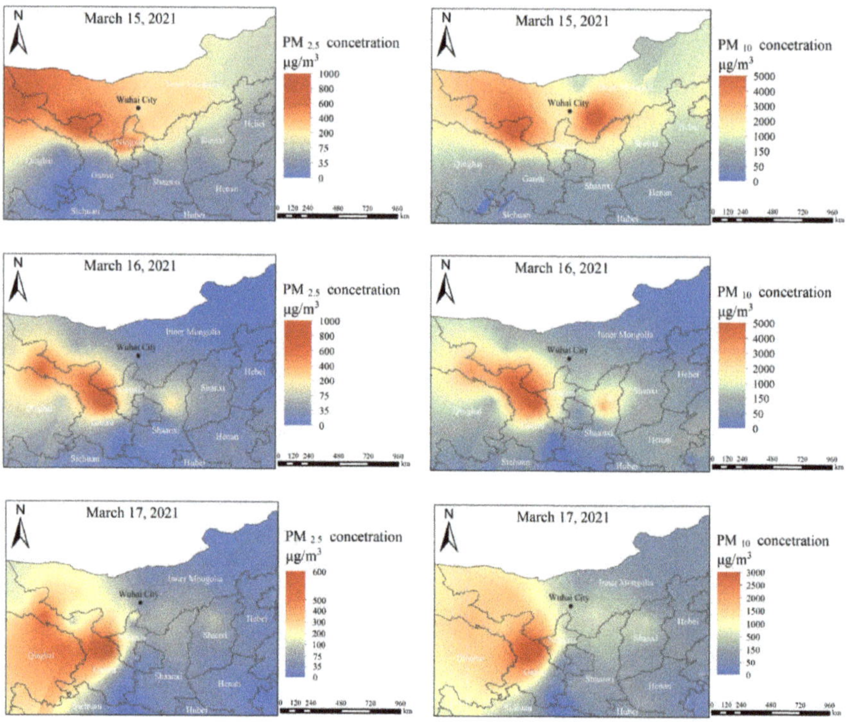

Figure 2. *Cont.*

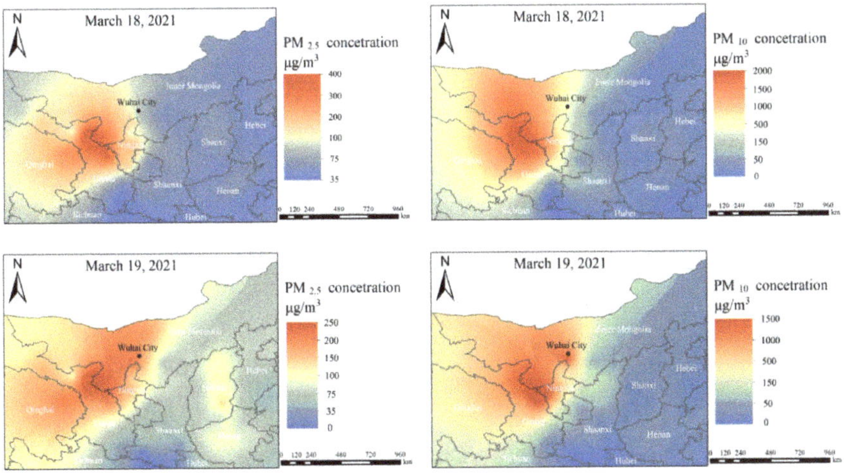

Figure 2. Spatial distribution of $PM_{2.5}$ and PM_{10} concentrations during the sandstorm.

3.2.2. Temporal Variation

The daily changes in PM concentrations before and during the sandstorm event in Wuhai (Figure 3a) showed that $PM_{2.5}$ and PM_{10} concentrations were less than the Level II limit values (75μg/m^3 and 150μg/m^3), with high temperature and humidity and low wind speed before the occurrence of the sandstorm (Figure 3b). The temperature and humidity decreased during the most severe hours of the sandstorm (15–16 March), while the wind speed increased. $PM_{2.5}$ concentration surged to 701μg/m^3 at 2:00 on March 15 and then continued to increase to 1585 μg/m^3 at 3:00. The following day, there was a continuous downward trend until the concentration dropped to 183 μg/m^3 at 15:00. $PM_{2.5}$ concentration remained between 100–300 μg/m^3 from 16:00 on the 15th to 23:00 the following day. PM_{10} concentration spiked to 3298 μg/m^3 at 2:00 on March 15 and remained at this level until 7:00; then decreased from 8:00 on the 15th to 6:00 on the 16th, but was still at a very high level between 1020–2528 μg/m^3. Then, the concentration decreased to less than 1000 μg/m^3 from 7:00 to 23:00 on the 16th, but the lowest concentration still reached 394 μg/m^3. Humidity increased and wind speed decreased on March 17 and 18. $PM_{2.5}$ and PM_{10} concentrations remained at low levels, below 200 μg/m^3 and 1000 μg/m^3,respectively during this period. $PM_{2.5}$ and PM_{10} concentrations surged again to 545 μg/m^3 and 3444 μg/m^3,respectively, at 10:00 on 19 March, with high concentrations (97 μg/m^3–545 μg/m^3, 566 μg/m^3–2602 μg/m^3) lasting until 23:00, during which time the temperature and humidity dropped while the wind speed increased. Air quality improved significantly from March 20, with both $PM_{2.5}$ and PM_{10} concentrations dropping significantly to below the Level II limit, and even below the Level I limit (35 μg/m^3, 50 μg/m^3) at certain times.

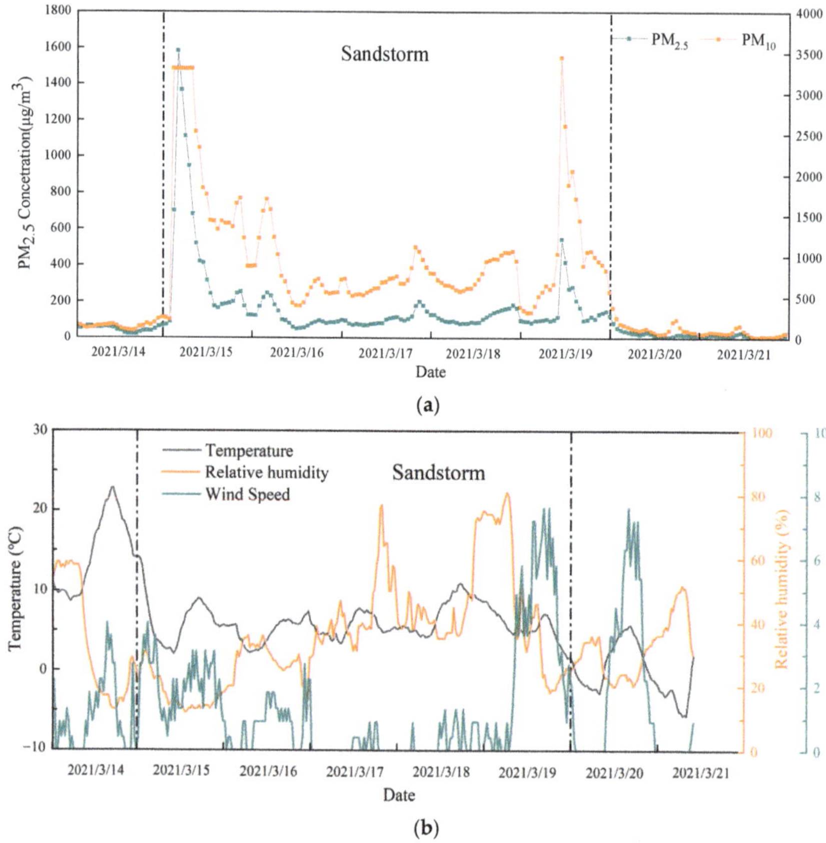

Figure 3. (**a**) Daily change of $PM_{2.5}$ and PM_{10} concentrations in Wuhai during the sandstorm; (**b**) changes of meteorological conditions in Wuhai during the sandstorm.

3.2.3. Chemical Composition of Particulate Matter

The content of each chemical component increased with increasing particle size (Supplementary Table S3). Compared with non-sandstorm days, trace metal concentrations in particles of different particle sizes generally increased on sandstorm days, and the patterns were similar for different trace metals. Al and Zn were the most abundant trace metals. Moderate concentrations of Ba, Fe, Sr and Mg were found, followed by Ti, Cr, Cu, Pb, Mn and Sn. Water-soluble ion concentrations in particles generally decreased on sandstorm days. Na^+ content in different particle sizes differed significantly, with concentrations of only 1.39 μg/m^3 in PM_1 and up to 70.88 μg/m^3 in TSP. Meanwhile, the concentrations of $SO_4{}^{2-}$ in different particle sizes, Ca^{2+} and Mg^{2+} in PM_{10} and Ca^{2+}, K^+, Mg^{2+} and Cl^- in TSP during the sandstorm were also higher than on non-sandstorm days. The patterns of the content of water-soluble ions were also similar during sandstorm and non-sandstorm. Ca^{2+} was the most abundant water-soluble ion, followed by Na^+, $SO_4{}^{2-}$, K^+ and Mg^{2+}. F^-, Cl^- and $NO_3{}^-$ were the least abundant water-soluble ions.

3.3. Analysis of Pollutants Sources

3.3.1. PMF

In this study, we used a PMF model combined with an analysis of emission sources of different chemical components in particulate matter to apportion the pollution sources of eleven metal elements and seven water-soluble ions, during non-sandstorm (Figure 4a) and sandstorm days (Figure 4b). Four main sources were identified for the metals and water-soluble ions for PM_1, $PM_{2.5}$, PM_{10} and TSP, combined with analysis of emission sources of different chemical components in particulate matter (Supplementary Table S4). The characteristic elements of each source of PM_1, $PM_{2.5}$, PM_{10} and TSP during both sandstorm and non-sandstorm days are shown in Supplementary Table S5 and Supplementary Figure S1. During non-sandstorm days, the main sources of pollutants for PM_1 were from industrial emissions (35.2%), dust and coal combustion (27.7%), traffic emissions (21.8%) and biomass burning (15.3%). Similar to PM_1, the largest sources of $PM_{2.5}$ and PM_{10} were industrial emissions (content 33.6% and 42.7%, respectively) with the other portion came from dust (content 27.9% and 20.5%, respectively), traffic emissions (content 21.5% and 24.3%, respectively) and combustion sources (content 17.0% and 12.5%, respectively). Traffic emissions (70.1%) was the main source of TSP, followed by industrial emissions (26.2%), dust (3.4%) and combustion sources (0.3%).

The sources of pollutants during sandstorm days were essentially similar to those during non-sandstorm days. The proportion of industrial emissions and combustion sources increased compared with non-sandstorm days, with combustion sources including coal combustion, biomass burning and vehicle exhaust emissions, while traffic emissions and dust from roads and construction decreased. During sandstorm days, the main sources of pollutants for PM_1 were from combustion sources (50.5%), industrial emissions (37.2%), dust (8.0%) and traffic emissions (4.3%). Both $PM_{2.5}$ and TSP were mainly from industrial emissions and biomass burning (content 40.1% and 87.4%, respectively), traffic emissions (content 8.4% and 3.2%, respectively) and vehicle exhaust (content 20.5% and 0.8%, respectively). The other portion of $PM_{2.5}$ came from combustion sources (31.0%), while TSP came from dust and coal combustion (8.6%). PM_{10} was mainly from industrial emissions (53.2%), vehicle exhaust (24.1%), dust and traffic emissions (14.1%) and coal combustion (8.6%).

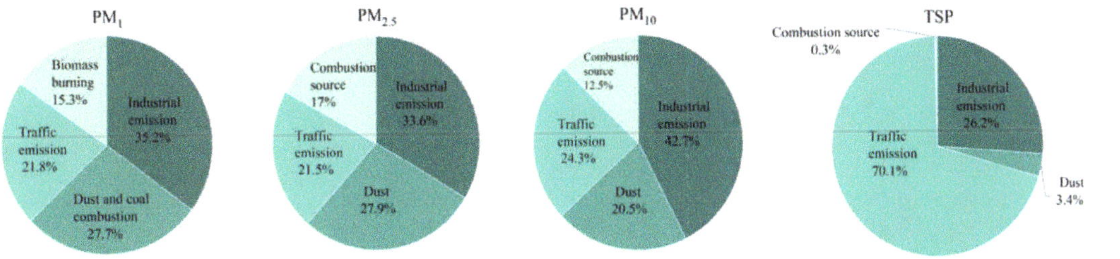

(a)

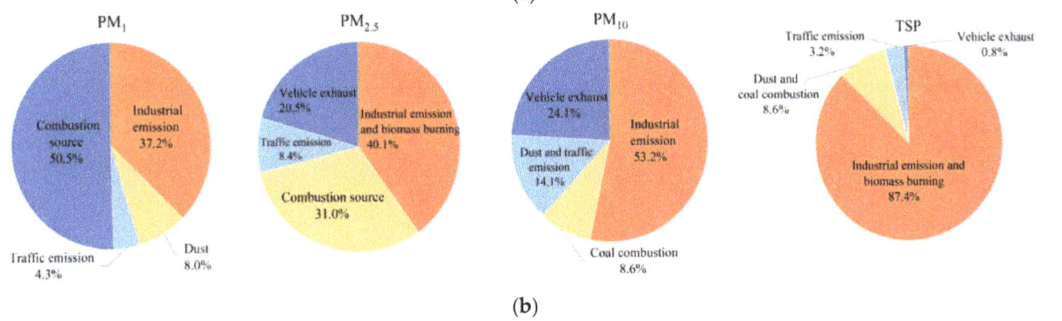

(b)

Figure 4. Source apportionment of PM_1, $PM_{2.5}$, PM_{10} and TSP during the (a) non-sandstorm and (b) sandstorm.

3.3.2. Analysis of Dust Transport Paths Based on MeteoInfo Trajectory Model

To further investigate the airflow transport characteristics in Wuhai during non-sandstorm and sandstorm days, 72-h backward airflow trajectories arriving at Wuhai were clustered and analyzed using MeteoInfo software (Figure 5). The results showed that two clusters during non-sandstorm periods were dominated by short-distance transport, from southern Ningxia (52.78%) and south-central Inner Mongolia (34.72%). Another cluster (12.50%) came from northern Gansu, passing through the southwest edge of Inner Mongolia to reach Wuhai with long-distance transport. However, two clusters were dominated by long-distance transport and were characterized by rapidly moving speeds during the sandstorm. One of the clusters (30.83%) came from the south of Russia and reached Wuhai via central Mongolia and northern Inner Mongolia. Another cluster (17.50%) came from the northern border of Xinjiang, passing through north and central Xinjiang in a northwesterly direction, crossing a small part of southwestern Mongolia and northwestern Inner Mongolia to reach Wuhai. The highest percentage (51.67%) of the cluster from south-central Inner Mongolia was dominated by short-distance transport, and the cluster moved slowly, passing through Ertokqi and Hainan district of Wuhai city.

By assigning the corresponding $PM_{2.5}$ and PM_{10} hourly concentration data to each airflow, the potential pollution source areas in Wuhai were further analyzed (Figure 6). PSCF results showed that the potential pollution source areas for both $PM_{2.5}$ and PM_{10} during non-sandstorms were mainly from southern Ningxia, southeastern Gansu and western Shaanxi and the contribution values decreased gradually to the north. However, the potential pollution sources of $PM_{2.5}$ and PM_{10} expanded and the sources were located further during sandstorm days, but their distributions were basically similar. The potential pollution source areas in Wuhai were distributed along three clusters and the contribution values decreased gradually to both sides of the airflow, mainly in southern Russia, north-central and south-western Mongolia, northern Xinjiang, west-central and north-western Inner Mongolia. Northern Shaanxi also contributed to particulate matter in Wuhai.

Compared to the PSCF method, the CWT method is more intuitive to identify potential source areas with high concentration contributions (Figure 7). The results showed that the high CWT areas of $PM_{2.5}$ and PM_{10} (>120 $\mu g \cdot m^{-3}$, 600 $\mu g \cdot m^{-3}$, respectively) during non-sandstorm days were mainly in southern Ningxia, southeast Gansu and western Shaanxi. The concentration contribution values of particulate matter during sandstorm days were all much higher. The high CWT area (>240 $\mu g \cdot m^{-3}$) of $PM_{2.5}$ was mainly in northwestern Inner Mongolia. The contributions of southern Russia, northern and southwestern Mongolia, and northern Xinjiang to the $PM_{2.5}$ mass concentration in Wuhai were also in the range of 160–240 $\mu g \cdot m^{-3}$. All regions contributed more than 200 $\mu g \cdot m^{-3}$ to the PM_{10} concentration. Among them, the contribution of northwestern Inner Mongolia was as high as 1800 $\mu g \cdot m^{-3}$. The contribution to PM_{10} mass concentrations of southern and southwestern Russia, northern Mongolia and parts of northern Inner Mongolia also ranged from 1200–1800 $\mu g \cdot m^{-3}$.

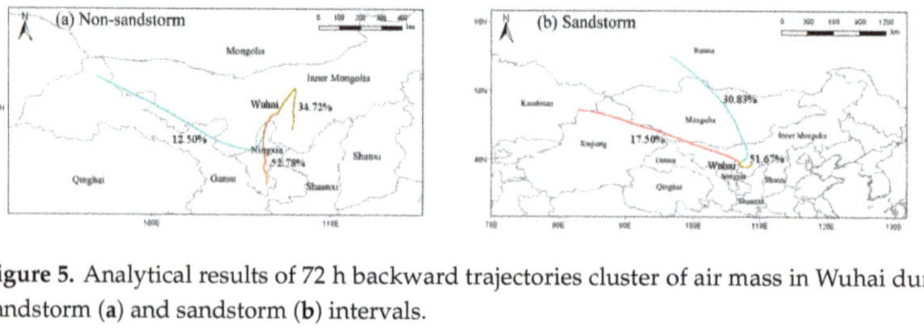

Figure 5. Analytical results of 72 h backward trajectories cluster of air mass in Wuhai during non-sandstorm (**a**) and sandstorm (**b**) intervals.

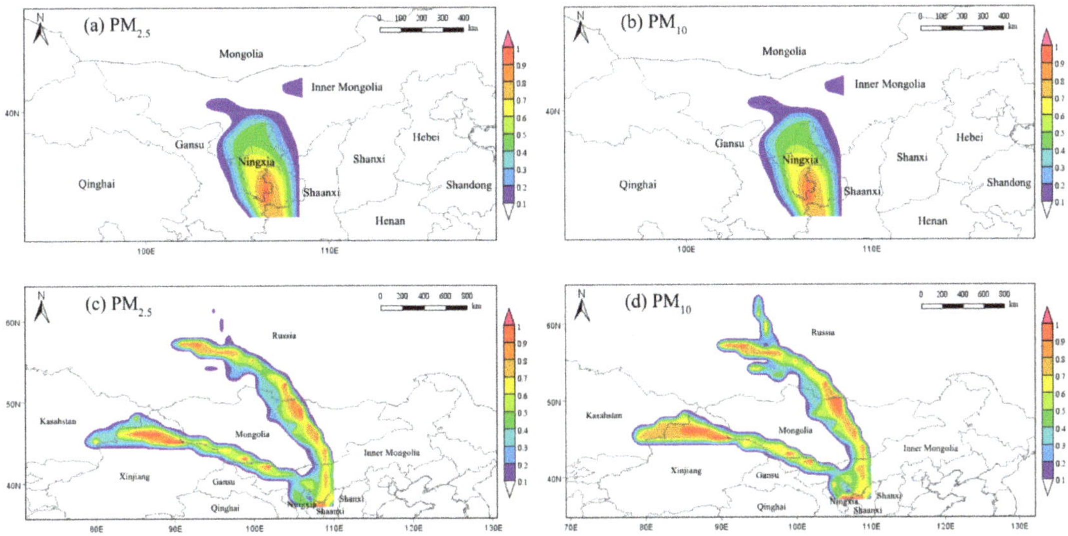

Figure 6. Weighted potential source contribution function (PSCF) for $PM_{2.5}$ and PM_{10} during non-sandstorm (**a**,**b**) and sandstorm (**c**,**d**) intervals.

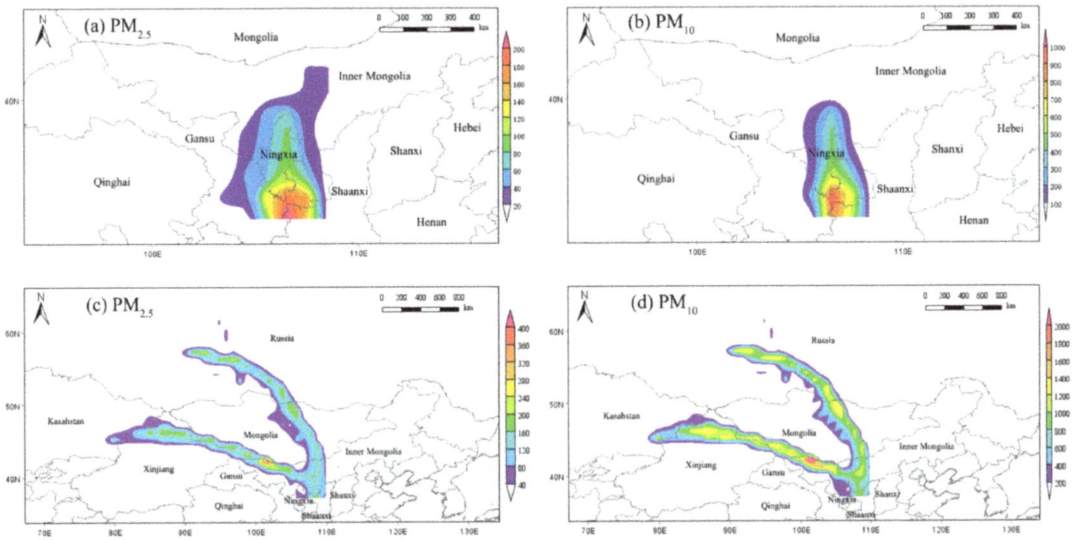

Figure 7. Weighted concentration weight trajectories (CWT) for $PM_{2.5}$ and PM_{10} during non-sandstorm (**a**,**b**) and sandstorm (**c**,**d**) intervals.

3.4. Health Risk Assessment

The assessment of health risk due to heavy metals in particles of different particle sizes entering the human body through the respiratory route is shown in Table 1. The concentrations of the heavy metal (Cr) in atmospheric particles all exceeded 10^{-4}, indicating a high risk of carcinogenicity in humans whether there is a sandstorm or not. Our study found that the human carcinogenic risk of heavy metals was similar during sandstorm and non-sandstorm days. The magnitude of the carcinogenic effect of Cr in different particle sizes on the human body showed that the coarse particles (PM_{10} and TSP) were larger than the fine particles (PM_1 and $PM_{2.5}$). Moreover, the risk coefficient showed that children were at greater risk than adult males and adult females, in that order. The carcinogenic risk of Cr in PM_1, $PM_{2.5}$, PM_{10} and TSP for children during sandstorm and non-sandstorm days was about two times higher than for adult males and adult females. The carcinogenic risk of Cr in PM_1, PM_{10} and TSP for all populations increased during sandstorm days. The risk was 1.2, 1.5 and 2.3 times higher than on non-sandstorm days for PM_1, PM_{10} and TSP, respectively, while Cr in $PM_{2.5}$ was less than on non-sandstorm days.

Table 1. Health risk assessment of toxic metal elements in atmospheric particulates to different populations.

		Element	CR			HQ		
			Adult Male	Adult Women	Children	Adult Male	Adult Women	Children
PM_1	Non-sandstorm	Cr	6.57×10^{-3}	6.25×10^{-3}	1.53×10^{-2}	2.31	2.06	12.98
		Cu	ne	ne	ne	3.19×10^{-5}	2.84×10^{-5}	1.79×10^{-4}
		Mn	ne	ne	ne	1.72	1.53	9.66
		Pb	ne	ne	ne	1.82×10^{-2}	1.62×10^{-2}	1.02×10^{-1}
		Zn	ne	ne	ne	7.69×10^{-3}	6.84×10^{-3}	4.3×10^{-2}
	Sandstorm	Cr	8.17×10^{-3}	7.77×10^{-3}	1.90×10^{-2}	2.80	2.56	16.14
		Cu	ne	ne	ne	2.07×10^{-5}	1.84×10^{-5}	1.16×10^{-4}
		Mn	ne	ne	ne	3.25	2.89	18.22
		Pb	ne	ne	ne	3.64×10^{-5}	3.25×10^{-5}	2.04×10^{-4}
		Zn	ne	ne	ne	7.67×10^{-3}	6.83×10^{-3}	4.3×10^{-2}
$PM_{2.5}$	Non-sandstorm	Cr	8.77×10^{-3}	8.35×10^{-3}	2.04×10^{-2}	3.09	2.75	17.33
		Cu	ne	ne	ne	5.94×10^{-5}	5.29×10^{-5}	3.33×10^{-4}
		Mn	ne	ne	ne	0.58	1.41	8.87
		Pb	ne	ne	ne	1.56×10^{-2}	1.39×10^{-2}	8.78×10^{-2}
		Zn	ne	ne	ne	7.87×10^{-3}	7.01×10^{-3}	4.41×10^{-2}
	Sandstorm	Cr	5.91×10^{-3}	5.63×10^{-3}	1.37×10^{-2}	2.08	1.86	11.69
		Cu	ne	ne	ne	4.84×10^{-5}	4.31×10^{-5}	2.71×10^{-4}
		Mn	ne	ne	ne	5.89	5.24	33.02
		Pb	ne	ne	ne	7.44×10^{-3}	6.62×10^{-4}	4.17×10^{-4}
		Zn	ne	ne	ne	6.2×10^{-3}	5.52×10^{-3}	3.48×10^{-2}
PM_{10}	Non-sandstorm	Cr	6.08×10^{-3}	5.79×10^{-3}	1.41×10^{-2}	2.14	1.91	12.02
		Cu	ne	ne	ne	6.13×10^{-5}	5.45×10^{-5}	3.44×10^{-4}
		Mn	ne	ne	ne	1.29	1.15	7.25
		Pb	ne	ne	ne	8.95×10^{-3}	7.97×10^{-3}	5.02×10^{-2}
		Zn	ne	ne	ne	7.75×10^{-3}	6.90×10^{-3}	4.34×10^{-2}
	Sandstorm	Cr	9.12×10^{-3}	8.68×10^{-3}	2.12×10^{-2}	3.21	2.86	18.03
		Cu	ne	ne	ne	9.92×10^{-5}	8.83×10^{-5}	5.57×10^{-4}
		Mn	ne	ne	ne	12.80	11.37	71.62
		Pb	ne	ne	ne	3.24×10^{-3}	2.89×10^{-3}	1.82×10^{-2}
		Zn	ne	ne	ne	6.90×10^{-3}	6.14×10^{-3}	3.87×10^{-2}
TSP	Non-sandstorm	Cr	4.61×10^{-3}	4.39×10^{-3}	1.07×10^{-2}	1.62	1.45	9.11
		Cu	ne	ne	ne	2.54×10^{-5}	2.27×10^{-5}	1.43×10^{-4}
		Mn	ne	ne	ne	1.23	1.10	6.90
		Pb	ne	ne	ne	6.68×10^{-3}	5.95×10^{-3}	3.75×10^{-2}
		Zn	ne	ne	ne	7.82×10^{-3}	6.96×10^{-3}	4.38×10^{-2}
	Sandstorm	Cr	1.05×10^{-2}	1.00×10^{-2}	2.45×10^{-2}	3.71	3.31	20.83
		Cu	ne	ne	ne	9.14×10^{-5}	8.13×10^{-5}	5.13×10^{-4}
		Mn	ne	ne	ne	16.08	14.32	90.20
		Pb	ne	ne	ne	6.39×10^{-3}	5.69×10^{-3}	3.59×10^{-2}
		Zn	ne	ne	ne	9.17×10^{-3}	8.17×10^{-3}	5.14×10^{-2}

CR: carcinogenic risk; HQ: non-carcinogenic risk; ne: no effect. CR $< 1 \times 10^{-6}$, no carcinogenic risk; $1 \times 10^{-6} <$ CR $< 1 \times 10^{-4}$, acceptable range; CR $> 1 \times 10^{-4}$, high risk of causing cancer. HQ ≥ 1, a possible non-cancer risk, with higher values increasing the risk.

Of all the heavy metal elements (Cr, Cu, Mn, Pb and Zn), only Cr and Mn pose a non-carcinogenic risk to humans. The effects of these elements on human non-carcinogenic risk were also similar under both weather conditions. The effects of Cr and Mn on non-carcinogenic risk were manifested mostly in children, then in adult males and females, in that order. The non-carcinogenic risks of Cr and Mn in PM_1, $PM_{2.5}$, PM_{10} and TSP for children were about 5.6 and 6.3 times higher than for adult males and females, respectively. The magnitude of the carcinogenic effect of Cr and Mn in different particle sizes on the human body was in this order: TSP>PM_{10}>$PM_{2.5}$>PM_1. The non-carcinogenic risks of Cr and Mn related to different particle sizes during sandstorm days were generally greater than on non-sandstorm days, except for Cr in $PM_{2.5}$. The non-carcinogenic risk of Cr in PM_1, PM_{10} and TSP for all populations was 1.2, 1.5 and 2.3 times higher than on non-sandstorm days, while the risk of Mn in PM_1, PM_{10} and TSP for all populations was 1.9, 10.0 and 13.1 times higher than on non-sandstorm days, respectively.

4. Discussion

4.1. Comparison of Pollutant Characteristics

Due to the coupling effect of local dust and regional transportation, the concentrations of PM_1, $PM_{2.5}$, PM_{10} and TSP in Wuhai City during the sandstorm period significantly increased, and were 2.2, 2.6, 4.8 and 6.0 times higher than during the non-sandstorm period, respectively. This can be attributed to the fact that the study area is located in the arid desert area of northwest China, with dry weather, low precipitation and severe wind erosion, coupled with a large number of minerals distributed in the region. The process of mining leads to looser soil and higher accumulation of ground dust resulting in reduced air quality. Moreover, particle concentration is also related to meteorological factors. Meteorological conditions in Wuhai during the sandstorm event (Figure 3b) showed that the low humidity on March 15 led to dry ground, while higher wind speed was more likely to carry ground dust, leading to an increase in the concentration of particulate matter in the atmosphere. PM concentrations were still at a high level on March 16–18, caused by high humidity and low wind speed conditions unfavorable to the diffusion of dust. As the wind speed increased on March 20, it accelerated the dispersion of particulate matter and led to a rapid decrease in PM concentration.

Compared with the non-sandstorm period, the occurrence of sandstorm caused the accumulation of large amounts of particulate matter in the atmosphere, resulting in an increase in the content of metallic elements in particulate matter. Liu et al. [17] also found that the sum of dust tracers (i.e., Al, Ca, Fe, Ti) concentrations were increased during sandstorm days. Conversely, water-soluble ions content generally decreased, probably due to the fact that these ions are mainly from local pollution sources. It is also possible that the enrichment of water-soluble ions varies between particle sizes [18], leading to an increase of water-soluble ions in some particle sizes. For instance, the concentration of Na^+ during sandstorm days were much higher than during non-sandstorm days because Na^+ may come from crustal sources such as construction and road dust. Ca^{2+} is mainly concentrated in coarse particles [35]. Windy and dusty weather can lead to higher Ca^{2+} concentrations in atmospheric particles. The dry and sandy environment around Wuhai is also one of the reasons for higher Ca^{2+} concentrations in PM_{10} and TSP during sandstorm events. Mg^{2+} content which mainly comes from soil is normally relatively low in atmospheric particulate matter. The large accumulation of ground dust during sandstorm days may increase the Mg^{2+} content in PM_{10} and TSP. $SO_4{}^{2-}$ is mainly derived from the combustion of fossil fuels such as coal, and the superposition of high concentrations of dust during long-distance transport with anthropogenic emissions of typical sulphur-containing pollutants may be responsible for its increased concentration. Due to the different elements had different degrees of enrichment in the particles, the content of each chemical component in different particles increased with the increase of particle size during both sandstorm and non-sandstorm days. The composition testing showed that PM_1, $PM_{2.5}$, PM_{10}, and TSP in the study area contained large amounts of Al, Zn, and Ca^{2+} during both sandstorm and non-sandstorm days, tentatively indicating that the sources of pollutants may be similar, dust and industrial emissions had a certain contribution.

4.2. Comparison of Pollutants Sources

During non-sandstorm days, the airflow was mainly transported over short distances, and potential pollution sources were mainly from southern Ningxia, southeast Gansu and western Shaanxi. These areas are rich in mineral resources and have developed numerous industries such as coal mining, chemical industries and metal smelting. The exhaust gas and particulate matter emitted from these industries seriously pollute the local environment and thus become a potential source of transported pollution in Wuhai area. The predominance of long-distance transport of airflow during sandstorms is due to the influence of meteorological conditions, coupled with the combined effects of local dust and regional transport. The potential source areas that had a large impact on $PM_{2.5}$ and PM_{10} concentrations in Wuhai during sandstorm days were basically similar and were mainly distributed in three directions. The first direction was southern Russia, north-central and southwestern Mongolia. Influenced by Mongolian cyclonic activity, this was the main direction for the occurrence of the sandstorm, passing through the Mongolian plateau, carrying a large concentration of particulate matter affecting the Wuhai area. The second direction was northern Xinjiang, west-central and northwestern Inner Mongolia. Airflow transport went through four of the eight major deserts in China, in order Gurbantunggut Desert, Batangilin Desert, Tengri Desert and Ulan Buh Desert. Furthermore, a large number of minerals are distributed within these source areas, accompanied by massive pollutants generated during industrial processes. Coupled with factors such as drought and little rain, low vegetation cover and severe desertification, airflow passing through these areas is likely to carry large amounts of particulate matter and is transported over long distance from the northwest to bring particulate matter to Wuhai City, affecting its air quality. The third direction was northern Shaanxi. The airflow trajectory showed the highest percentage of air mass trajectory from south-central Inner Mongolia (51.67%), which was dominated by short-distance transport. This indicated that the potential source area in this direction was mainly from local emissions. At the same time, the Kubuqi Desert is in the south-central part of Inner Mongolia, and cities in the northwest region have centralized heating in the form of coal combustion, which leads to high soot content in the atmosphere and affects the air quality.

The results showed that the sources of pollutants in different particle sizes during the sandstorm event were mainly from industrial and traffic emissions, combustion sources and dust, of which industrial emissions and combustion sources account for a large proportion. As a typical coal resource-based city, Wuhai area has many exposed mines after mineral extraction, accompanied by a large number of industrial and mining emissions. The occurrence of sandstorms accompanied by the transport of regional pollutants further aggravates local pollution. Therefore the proportion of industrial emissions and combustion sources of particulate matter increased compared with non-sandstorm days. However, the sources of traffic emissions and dust from roads and construction decreased during sandstorm days. This may indicate the impact of human activities, especially the industrialization process on the environment. Accompanied by extreme weather conditions, pollutants emitted by industrial production are more likely to diffuse and thus affect the regional eco-environmental. Combined with the results of CWT, the maximum high concentration contribution source areas of $PM_{2.5}$ and PM_{10} during sandstorm days were from northwestern Inner Mongolia, which further indicated that air pollution was affected by both local pollutant emissions and cross-regional transport [36,37]. Therefore, the control of atmospheric environment in the study area should strengthen the control of mineral exploitation in local areas, but one should also pay attention to pollutants transported from external areas. The implementation of afforestation, windbreak and sand fixation and mitigation of land desertification policies, improved vegetation coverage in arid desert areas of northwest China and reduced desertification area can impede the long-distance transport of particulate matter. In addition, the development of clean energy, green production and other technologies to reduce emissions of SO_2, NO_x and soot from coal combustion in northwest China can help to improve local air quality.

4.3. Impact of Pollutants on Human Health

The risk evaluation results showed that among the heavy metals selected for evaluation, only Cr pose a carcinogenic risk. Cr and Mn also have respiratory toxicity, which poses non-carcinogenic risks to human health [38]. Cr can precipitate some proteins in the blood, and long-term contact with this substance can cause respiratory tract inflammation and induce lung cancer [39]. Industrial production and coal mining activities are intense in Wuhai, which is a coal resource-based city, and thus the atmosphere continuously contains Cr and Mn. These elements can have an impact on human health in both sandstorm and non-sandstorm days, and pose greater risks to children's health than adults' health [40]. Masks should also be worn during non-sandstorm days, especially by infants and coal miners. Dust production in open coal mines can be limited by optimizing the coal mining process and deploying dust suppression measures to reduce the impact on human health. The risk of these elements to human health in sandstorm weather was generally higher than in non-sandstorm days because of the increased concentration of particulate matter in the atmosphere. Dusty weather can lead to allergies and respiratory diseases. It is recommended to minimize outdoor activities during sandstorms and to avoid outdoor activities for a few days after the sandstorm has passed, especially on the second day. Protective measures should be taken outdoors, such as wearing masks, hats, wind goggles and other protective devices to prevent dust from entering the respiratory tract. In addition, due to the different enrichment of elements in different particle sizes, heavy metal elements in different particle sizes had different effects on human health. The chemical content of heavy metals generally increased with the increase in particle size. However, due to the larger specific surface area of small particles, they are easier to adsorb into the human body, resulting in greater toxicity [41,42]. Therefore, the study of the enrichment of PM_1 during sandstorm events is essential to further understand the influence mechanism of small particle size particles on human health.

5. Conclusions

The concentrations of PM_1, $PM_{2.5}$, PM_{10} and TSP in Wuhai City during sandstorm days were significantly increased due to the coupling effects of local dust and regional transportation. Compared with non-sandstorm days, trace metals concentrations in particles of different particle sizes generally increased during sandstorm days, while water-soluble ions decreased. PMF results showed that the sources of pollutants during both sandstorm days and non-sandstorm days mainly came from industrial and traffic emissions, combustion sources and dust. The proportion of industrial emissions and combustion sources increased compared with non-sandstorm days, while traffic emissions and dust decreased. The airflows were mainly transported over short distances during non-sandstorm days, and high concentration contribution source areas were from southern Ningxia, southeast Gansu and western Shaanxi. While the airflows were mainly transported over long distances during sandstorm days, and high concentration contribution source areas were from northwestern Inner Mongolia, southern Russia, northern and southwestern Mongolia, and northern Xinjiang. Therefore, the management and prevention to the local pollutant emissions and the cross-regional pollutant transport should be paid more attention at the same time. Moreover, the risk of elements to human health during sandstorm days was generally higher than during non-sandstorm days. It is recommended that protective measures should be taken during both sandstorm and non-sandstorm days, especially for the infants and coal miners.

Supplementary Materials: The following supporting information can be downloaded at: https://www.mdpi.com/article/10.3390/atmos13101629/s1, Supplement S1: PMF; Supplement S2: Backward clustering trajectory; Supplement S3: Health risk assessment; Table S1: The meaning and value of each parameter in the exposure formula; Table S2: Values of SF and RfD for related elements; Table S3: Mean concentrations and standard deviations of elements during the sandstorm and non-sandstorm; Table S4: Emission sources of different chemical components in particulate matters; Table S5: The characterizing elements of each source for PM_1, $PM_{2.5}$, PM_{10} and TSP based on PMF model; Figure S1: Profiles of four sources identified from the PMF model for PM_1, $PM_{2.5}$, PM_{10} and TSP during the (a) non-sandstorm and (b) sandstorm. References [43–63] are cited in Supplementary Materials.

Author Contributions: Conceptualization, Y.L., T.Z. and R.W.; Data curation, Y.L.; Formal analysis, Y.L.; Funding acquisition, T.Z. and R.W.; Investigation, Y.L. and M.W.; Methodology, Y.L. and Q.J.; Project administration, R.W.; Resources, T.S.; Software, X.A.; Writing—original draft, Y.L.; Writing—review & editing, R.W. All authors have read and agreed to the published version of the manuscript.

Acknowledgments: This work was supported by the National Key Research and Development Plan Project of China (2017YFC0504403). We thank our colleagues for their help with the experiments.

Conflicts of Interest: The authors declare no conflict of interest.

References

1. Yassin, M.F.; Almutairi, S.K.; Al-Hemoud, A. Dust storms backward Trajectories' and source identification over Kuwait. *Atmos. Res.* **2018**, *212*, 158–171. [CrossRef]
2. Kang, J.H.; Keller, J.J.; Chen, C.S.; Lin, H.C. Asian dust storm events are associated with an acute increase in pneumonia hospitalization. *Ann. Epidemiol.* **2012**, *22*, 257–263. [CrossRef] [PubMed]
3. Trianti, S.M.; Samoli, E.; Rodopoulou, S.; Katsouyanni, K.; Papiris, S.A.; Karakatsani, A. Desert dust outbreaks and respiratory morbidity in Athens, Greece. *Environ. Health* **2017**, *16*, 72. [CrossRef]
4. Vodonos, A.; Friger, M.; Katra, I.; Avnon, L.; Krasnov, H.; Koutrakis, P.; Schwartz, J.; Lior, O.; Novack, V. The impact of desert dust exposures on hospitalizations due to exacerbation of chronic obstructive pulmonary disease. *Air Qual. Atmos. Health* **2014**, *7*, 433–439. [CrossRef]
5. Akinwumiju, A.S.; Ajisafe, T.; Adelodun, A.A. Airborne particulate matter pollution in Akure Metro city, southwestern Nigeria, West Africa: Attribution and meteorological influence. *J. Geovis. Spat. Anal.* **2021**, *5*, 11. [CrossRef]
6. Fernández, A.J.; Sicard, M.; Costa, M.J.; Guerrero-Rascado, J.L.; Gómez-Amo, J.L.; Molero, F.; Barragán, R.; Basart, S.; Bortoli, D.; Bedoya-Velásquez, A.E.; et al. Extreme, wintertime Saharan dust intrusion in the Iberian Peninsula: Lidar monitoring and evaluation of dust forecast models during the February 2017 event. *Atmos. Res.* **2019**, *228*, 223–241. [CrossRef]
7. Filonchyk, M.; Peterson, M.; Hurynovich, V. Air pollution in the Gobi Desert region: Analysis of dust-storm events. *Q. J. R. Meteorol. Soc.* **2021**, *147*, 1097–1111. [CrossRef]
8. Liu, L.; Guo, J.; Gong, H.; Li, Z.; Chen, W.; Wu, R.; Wang, L.; Xu, H.; Li, J.; Chen, D.; et al. Contrasting influence of Gobi and Taklimakan deserts on the dust aerosols in western North America. *Geophys. Res. Lett.* **2019**, *46*, 9064–9071. [CrossRef]
9. An, L.; Che, H.; Min, X.; Zhang, T.; Wang, H.; Wang, Y.; Zhou, C.; Zhao, H.; Gui, K.; Zheng, Y.; et al. Temporal and spatial variations in sand and dust storm events in East Asia from 2007 to 2016: Relationships with surface conditions and climate change. *Sci. Total Environ.* **2018**, *633*, 452–462. [CrossRef] [PubMed]
10. Barbulescu, A.; Nazzal, Y. Statistical analysis of dust storms in the United Arab Emirates. *Atmos. Res.* **2020**, *231*, 104669. [CrossRef]
11. Chen, Q.C.; Wang, M.M.; Sun, H.Y.; Wang, X.; Wang, Y.Q.; Lim, Y.G.; Zhang, L.X.; Mu, Z. Enhanced health risks from exposure to environmentally persistent free radicals and the oxidative stress of $PM_{2.5}$ from Asian dust storms in Erenhot, Zhangbei and Jinan, China. *Environ. Int.* **2018**, *121*, 260–268. [CrossRef] [PubMed]
12. Ismail, S.; Jean-François, L.; Mar, S.; Benjamin, G.; Carmen, C.J.; Amaury, D.S.; Faisal, A.S. Dust and dust storms over Kuwait: GROUND-BASED and satellite observations. *J. Atmos. Sol.-Terr. Phys.* **2018**, *179*, 105–113.
13. Soleimani, Z.; Teymouri, P.; Boloorani, A.D.; Mesdaghinia, A.; Middleton, N.; Griffin, D.W. An overview of bioaerosol load and health impacts associated with dust storms: A focus on the Middle East. *Atmos. Environ.* **2020**, *223*, 117187. [CrossRef]
14. Tiwari, S.; Kumar, A.; Pratap, V.; Singh, A.K. Assessment of two intense dust storm characteristics over indo-gangetic basin and their radiative impacts: A case study. *Atmos. Res.* **2019**, *228*, 23–40. [CrossRef]
15. Tang, G.L.; Chao, Q.C. Analysis on Sandstorm Variation in China for Last 49 Years. *Meteorol. Mon. (China)* **2005**, *5*, 8–11.
16. Filonchyk, M. Characteristics of the severe March 2021 Gobi Desert dust storm and its impact on air pollution in China. *Chemosphere* **2021**, *287*, 132219. [CrossRef] [PubMed]
17. Liu, Q.Y.; Liu, Y.J.; Zhao, Q.C.; Zhang, T.T.; Schauer, J.J. Increases in the formation of water soluble organic nitrogen during asian dust storm episodes. *Atmos. Res.* **2021**, *253*, 105486. [CrossRef]

18. Wang, G.H.; Cheng, C.L.; Huang, Y.; Tao, J.; Ren, Y.Q.; Wu, F.; Meng, J.J.; Li, J.J.; Cheng, Y.T.; Cao, J.J.; et al. Evolution of aerosol chemistry in Xi'an, inland China, during the dust storm period of 2013—Part 1: Sources, chemical forms and formation mechanisms of nitrate and sulfate. *Atmos. Chem. Phys.* **2014**, *14*, 11571–11585. [CrossRef]
19. Tao, M.; Gui, L.; Li, R.; Wang, L.; Liang, S.; Li, Q.; Wang, L.; Yu, C.; Chen, L. Tracking prevailing dust aerosol over the air pollution in central China with integrated satellite and ground observations. *Atmos. Environ.* **2021**, *253*, 118369. [CrossRef]
20. Zhou, L.; Dang, X.; Mu, H.; Wang, B.; Wang, S. Cities are going uphill: Slope gradient analysis of urban expansion and its driving factors in China. *Sci. Total Environ.* **2021**, *775*, 145836. [CrossRef] [PubMed]
21. Titos, G.; Ealo, M.; Pandolfi, M.; Perez, N.; Sola, Y.; Sicard, M.; Comerón, A.; Querol, X.; Alastuey, A. Spatiotemporal evolution of a severe winter dust event in the western Mediterranean: Aerosol optical and physical properties. *J. Geophys. Res. Atmos.* **2017**, *122*, 4052–4069. [CrossRef]
22. Wang, Q.; Dong, X.; Fu, J.S.; Xu, J.; Deng, C.; Jiang, Y.; Fu, Q.; Lin, Y.; Huang, K.; Zhuang, G. Environmentally dependent dust chemistry of a super Asian dust storm in March 2010: Observation and simulation. *Atmos. Chem. Phys.* **2018**, *18*, 3505–3521. [CrossRef]
23. Zheng, S.; Singh, R.P. Aerosol and meteorological parameters associated with the intense dust event of 15 April 2015 over Beijing, China. *Remote Sens.* **2018**, *10*, 957. [CrossRef]
24. Francis, D.; Eayrs, C.; Chaboureau, J.P.; Mote, T.; Holland, D.M. Polar jet associated circulation triggered a Saharan cyclone and derived the poleward transport of the African dust generated by the cyclone. *J. Geophys. Res. Atmos.* **2018**, *123*, 11-899. [CrossRef]
25. She, L.; Xue, Y.; Guang, J.; Che, Y.; Fan, C.; Li, Y.; Xie, Y. Towards a comprehensive view of dust events from multiple satellite and ground measurements: Exemplified by the May 2017 East Asian dust storm. *Nat. Hazards Earth Syst. Sci.* **2018**, *18*, 3187–3201. [CrossRef]
26. Tan, S.C.; Li, J.; Che, H.; Chen, B.; Wang, H. Transport of East Asian dust storms to the marginal seas of China and the southern North Pacific in spring 2010. *Atmos. Environ.* **2017**, *148*, 316–328. [CrossRef]
27. Yuan, T.; Chen, S.; Huang, J.; Zhang, X.; Luo, Y.; Ma, X.; Zhang, G. Sensitivity of simulating a dust storm over Central Asia to different dust schemes using the WRF-Chem model. *Atmos. Environ.* **2019**, *207*, 16–29. [CrossRef]
28. Li, G.; Lu, D.; Yang, X.; Zhang, H.; Guo, Y.; Qu, G.; Wang, P.; Chen, L.; Ruan, T.; Hou, X.; et al. Resurgence of sandstorms complicates China's air pollution situation. *Environ. Sci. Technol.* **2021**, *55*, 11467–11469. [CrossRef]
29. Liu, Y.; Wang, R.S.; Zhang, Y.; Zhao, T.N.; Wang, J.H.; Wu, H.X.; Hu, P. Temporal and spatial distributions of particulate matters around mining areas under two coal mining methods in arid desert region of northwest China. *Environ. Technol. Innov.* **2020**, *19*, 101029. [CrossRef]
30. Cheng, M.T.; Tang, G.Q.; Lv, B.; Li, X.R.; Wu, X.R.; Wang, Y.M.; Wang, Y.S. Source apportionment of $PM_{2.5}$ and visibility in Jinan, China. *J. Environ. Sci. (China)* **2021**, *102*, 207–215. [CrossRef]
31. Wang, Y.Q. MeteoInfo: GIS software for meteorological data visualization and analysis. *Meteorol. Appl.* **2012**, *21*, 360–368. [CrossRef]
32. Lee, S.; Ashbaugh, L. The impact of trajectory starting heights on the MURA trajectory source apportionment (TSA) method. *Atmos. Environ.* **2007**, *41*, 7022–7036. [CrossRef]
33. Sapkota, A.; Symons, J.M.; Kleissl, J.; Wang, L.; Parlange, M.B.; Ondov, J.; Breysse, P.N.; Diette, G.B.; Eggleston, P.A.; Buckley, T.J. Impact of the 2002 Canadian forest fires on particulate matter air quality in Baltimore City. *Environ. Sci. Technol.* **2005**, *39*, 24–32. [CrossRef] [PubMed]
34. *GB 3095-2012*; CNRAES (Chinese Research Academy Environmental Sciences). Ambient Air Quality Standards. Chinese Research Academy Environmental Sciences: Beijing, China, 2012.
35. Yang, Y.J.; Zhou, R.; Yan, Y.; Yu, Y.; Liu, J.; Di, Y.; Du, Z.; Wu, D. Seasonal variations and size distributions of water-soluble ions of atmospheric particulate matter at Shigatse, Tibetan Plateau. *Chemosphere* **2016**, *145*, 560–567. [CrossRef] [PubMed]
36. Wang, W.D.; Chen, H.S.; Wu, Q.Z.; Wei, L.F.; Wang, Z.F.; Li, C.; Chen, D.H.; Jiang, Z.M.; Wu, W.W. Numerical study of $PM_{2.5}$ regional transport over Pearl River Delta during a winter heavy haze event. *Acta Sci. Circumstantiae (China)* **2016**, *36*, 2741–2751.
37. Yu, G.M.; Xu, J.Z.; Kang, S.C.; Ren, J.W.; Cui, X.Q. Trajectory analysis of atmospheric transport of particles in Laohugou area, Western Qilian Mountains. *Arid Zone Res. (China)* **2020**, *37*, 671–679.
38. Xie, J.W.; Jin, L.; Cui, J.L.; Luo, X.S.; Li, J.; Zhang, G.; Li, X.D. Health risk oriented source apportionment of $PM_{2.5}$-associated trace metals. *Environ. Pollut.* **2020**, *262*, 114655. [CrossRef]
39. Lin, Y.C.; Li, Y.C.; Amesho, K.T.T.; Chou, F.C.; Cheng, P.C. Filterable $PM_{2.5}$, metallic elements, and organic carbon emissions from the exhausts of diesel vehicles. *Aerosol Air Qual. Res.* **2020**, *20*, 1319–1328. [CrossRef]
40. Elif, Y.; Akif, A.; Ahmet, A. Land application of municipal sewage sludge: Human health risk assessment of heavy metals. *J. Clean. Prod.* **2021**, *319*, 128568.
41. Guo, G.; Zhang, D.; Wang, Y. Characteristics of heavy metals in size-fractionated atmospheric particulate matters and associated health risk assessment based on the respiratory deposition. *Environ. Geochem. Health* **2021**, *43*, 285–299. [CrossRef]
42. Penttinen, P.K.; Timonen, K.L.; Tiittanen, P.; Mirme, A.; Ruuskanen, J.; Pekkanen, J. Ultrafine particles in urban air and respiratory health among adult asthmatics. *Eur. Respir. J.* **2001**, *17*, 428–435. [CrossRef]
43. He, R.-D.; Zhang, Y.-S.; Chen, Y.-Y.; Jin, M.-J.; Han, S.-J.; Zhao, J.-S.; Zhang, R.-Q.; Yan, Q.-S. Heavy Metal Pollution Characteristics and Ecological and Health Risk Assessment of Atmospheric PM2. *5 in a Living Area of Zhengzhou City. Huan Jing ke Xue= Huanjing Kexue.* **2019**, *40*, 4774–4782.

44. Beijing Quality and Technology Supervision Bureau. *Environmental Site Assessment Guideline*; DB11/T 656-2009; Beijing Quality and Technology Supervision Bureau: Beijing, China, 2009.
45. Kuo, S.-C.; Hsieh, L.-Y.; Tsai, C.-H.; Tsai, Y.I. Characterization of $PM_{2.5}$ fugitive metal in the workplaces and the surrounding environment of a secondary aluminum smelter. *Atmos. Environ.* **2007**, *41*, 6884–6900. [CrossRef]
46. Li, S.L. *Characteristics and Source Analysis of Dust Pollution in Jinan City*; Shandong Jianzhu University: Jinan, China, 2019.
47. Amato, F.; Viana, M.; Richard, A.; Furger, M.; Prévôt, A.S.H.; Nava, S.; Lucarelli, F.; Bukowiecki, N.; Alastuey, A.; Reche, C.; et al. Size and time-resolved roadside enrichment of atmospheric particulate pollutants. *Atmos. Chem. Phys.* **2011**, *11*, 2917–2931. [CrossRef]
48. Grigoratos, T.; Martini, G. Brake wear particle emissions: A review. *Environ. Sci. Pollut. Res.* **2014**, *22*, 2491–2504. [CrossRef] [PubMed]
49. Lin, Y.C.; Zhang, Y.L.; Song, W.; Yang, X.; Fan, M.Y. Specific sources of health risks caused by size-resolved pm-bound metals in a typical coal-burning city of northern China plain during the winter haze event. *Sci. Total Environ.* **2020**, *734*, 138651. [CrossRef] [PubMed]
50. Altıntas, Y.; Kaygısız, Y.; Öztürk, E.; Aksöz, S.; Keşlioğlu, K.; Maraşlı, N. The measurements of electrical and thermal conductivity variations with temperature and phonon component of the thermal conductivity in Sn-Cd-Sb, Sn-In-Cu, Sn-Ag-Bi and Sn-Bi-Zn alloys. *Int. J. Therm. Sci.* **2016**, *100*, 1–9. [CrossRef]
51. Bozkurt, Z.; Gaga, E.O.; Taşpınar, F.; Arı, A.; Pekey, B.; Pekey, H.; Döğeroğlu, T.; Üzmez, Ö.Ö. Atmospheric ambient trace element concentrations of PM_{10} at urban and sub-urban sites: Source apportionment and health risk estimation. *Environ. Monit. Assess.* **2018**, *190*, 1–17. [CrossRef]
52. Ledoux, F.; Kfoury, A.; Delmaire, G.; Roussel, G.; El Zein, A.; Courcot, D. Contributions of local and regional anthropogenic sources of metals in $PM_{2.5}$ at an urban site in northern France. *Chemosphere* **2017**, *181*, 713–724. [CrossRef]
53. Zhang, X.; Chen, W.; Ma, C.; Zhan, S. Modeling particulate matter emissions during mineral loading process under weak wind simulation. *Sci. Total Environ.* **2013**, *449*, 168–173. [CrossRef]
54. Zhang, R.J.; Jing, J.; Tao, J.; Hsu, S.C.; Wang, G.; Cao, J. Chemical characterization and source apportionment of $PM_{2.5}$ in Beijing: Seasonal perspective. *Atmos. Chem. Phys.* **2013**, *13*, 7053–7074. [CrossRef]
55. Lin, Y.-C.; Tsai, C.-J.; Wu, Y.-C.; Zhang, R.; Chi, K.-H.; Huang, Y.-T.; Lin, S.-H.; Hsu, S.-C. Characteristics of trace metals in traffic-derived particles in Hsuehshan Tunnel, Taiwan: Size distribution, potential source, and fingerprinting metal ratio. *Atmos. Chem. Phys.* **2015**, *15*, 4117–4130. [CrossRef]
56. Minguillón, M.; Monfort, E.; Escrig, A.; Celades, I.; Guerra, L.; Busani, G.; Sterni, A.; Querol, X. Air quality comparison between two European ceramic tile clusters. *Atmos. Environ.* **2013**, *74*, 311–319. [CrossRef]
57. Xue, G.-Q.; Zhu, B.; Wang, H.-L. Size distributions and source apportionment of soluble ions in aerosol in Nanjing. *Huan jing ke xue=Huanjing kexue* **2014**, *35*, 1633–1643. [PubMed]
58. Kong, S.; Han, B.; Bai, Z.; Chen, L.; Shi, J.; Xu, Z. Receptor modeling of $PM_{2.5}$, PM_{10} and TSP in different seasons and long-range transport analysis at a coastal site of Tianjin, China. *Sci. Total Environ.* **2010**, *408*, 4681–4694. [CrossRef]
59. Wang, Y.; Zhuang, G.; Sun, Y.; An, Z. The variation of characteristics and formation mechanisms of aerosols in dust, haze, and clear days in Beijing. *Atmos. Environ.* **2006**, *40*, 6579–6591. [CrossRef]
60. Guo, W.; Zhang, Z.-Y.; Zhang, N.J.; Luo, L.; Xiao, H.Y.; Xiao, H.W. Chemical characterization and source analysis of water-soluble inorganic ions in $PM_{2.5}$ from a plateau city of Kunming at different seasons. *Atmos. Res.* **2020**, *234*, 104687. [CrossRef]
61. Zhao, J.; Zhang, F.; Xu, Y.; Chen, J. Characterization of water-soluble inorganic ions in size-segregated aerosols in coastal city, Xiamen. *Atmos. Res.* **2011**, *99*, 546–562. [CrossRef]
62. Eliani, E.; Nicola, M.; Jonathan, G.; Tara, K.B.; Zoe, K.S.; Iain, J.B. Measurement of diesel combustion-related air pollution downwind of an experimental unconventional natural gas operations site. *Atmos. Environ.* **2018**, *189*, 30–40.
63. McCulloch, A.; Aucott, M.L.; Benkovitz, C.M.; E Graedel, T.; Kleiman, G.; Midgley, P.M.; Li, Y.-F. Global emissions of hydrogen chloride and chloromethane from coal combustion, incineration and industrial activities: Reactive Chlorine Emissions Inventory. *J. Geophys. Res. Earth Surf.* **1999**, *104*, 8391–8403. [CrossRef]

atmosphere

Article

Bioaerosol Concentration in a Cattle Feedlot in Neuquén, Argentina

Marisa Gloria Cogliati [1,*], Paula Andrea Paez [2,*], Luis Alfredo Pianciola [3], Marcelo Alejandro Caputo [2] and Paula Natalia Mut [2]

1 Faculty of Environmental and Health Sciences, National University of Comahue, 8300 Neuquén, Argentina
2 Río Negro Research and Transfer Center, National University of Río Negro, 8332 Viedma, Argentina
3 Central Laboratory, Ministry of Health of the Province of Neuquén, 8300 Neuquén, Argentina
* Correspondence: marisa.cogliati@fahu.uncoma.edu.ar (M.G.C.); ppaez@unrn.edu.ar (P.A.P.)

Abstract: There is a global trend toward intensive livestock breeding, which tends to increase the microbial load in the environment as well as the presence of volatile compounds and dust that can cause health issues. Cattle is the major producer of *Escherichia coli* (*E. coli*), a group of foodborne bacteria associated with severe human diseases, and Neuquén province in Argentina has one of the highest rates of uremic hemolytic syndrome incidence in the world. This paper presents the results of two sampling events of *E. coli* bacteria at 39 sites in La Paisana ranch (LPR), in Añelo (Neuquén), considering locations inside the pens, upwind, and downwind of the feedlot with different time steps, using a Microflow α equipment. The ranch has approximately 600 heads and clean and controlled installations. The field experiment included sampling airborne aerosol deposition and concentration using passive and active methods. Concentrations were also estimated using an atmospheric dispersion model. During the field experiment, counts of up to 2970 CFU/m^3 were obtained in the cattle stockyards and up to 111 CFU/m^3 at a distance of 100 m.

Keywords: cattle; intensive livestock farming; particulate matter; *Escherichia coli*

Citation: Cogliati, M.G.; Paez, P.A.; Pianciola, L.A.; Caputo, M.A.; Mut, P.N. Bioaerosol Concentration in a Cattle Feedlot in Neuquén, Argentina. *Atmosphere* **2022**, *13*, 1761. https://doi.org/10.3390/atmos13111761

Academic Editor: Daniel Viúdez-Moreiras

Received: 16 September 2022
Accepted: 17 October 2022
Published: 26 October 2022

1. Introduction

Currently, there is a global trend toward the intensification and industrialization of animal breeding, especially cattle, in order to enhance efficiency and reduce production costs.

Bioaerosols are airborne particles of biological origin which include bacteria, fungi, viruses, microbial toxins, pollen, proteins, and enzymes; they can be individual organisms or can be attached to dust particles or small water droplets [1]). Matthias-Maser & Janicke [2] reported that bioaerosols might contribute to almost 25% of atmospheric aerosols both in dry air and in cloud water from data of the field campaign FELDEX 95 in a rural/urban environment and Fröhlich-Nowoisky et al. [3] in a review concerning the sources, abundance, composition, and effects of biological aerosols, pointed that for diameters larger than ~1 μm, bioaerosols typically account for around 30% in urban and rural air.

Cattle feedlot pens are effective production systems because the feeding is highly managed, and animals gain weight rapidly. However, because of the density of cattle heads, pens normally generate high concentrations of ammonia [4], methane [5], and a great variety of bioaerosols (bacteria, endotoxins, viruses, fungi, parasites, etc.) [6] and dust [7] changing nearby air quality

In livestock, bioaerosols are produced throughout the primary production environment due to the increased volumes of animals and organic waste present: accumulation of manure in pens and collection areas, in manure storage lagoons, and in the soil [8–10]. Without adequate treatment, these areas can also become sources of contamination of surface and groundwater [11]. Post-processing such as slaughter, application of manure in fields as fertilizer [12], and treatment of wastewater from livestock facilities [13] are sources of bioaerosols, which are transported toward the environment and residential areas [5]. Bioaerosols, together with other pollutants in the air, can negatively affect the health of people working in agricultural operations [14].

There is growing evidence that bioaerosol emissions also have a negative impact on the general population, especially in areas with a high density of livestock and in the vicinity of intensive livestock operations [14–16]. This problem has been markedly exacerbated by the increasing number of animals kept in close proximity to the human population [17,18].

Feed management regimes affect dust emissions in the feedlot pens. Cattle activity in the feedlots, particularly when surface moisture is low, may contribute to increased particulate matter (PM) emissions [19,20]. Particulate matter concentrations in the feedlots are usually higher at dawn and nightfall when animal activity is more intense and meteorological conditions more stable [21]. Changes in the feeding regime during those times can reduce dust emissions by replacing active periods with periods of eating and chewing the cud [19].

Cattle are the main carrier of the zoonotic pathogen *Escherichia coli* O157:H7. This pathogen has been related to numerous infectious outbreaks around the world [22] that can progress to Hemolytic Uremic Syndrome (HUS). HUS is a life-threatening disease, mainly in children, that can lead to blood transfusions and dialysis [23]. The province of Neuquén has one of the highest HUS incidence rates in the world, which is connected to the high proportion of *E. coli* O157:H7 (clade 8 strain of Shiga Toxin-producing *E. coli* (STEC) 0157) in Argentine cattle [24,25]. Hence, although this study focuses on aerosol behavior in feedlots, the microbiological analysis concentrates on the bacterium *E. coli*.

The magnitude of the risk of airborne transport of *E. coli* O157:H7 is not fully known. Berry et al. [26] studied the impact of proximity to a beef cattle feedlot on *E. coli* O157:H7 contamination of leafy greens finding that distance guidelines of 120 m may not be adequate to limit the transmission. Studies about emission rates in livestock or transport distances of bacteria are very limited because bioaerosol concentrations are assumed to be insignificant compared to non-biological particles [27]. As a consequence, given that outdoor cattle feedlots emit several pollutants into the air, including PM with an equivalent aerodynamic diameter of less than 10 μm (PM10), under different environmental and growing conditions, the approach of such studies consists of determining PM10 emission rates under different weather conditions. For instance, Bonifacio et al. [28] determined the PM10 emission rates from two large feedlot pens in Kansas under different weather conditions; McGinn et al. [6] modeled the PM10 emission rates in two feedlot pens in Australia.

Millner & Suslow [29] collected bioaerosol samples downwind of a cattle feedlot; and found higher concentrations of *E. coli* in air at 9 m and lower at 30 m, without any detection at distances greater than 60 m. The samples were negative for *E. coli* O157:H7. The importance of this pathogen in public health deserves more research to determine set-back distances or buffer zones to effectively reduce the risk of airborne *E. coli* O157:H7 contamination of produced crops and population.

Although some studies address the relationship between weather conditions and bioaerosol concentrations or the size distribution of bacterial aerosols, their relationship with meteorological factors has been recognized only preliminarily [30]. The air temperature, relative humidity, and wind speed affect the concentration and viability of bioaerosols [31]. Inert materials are raised into the air when the surface dries and the bonding forces weaken, when air motion is strong enough, or due to mechanical perturbation [7].

This study aims to assess the presence and airborne dispersion of *E. coli* from intensive confined cattle farming through measurements of bioaerosols in situ at the Loma de la Paisana ranch (LPR) in Añelo, Neuquén, with analysis of bacterium *E. coli*, total bacteria, and estimations of bioaerosol concentration and deposition using an atmospheric dispersion model; it also analyzes the incidence of meteorological conditions in the region on the dispersion of *E. coli*.

2. Study Area Characteristics and Climate

The study area includes a feedlot located at LPR, close to the town of Añelo in the valley of Neuquén River, northern Argentine Patagonia (Figure 1). Figure 1 also shows the distribution of the sampling points. Table 1 connects the sites marked in Figure 1 with the active (Ai) and passive (Pi) measurements made during each of the field experiments. The pens are distributed in the ranch, as shown in Figure 1b. The ranch has 600 heads of confined cattle only. Feedlot manure is collected and stored in a remote area. Drinking troughs with clean water, shade areas, and enough room for short walks are available to cattle in the feedlot (Figure 1b).

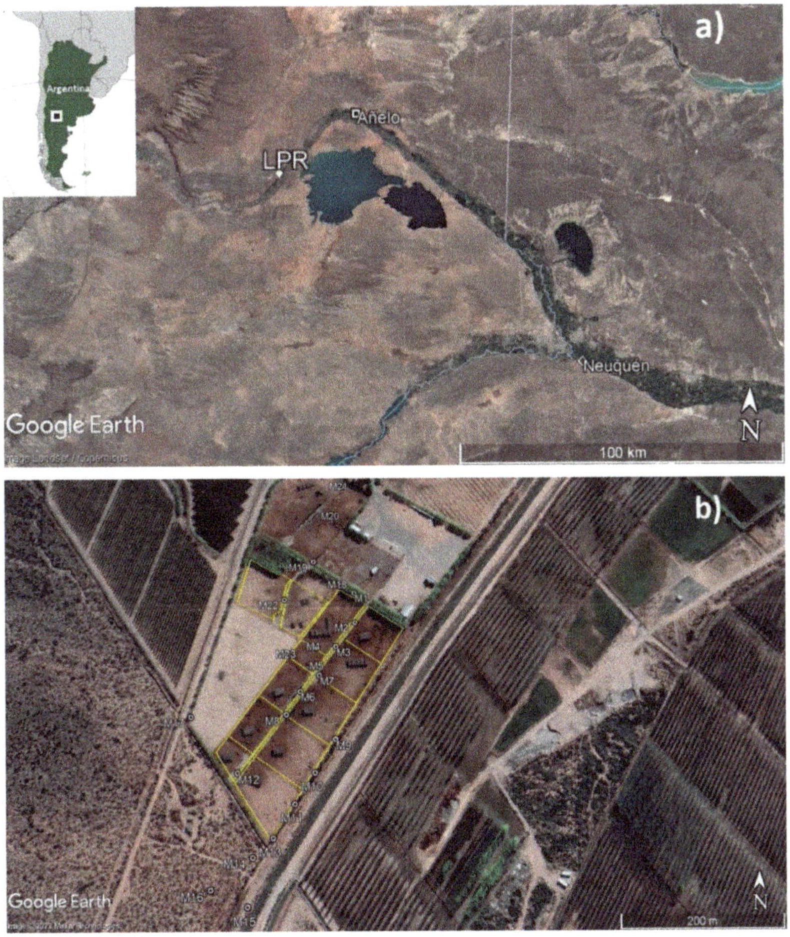

Figure 1. (**a**) Location of La Paisana ranch (LPR) in Neuquén. (**b**) Enlarged satellite image of the pens at LPR (yellow lines) and location of the sampling points (Mi) described in Table 1.

The main characteristic of the climate in the area is the constancy and intensity of wind [32] and corresponds to type BWk in Köeppen's classification: cold desert with warm summer [33]. According to Prohaska [32] the annual distribution of wind directions presents a peak between 50–70% of westerly winds (including calm winds). The mean monthly wind speed presents minimum values in July and increases to reach a peak in October. It remains high during the summer and decreases as of March. The mean wind speed is 3.8 m/s reaching 6 and 5 m/s from the southwest to west sectors, respectively, and calm wind represents 18% [34]. The daily thermal amplitude reaches values between 17.0 °C and 19.0 °C at the end of the summer in Neuquén [32]. The mean monthly temperature in the hottest months is above 30 °C and the difference between the maximum temperature of the warmest month and the minimum temperature in the coldest month is greater than 30 °C in northern Patagonia [32].

Table 1. Correspondence between sampling points (Mi) and passive (Pi) and active (Ai) measurements of the samplings of days 20 February 2020 and 5 April 2022.

Mi	20 February 2020	5 April 2022	
	Ai	Pi	Ai
M1	A2	P11–P12	A14
M2	A1	P9–P10	A15
M3	A3		
M4	A4		
M5	A5	P7–P8	A16–A17
M6	A6		
M7		P1–P2	A18
M8	A7		
M9	A8		
M10	A9		
M11		P5–P6	A19
M12	A10		
M13	A11		
M14	A12		
M15	A13		
M16			A20
M17		P17–P18	
M18		P19	
M19		P20	
M20		P15–P16	A21
M21		P13–P14	A22
M22			A24
M23			A25 (manure storage)

Atmospheric pollution studies should include the analysis of source emissions and the influence of meteorological conditions on the dispersion of pollutants toward the receptors. High pollutant concentrations at a reception site might be related to atmospheric conditions rather than to the exceeding of emission thresholds at the source [35]. One of the variables that account for the atmospheric situation is the atmospheric pollution potential [35] which connects two atmospheric parameters in a region: the thickness of the boundary layer and the mean wind speed within the boundary layer or transport wind. Gassmann & Mazzeo [35] calculated the ventilation potential in Argentina considering the transport wind and the height of the mixed layer in the period 1972–1982. They found that the transport wind in Neuquén was 6.9 ± 5.3 m/s in autumn and 10.5 ± 7.1 m/s in spring, and the maximum average height of the mixed layer was 1100 m in winter and 2700 m in summer with standard deviation peaks of 543 m and 757 m in winter and spring, respectively [35].

As to atmospheric stability and taking into account the months when field experiments were made, i.e., February and April, the most frequent class in Neuquén was neutral stability (D) during measurement hours (15:00 HL in Figure 2) in the period 2015–2021. These data were obtained from the meteorological data archive of the Real-time Environmental Applications and Display system (READY, http://www.ready.noaa.gov (accessed on 10 May 2022)). In the same period and time, the mean mixing height was 662 ± 289 m and 630 ± 578 m, reaching peaks of 1870 m and 2610 m in April and February, respectively.

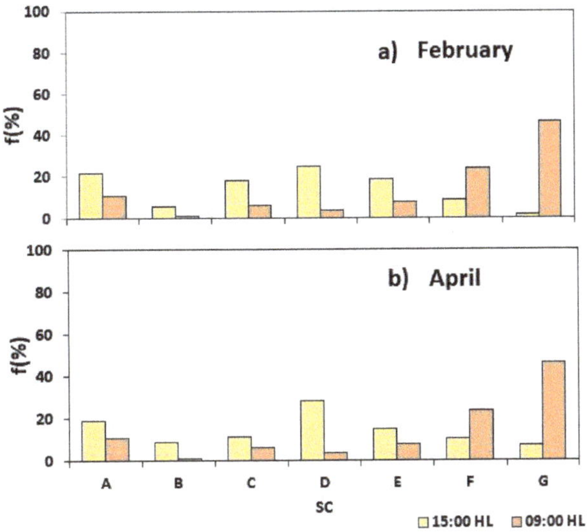

Figure 2. Relative frequency (f [%]) of atmospheric stability (SC, [36]) in Neuquén as a function of the month ((**a**) February, (**b**) April) at 09:00 LT (light yellow) and at 15:00 LT (dark yellow), for the period 2015–2021. Classes: A: extremely unstable; B: moderately unstable; C: lightly unstable; D: neutral; E: lightly stable; F: moderately stable, G: extremely stable. Data from the Real-time Environmental Applications and Display sYstem (READY, http://www.ready.noaa.gov (accessed on 10 May 2022)).

3. Materials and Methods

3.1. Sampling and Detection of Airborne E. coli Concentration

3.1.1. Sampling of Airborne *E. coli*

There are different methods for sampling bioaerosols, of which deposition and impaction belong to the relatively simple and most used collecting techniques [5,37] We included two types of sampling in our experiment: active and sedimentation samples. Airborne microorganisms were collected directly on Petri dishes prepared with a culture medium or nutrient [30,31,37]. Active measurements at LPR were made on 20 February 2020 and 5 April 2022 with an impaction microflow sampler (Microflow α Aquaria version 3.0.0 cod. G.1015) disposable 90 mm Petri dishes and a sampling speed of 30 L/min. The passive deposition samplings were all made on 5 April 2022. The impaction samplers were disinfected after each sampling with tissue paper soaked with 70% ethyl alcohol. The chromogenic culture media was CHROMagar™ (CHROMagar™ Orientation of CHROMagar).

Commonly, the concentrations of living microorganisms present in the air is the number of colony-forming units in the volume of air, the results of deposition sampling methods are expressed as bacterial colony-forming units (CFU) per unit area and unit time, so they cannot be compared directly with the results from volumetric measurements [5]. Samples were taken at a height of approximately 1.1 m and were georeferenced in situ, and the Petri dishes were distributed among the pens and windward and downwind from the emission points at the sites shown in Figure 1b (Mi). An additional measurement was made in the place where manure is stored (M23, not shown in Figure 1). The samples were kept cool until they arrived at the laboratory. Sampling times were set from 10 a.m. to 5 p.m. (LT, local time), which is when the greatest PM10 emissions take place, according to Bonifacio et al. [28].

The sampling method was designed during a prior experiment in a feedlot in Chel Cura (Choele Choel, Río Negro Argentina), where measurements were made with exposure times of 3, 5, and 10 min. Three-minute exposures within the pens resulted in saturated dishes, which hindered the measurement of CFU [38]. Based on these results, and considering that there were more cattle heads at LPR than in Chel Cura, as well as the studies by Bragoszewska et al. [30,39] and Environment Agency [40] the sampling speed was set at 30 L/min for active sampling (sites called Ai in Table 1). The locations of those sites are indicated as Mi in Figure 1b.

The sampling volume in the pens was 30 L and was augmented to 300 L with increasing distance from the more crowded areas. Deposition samples were called Pi (see Table 1) and were obtained from exposures of 10 and 30 min at the sites shown in Figure 1a (puntos Mi).

3.1.2. Microbial Analysis

The samples were incubated at 37 ± 1 °C for 48 h with readings at 24 h. Under the same conditions, a control measurement was made in CHROMagar medium using the reference strain *E. coli* ATCC 25922. The total count of bacteria present was also carried out, identifying them according to morphology, color and appearance.

3.2. Emission and Dispersion

USEPA [40] establishes a PM10 emission factor of 17 ton/1000 hd-yr as a methodology for the estimation of feedlot emissions. We estimated PM10 emissions based on these results adjusted by the number of heads in the pens to assess the emission; and Li et al. [37] we estimated bioaerosols as 25% of PM10 mass fraction.

The feedlot dispersion plumes were simulated using the HYSPLIT model, developed by the Air Resources Laboratory (ARL) [41], with meteorological observations from Neuquén station (see Figure 1a)—which is the closest one to the study area—and Global Data Assimilation System (GDAS) data. For modeling purposes, particles were assumed to be lifted by cattle activity and wind. The dry deposition rate was calculated assuming a particle density of 1.0 g·cm^{-3} [12]. The values of the highest concentrations at the breathing level were estimated with a screening model.

4. Results and Discussion

4.1. Weather Parameters and Airborne E. coli Concentration

The weather conditions during the sampling of 20 February 2020 and 5 April 2022 are informed in Table 2 and were estimated from GDAS data for the study area. The results of the active samplings in the field experiments showed that in-pen *E-coli* concentrations were greatest at sites M6 and M7 (see Table 3, Table 1). The maximum concentration of airborne *E. coli* in the two events under analysis was 2967 CFU/m^3 and took place on 20 February 2020 (Table 3). On 5 April 2022, the maximum concentration was 33 CFU/m^3. The average concentration was 1050 ± 1090 CFU/m^3 on the 20 February 2020 and 8 ± 17 CFU/m^3 on 5 April 2022 (Table 3), which amounts to a difference in concentration of 76.2%. According to the literature, concentration differences can be expected in connection to weather conditions and cattle activity [31].

The high temperatures that are prevalent in summer support the growth and physiological activity of bioaerosols [42]. The maximum growth rate of *E. coli* bacteria occurs in the range of 21–42 °C, and optimum growth is at 37 °C [43].The greatest count of *E. coli* was registered on 20 February 2020, after several days of air temperature ranging from 15.3 °C to 31.2 °C and an average ground temperature of 39.5 °C with low rainfall and low relative humidity. These ambient conditions favored dust and manure particle suspension by the light breeze (see Table 2). On 5 April 2022, the mean ground temperature was 30.3 °C at the time of sampling (see Table 2), and the air temperature range on the three previous days was 8.9–27.7 °C (see Table 2), with weak wind, which would limit bacteria growth and subsequent emission of particles from the surface into the air.

Table 2. Mean meteorological parameters during the field experiments at LPR (15 UTC to 18 UTC). Rainfall (PP, mm), relative humidity (RH, %), evolution of atmospheric stability (SC, [36], wind speed (v, m/s), wind direction (DD), mean air temperature (T, °C), height of the mixed layer (H, m), atmospheric pressure (P, hPa), downward shortwave radiation range (I, W/m^2), temperature range on the 3 days prior to the sampling (RT, °C), 3-hourly soil moisture during the sampling period (RTS, °C) (source: NOAA Air Resources Laboratory). Rainfall corresponds to days 17 February 2020 and 3 April 2022.

	20 February 2020	**5 April 2022**
DD	ENE–E	SW–NE
V (m/s)	2.6–3.9 3.9–6.7	<0.5 3.3
SC	E–D–C	E–E–D
PP (mm)	0.5 (17 February 2020)	2.1 (3 April 2022)
T(°C)	21.8	21.2
RH (%)	20.0	23.6
P (hPa)	1021	1011
H (m)	<2069.8	<1470
I (W/m^2)	523–605	397.7–450
RT (°C)	15.3–31.2	8.9–27.7
TSm (°C)	39.5	30.3

Table 3. Average *E. coli* concentration per m^3 of air (Cm, CFU/m^3), concentration standard deviation (ST_C), minimum concentration *E. coli* per m^3 of air (C_{min}, CFU/m^3), maximum concentration per m^3 of air (C_{max}, CFU/m^3), number of samples (N) within the LPR pens. Sampling dates: 20 February 2020 and 5 April 2022. *w/c*: Petri dishes without count.

	20 February 2020	**5 April 2022**
C_m (CFU/m^3)	1050	8
ST_C	1090	17
C_{min} (CFU/m^3)	w/c	w/c
C_{max} (CFU/m^3)	2967	33
N	5	5

Factors affecting the abundance of microbial activity are quite complex and include meteorological parameters, weather conditions, the intensity of the source, and the geographical environment [30]. Zhong et al. [42] and Li et al. [31] analyzed the correlations between meteorological variables and bacteria concentration in bioaerosols and found that atmospheric temperature and wind speed have, respectively, positive and negative effects on bacterial concentrations; while relative humidity and wind direction would have no significant influence. The sampling results agree with those of Zhong et al. [42] who found that the seasonal distribution of bacterial concentration in bioaerosols was greatest in summer > autumn > winter > spring, with large fluctuations in summer and autumn.

To understand the airborne transport of *E. coli*, we collected air samples in situations and conditions that would allow describing the dissemination of these microorganisms around the source. In an exploratory analysis, we estimated the most unfavorable values associated with the highest concentrations at the breathing height, considering weather information and pen size (322 m × 176 m). The most unfavorable situations downwind of the feedlot would occur under a moderately stable atmosphere (SC (stability class): F), refs [36,44] at 183 m from the source, with the receptors assumed to be 1.5 m height above the ground (human breathing height). For which the following results were obtained:

4.2. Detection of *E. coli* by Impaction with Distance

E. coli concentrations in the air samples collected by impaction on 20 February 2020 were much higher than those obtained on 5 April 2022 (Table 3) in all comparable situations.

At 180, 221, and 300 m from the feedlot pens, concentrations were 111 CFU/m^3, 30 CFU/m^3, and 20 CFU/m^3, respectively. The presence of *E. coli* was observed to decrease by 82% with increasing distance from the source of the greatest emissions. Such a decrease agrees with the changes in concentration within the bioaerosol dispersion plume (see Table 4). The measurements at 180 m distance agree with the distance where the concentration peak would be found under worse air quality conditions using the screening method.

Table 4. *E. coli* concentration per m^3 of air (R, CFU/m^3), sampling site (ID), wind direction (DD), wind speed (v, m/s), distance between the sampling site and the pen with the highest *E. coli* concentration (M6 on 20 February 2020 and M7 on 5 April 2022) (D, m), sampling duration at 30 L/min (T, min), sampling time (hh:mm, local time (LT) of samplings on 20 February 2020 and 5 April 2022. w/c: Petri dishes without count.

Date	hh:mm (LT)	ID	DD	v (m/s)	D (M6, M7–Mi) (m)	T (min)	R (CFU/m^3)
20 February 2020	13:26	M8 (A7)	E	2.0–3.6	72	1	67
	13:45	M10 (A9)	E	3.6–<5.6	182	3	111
	13:55	M12 (A10)	NE	3.6–<5.6	221	3	33
	14:15	M14 (A12)	NE	2.0–3.6	300	5	20
5 April 2022	13:03	M20 (A21)	SSE	<0.5	160	3	11
	13:15	M21 (A22)	SSE	<0.5	55	3	w/c
	13:45	M16 (A23)	NE	<0.5	174	3	w/c
	14:08	M22 (A24)	NE	<0.5	210	10	w/c

On 5 April 2022, the peak measured concentration was found at 160 m from the pen, with a value of 11 CFU/m^3 (see Table 4) and a sampling period of 3 min.

4.3. Detection of *E. coli* by Deposition with Distance

The greatest deposition rate (DR) in the samples was 3.14 CFU/m^2 s in the pen with the greatest number of animals (see Table 5). The second greatest value was 2.6 $\pm$ 0.5 CFU/m^2 s, and in M1, M2, and M21 with a DR of 0.52 CFU/m^2 s. The maximum DR in sites located far from the pens was 1.57 CFU/m^2 s measured at M20, decreasing to 0.7 CFU/m^2 s at M16 and to 0.52 CFU/m^2 s at M21. The sampling point located 210 m away from the pens presented a deposition rate of 0.18 CFU/m^2 s. In the area of manure collection (M23), the measure was 0.52 CFU/m^2 s. (see Table 5 and Revised Supplementary Materials Tables S1–S5).

Table 5. Total mean *E. coli* deposition rate (DRm, CFU/m^2 s) standard deviation (ST), minimum *E. coli* deposition rate (DR_{min}, CFU/m^2 s), maximum *E. coli* deposition rate (DR_{max}, CFU/m^2 s), number of samples (N). Sampling date: 5 April 2022, sampling site: withing the pens (INT), around the pens (OUT). w/c: saturated Petri dishes without count.

	INT	OUT
DRm (CFU/m^2 s)	1.35	0.43
ST	1.24	0.50
DR_{min} (CFU/m^2 s)	w/c	w/c
DR_{max} (CFU/m^2 s)	3.14	1.57
N	7	9

4.4. Detection of Total Bacteria with Distance on 5 April 2022

The active sampling total bacterial count on 5 April 2022 showed a greater presence of colonies inside the pens at M1 and M2 (see Table 6 and Figure 1). The maximum number of total bacteria colonies in air was 1467 CFU/m^3, with a mean count of 889 $\pm$ 566 CFU/m^3 on 5 April 2022 (see Table 6). Meanwhile, outside the pens, the mean bacteria count was 534 $\pm$ 586 CFU/m^3, and the maximum was measured at M21 (see Table 6 and Figure 1), with 1378 CFU/m^3 (see Table 6).

Table 6. Average concentration of total bacteria per m^3 of air (Cm, CFU/m^3), concentration standard deviation (ST_C), minimum concentration of total bacteria (C_{min}, CFU/m^3), maximum concentration (C_{max}, CFU/m^3), number of considered samples (N). Sampling date: 5 April 2022, sampling site: within the pens (INT), around the pens (OUT).

	INT	OUT
C_m (CFU/m^3)	889	534
ST_C	566	586
C_{min} (CFU/m^3)	267	23
C_{max} (CFU/m^3)	1467	1378
N	5	5

The passive sampling presented 51.35 CFU/m^2 s inside the pens at M1 (see Table 7 and Figure 1) with a mean deposition of 26.45 ± 17.91 CFU/m^2 s inside the pens (see Table 7) and a maximum of 60.78 CFU/m^2 s at M20 (see Table 7 and Figure 1) downwind. The average deposition was 18.84 ± 17.94 CFU/m^2 s (see Table 7).

Table 7. Mean total bacteria deposition rate (DRm, CFU/m^2 s) standard deviation (ST), minimum total bacteria deposition rate (DR_{min}, CFU/m^2 s), maximum total bacteria deposition rate (DR_{max}, CFU/m^2 s), number of samples (N). Sampling date: 5 April 2022, sampling site: within the pens (INT), downwind the pens (OUT).

	INT	OUT
DRm (CFU/m^2 s)	26.45	18.84
ST	17.91	17.94
DR_{min} (CFU/m^2 s)	9.96	2.1 6
DR_{max} (CFU/m^2 s)	51.35	60.78
N	7	9

4.5. *Bioaerosol Atmospheric Dispersion*

Feedlot PM10 emissions were estimated based on the recommendations of USEPA [40], corrected by the number of heads at LPR, which resulted in a concentration of 27,945.2 kg/day 600 hd. Bioaerosols were estimated following the method defined in [44] and the result for LPR was 2812.5 mg/min [28] found that the emission from the feedlot pens they studied presented a diurnal variability with a peak between 10 am and 4 pm. Therefore, we assume that the emissions obtained in our work are associated with the time of maximum emissions at LPR.

Following the results of Jones and Harrison [7], we considered that bioaerosols were the 25% of the biological material that adhered to PM10, then, dispersion was modeled using the HYSPLIT model, and the 25% of the mean [40] pointed PM10 emission following the characteristics of LPR feedlot, resulting in 27,945.2 kg/day 600 hd. Dispersion estimates were calculated at the middle point of the sampling period.

Model vertical estimates show that the particles emitted on 5 April 2022 would reach a maximum height close to 1000 m, with peak concentrations in the lowest 500 m, while on 20 February 2020, the particles would reach 2000 m altitude with concentration peaks up to 1000 m. The maximum PM10 concentration on 20 February 2020 was 1.8 mg/m^3, and 0.44 mg/m^3 if only bioaerosol contribution is considered (see Figure 3a). The maximum concentration was at 340 m southwest of the feedlot, with a peak PM10 deposition of 8.6 mg/m^2 and 2.1 mg/m^2 of bioaerosols (see Figure 3b).

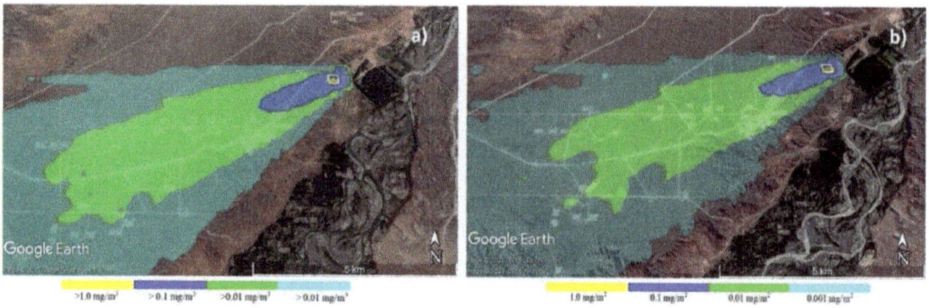

Figure 3. (**a**) Bioaerosol concentration downwind from the feedlot from HYSPLIT on 20 February 2020 at 13:00 UTC considering an emission source located from 0 to 1.5 m above the ground (**b**) deposition from HYSPLIT 20 February 2020 at 14:00 UTC.

On 5 April 2022, the peak PM10 concentration was 7.3 mg/m^3 or 1.7 mg/m^3 if only bioaerosols are considered (see Figure 4a,b). The maximum deposition was 430 mg/m^2 in the NE (M1–M4) and between 200 and 500 m downwind; bioaerosol concentration in a similar sector was 110 mg/m^2 with a peak between 17 and 300 m from the center of the feedlot, in the farmer working area. These values agreed with the sampling results shown in Table 4. *E coli* concentration peaks on 20 February 2020 and 5 April 2022 were at 300 m and 210 m from the center of the feedlot. The spatial distribution of airborne bioaerosol concentrations obtained from the dispersion model agrees with the measured values (see Figures 3 and 4 and Table 3). Site M6 had the greatest measured concentrations on 5 April 2022 and was located in the area of maximum concentration, as estimated by the model.

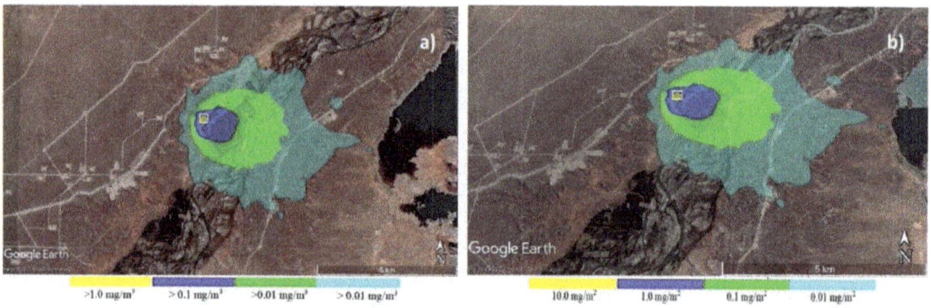

Figure 4. (**a**) Bioaerosol concentration downwind of the feedlot from HYSPLIT on the 5 April 2022 at 13:00 UTC considering an emission source between 0 and 1.5 m above the ground (**b**) deposition from HYSPLIT for the 5 April 2022 at 14:00 UTC.

5. Conclusions

This paper presents the results of two sampling events of bacteria at 39 spots in La Paisana ranch, in Añelo (Neuquén, Argentina), with emphasis on *E. coli*, considering spots inside the pens, upwind and downwind of the feedlot, with different time steps, using a Microflow α equipment and deposition sampling methods.

Cattle are the major reservoir of STEC O157, a group of foodborne bacteria associated with severe human diseases, such as the UHS. The province of Neuquén has one of the highest UHS incidence rates in the world, which is connected to the high proportion of *E. coli* (clade 8 strain of STEC O157:H7) in Argentine cattle. In our field experiments, we did not evaluate the presence of that serotype in the air. This is an issue that needs to be studied with further measurements.

From the methodological point of view, the analysis indicates that sampling times of 1 min in the pens and 3 min downwind are correct to measure bacteria concentrations in this feedlot. *E. coli* bacteria were detected in both field experiments in the center of the pens and downwind areas. The passive method measures of total bacteria and the modeled dispersion estimates presented good agreement in terms of the spatial distribution of bioaerosol concentrations. The model proved to be an important tool in field experiments planning and distribution of downwind concentration; however, the spatial resolution should be higher.

Airborne aerosol concentrations are likely affected by weather conditions. The greatest concentration was found on 20 February 2020, which seemed to be associated with a higher surface temperature. The mean surface temperature on 20 February 2020 was 39.5 °C, which would stimulate the growth of *E. coli* as this temperature is within the range of maximum growth of these bacteria. The maximum PM10 concentration on 20 February 2020 was 1.8 mg/m^3 and 0.44 mg/m^3 if only bioaerosol contribution was considered with maximum heights up to 2000 m according to the height of the mixed layer. The 20 February 2020 maximum estimated deposition was at 340 m downwind. This would indicate that the greatest concentration of bacteria would be associated with summertime conditions, in agreement with Zhong et al. [42]. However, it should be noted that viable bacteria can be found at temperatures as low as 5 °C or less. The concentration of total bacteria on 5 April 2022 showed a similar pattern of *E. coli* distribution.

The air temperature range on the days preceding the measurements was favorable for the surface microbial load to persist. After that, cattle behavior and the wind would lift the bacteria into the air. Because the study area is located in an arid environment, the relative humidity in summer is particularly low. This situation would also favor the suspension of dust and manure particles from the ground by even a light breeze.

Bioaerosol emissions have a negative impact on the population, especially in areas with a high density of livestock and in the vicinity of intensive livestock operations. This is in line with the global trend toward the intensification and industrialization of animal breeding. The understanding of the spatial and temporal dynamics of atmospheric bioaerosols as well as the pathways of emission and transport is important for future research, mainly in areas such as the province of Neuquén, which has one of the highest HUS incidence rates in the world, connected to the high proportion of *E. coli* in Argentine cattle.

6. Note

The measurement experiment was planned according to the equipment and resources available. The results are indeed variable because measurements were carried out at different times and dates. In the first instance, we planned to measure under different meteorological situations and seasons to explore the general behavior of *E.coli*. In the planning stage of the research project, we had the possibility of full collaboration with the Central Laboratory of Neuquén province for the analysis, but the COVID-19 pandemic changed our plans. The laboratory was fully dedicated to covid PCR studies, and we also had trouble entering the feedlot. The costs of chemical analysis and field experiments increased, which impeded us from performing as many experiments as we had planned. We are considering performing more experiments in similar conditions to obtain more reliable results.

Supplementary Materials: The following supporting information can be downloaded at: https://www.mdpi.com/article/10.3390/atmos13111761/s1, Table S1. Concentration of bacteria *E. coli* in active sampling in LPR feedlot pens on 20th February 2020. Table S2. Concentration of *E. coli* in active sampling around LPR feedlot pens on 20th February 2020. Table S3. Concentration of *E. coli* in active sampling in LPR feedlot pens on 5th April 2022. Table S4. Deposition of bacteria *E. coli* in passive sampling in LPR feedlot pens on 5th April 2022. Table S5. Deposition of *E. coli* in the passive sampling around the LPR feedlot pens on 5th April 2022.

Author Contributions: Conceptualization, M.G.C. and P.A.P.; methodology, M.G.C., P.A.P. and L.A.P.; formal analysis, M.G.C., P.A.P. and L.A.P.; investigation, M.G.C., P.A.P., L.A.P., M.A.C. and P.N.M.; resources, P.A.P., L.A.P., M.A.C. and P.N.M.; writing—original draft preparation, M.G.C. and P.A.P.; writing—review and editing, M.G.C., P.A.P. and L.A.P.; supervision, M.G.C.; project administration, P.A.P.; funding acquisition, M.G.C. and P.A.P. All authors have read and agreed to the published version of the manuscript.

Funding: This research was funded by Río Negro National University (PI 40-A-715 UNRN).

Institutional Review Board Statement: Not applicable.

Informed Consent Statement: Not applicable.

Data Availability Statement: Not applicable.

Acknowledgments: We acknowledge Julieta Sepúlveda, Melina Tello Sandoval, Luciana Di Sario, and Priscilla Lopez Maldonado for the technical support and P. Richter for the English translation and text editing.

Conflicts of Interest: The authors declare no conflict of interest.

References

1. Lighthart, B. The Ecology of Bacteria in the Alfresco Atmosphere. *FEMS Microbiol. Ecol.* **1997**, *23*, 263–274. [CrossRef]
2. Matthias-Maser, S.; Bogs, B.; Jaenicke, R. The Size Distribution of Primary Biological Aerosol Particles in Cloud Water on the Mountain Kleiner Feldberg/Taunus (FRG). *Atmos. Res.* **2000**, *54*, 1–13. [CrossRef]
3. Fröhlich-Nowoisky, J.; Kampf, C.J.; Weber, B.; Huffman, J.A.; Pöhlker, C.; Andreae, M.O.; Lang-Yona, N.; Burrows, S.M.; Gunthe, S.S.; Elbert, W.; et al. Bioaerosols in the Earth System: Climate, Health, and Ecosystem Interactions. *Atmos. Res.* **2016**, *182*, 346–376. [CrossRef]
4. McGinn, S.M.; Chen, D.; Loh, Z.; Hill, J.; Beauchemin, K.A.; Denmead, O.T. Methane Emissions from Feedlot Cattle in Australia and Canada. *Aust. J. Exp. Agric.* **2008**, *48*, 183–185. [CrossRef]
5. Clauß, M. *Emission of Bioaerosols from Livestock Facilities: Methods and Results from Available Bioaerosol Investigations in and around Agricultural Livestock Farming*; Johann Heinrich von Thünen-Institut: Braunschweig, Germany, 2020.
6. McGinn, S.M.; Flesch, T.K.; Chen, D.; Crenna, B.; Denmead, O.T.; Naylor, T.; Rowell, D. Coarse Particulate Matter Emissions from Cattle Feedlots in Australia. *J. Environ. Qual.* **2010**, *39*, 791–798. [CrossRef]
7. Jones, A.M.; Harrison, R.M. The Effects of Meteorological Factors on Atmospheric Bioaerosol Concentrations—A Review. *Sci. Total Environ.* **2004**, *326*, 151–180. [CrossRef]
8. Clark, S.; Rylander, R.; Larsson, L. Airborne Bacteria, Endotoxin and Fungi in Dust in Poultry and Swine Confinement Buildings. *Am. Ind. Hyg. Assoc. J.* **1983**, *44*, 537–541. [CrossRef]
9. Post, P.M.; Hogerwerf, L.; Bokkers, E.A.M.; Baumann, B.; Fischer, P.; Rutledge-Jonker, S.; Hilderink, H.; Hollander, A.; Hoogsteen, M.J.J.; Liebman, A.; et al. Effects of Dutch Livestock Production on Human Health and the Environment. *Sci. Total Environ.* **2020**, *737*, 139702. [CrossRef]
10. Straumfors, A.; Heldal, K.K.; Eduard, W.; Wouters, I.M.; Ellingsen, D.G.; Skogstad, M. Cross-Shift Study of Exposure response Relationships between Bioaerosol Exposure and Respiratory Effects in the Norwegian Grain and Animal Feed Production Industry. *Occup. Environ. Med.* **2016**, *73*, 685–693. [CrossRef]
11. Millner, P.D. Bioaerosols Associated with Animal Production Operations. *Bioresour. Technol.* **2009**, *100*, 5379–5385. [CrossRef]
12. Jahne, M.A.; Rogers, S.W.; Holsen, T.M.; Grimberg, S.J.; Ramler, I.P. Emission and Dispersion of Bioaerosols from Dairy Manure Application Sites: Human Health Risk Assessment. *Environ. Sci. Technol.* **2015**, *49*, 9842–9849. [CrossRef] [PubMed]
13. Kim, K.-Y.; Ko, H.-J.; Kim, D. Assessment of Airborne Microorganisms in a Swine Wastewater Treatment Plant. *Environ. Eng. Res.* **2012**, *17*, 211–216. [CrossRef]
14. Douwes, J.; Wouters, I.; Dubbeld, H.; van Zwieten, L.; Steerenberg, P.; Doekes, G.; Heederik, D. Upper Airway Inflammation Assessed by Nasal Lavage in Compost Workers: A Relation with Bio-Aerosol Exposure. *Am. J. Ind. Med.* **2000**, *37*, 459–468. [CrossRef]
15. Radon, K.; Schulze, A.; Ehrenstein, V.; van Strien, R.T.; Praml, G.; Nowak, D. Environmental Exposure to Confined Animal Feeding Operations and Respiratory Health of Neighboring Residents. *Epidemiology* **2007**, *18*, 300–308. [CrossRef]
16. Sneeringer, S. Does Animal Feeding Operation Pollution Hurt Public Health? A National Longitudinal Study of Health Externalities Identified by Geographic Shifts in Livestock Production. *Am. J. Agric. Econ.* **2009**, *91*, 124–137. [CrossRef]
17. Beran, G.W. Disease and Destiny–Mystery and Mastery. *Prev. Vet. Med.* **2008**, *86*, 198–207. [CrossRef] [PubMed]
18. Klous, G.; Huss, A.; Heederik, D.J.J.; Coutinho, R.A. Human–Livestock Contacts and Their Relationship to Transmission of Zoonotic Pathogens, a Systematic Review of Literature. *One Health* **2016**, *2*, 65–76. [CrossRef]

19. Rahman, S.; Mukhtar, S.; Wiederholt, R. Managing Odor Nuisance and Dust from Cattle Feedlots. 2008. Available online: https://www.semanticscholar.org/paper/Managing-Odor-Nuisance-and-Dust-from-Cattle-Rahman-Mukhtar/7e8ebd1f8c837fa7f4474ba509a9878b1b24a0f1 (accessed on 19 October 2022).
20. Bush, J.; Heflin, K.; Marek, G.; Bryant, T.; Auvermann, B. Increasing Stocking Density Reduces Emissions of Fugitive Dust from Cattle Feedyards. *Appl Eng Agric* **2014**, *30*, 815–824. [CrossRef]
21. Sweeten, J. Cattle Feedlot Manure and Wastewater Management Practices. In *Animal Waste Utilization: Effective Use of Manure as a Soil Resource*; Hatfield, J.L., Stewart, B.A., Eds.; CRC Press LLC: Boca Raton, FL, USA, 1998.
22. Greig, J.D.; Ravel, A. Analysis of Foodborne Outbreak Data Reported Internationally for Source Attribution. *Int. J. Food Microbiol.* **2009**, *130*, 77–87. [CrossRef]
23. Smith, J.L.; Fratamico, P.M.; Gunther, N.W. Shiga Toxin-Producing *Escherichia coli*. *Adv. Appl. Microbiol.* **2014**, *86*, 145–197. [CrossRef]
24. Pianciola, L.; Chinen, I.; Mazzeo, M.; Miliwebsky, E.; González, G.; Müller, C.; Carbonari, C.; Navello, M.; Zitta, E.; Rivas, M. Genotypic Characterization of *Escherichia coli* O157:H7 Strains That Cause Diarrhea and Hemolytic Uremic Syndrome in Neuquén, Argentina. *Int. J. Med. Microbiol.* **2014**, *304*, 499–504. [CrossRef] [PubMed]
25. Pianciola, L.; D'Astek, B.A.; Mazzeo, M.; Chinen, I.; Masana, M.; Rivas, M. Genetic Features of Human and Bovine *Escherichia coli* O157:H7 Strains Isolated in Argentina. *Int. J. Med. Microbiol.* **2016**, *306*, 123–130. [CrossRef] [PubMed]
26. Berry, E.D.; Wells, J.E.; Bono, J.L.; Woodbury, B.L.; Kalchayanand, N.; Norman, K.N.; Suslow, T.V.; López-Velasco, G.; Millner, P.D. Effect of Proximity to a Cattle Feedlot on *Escherichia coli* O157:H7 Contamination of Leafy Greens and Evaluation of the Potential for Airborne Transmission. *Appl. Environ. Microbiol.* **2015**, *81*, 1101–1110. [CrossRef] [PubMed]
27. Penner, J.E.; Andreae, M.O.; Annegarn, H.; Barrie, L.; Feichter, J.; Hegg, D.; Achuthan, J.; Leaitch, R.; Murphy, D.; Nganga, J.; et al. *Aerosols: Their Direct and Indirect Effects, The Scientific Basis, Contribution of Working Group I to the Third Assessment Report of the Intergovernmental Panel on Climate Change*; Dai, X., Maskell, K., Johnson, C.A., Eds.; Cambridge University Press: Cambridge, UK, 2001.
28. Bonifacio, H.F.; Maghirang, R.G.; Auvermann, B.W.; Razote, E.B.; Murphy, J.P.; Harner, J.P. Particulate Matter Emission Rates from Beef Cattle Feedlots in Kansas—Reverse Dispersion Modeling. *J. Air Waste Manag. Assoc.* **2012**, *62*, 350–361. [CrossRef]
29. Millner, P.; Suslow, T. *CA Lettuce Research Board 2007-08: Concentration and Deposition of Viable E. coli in Airborne Particulates from Composting and Livestock Operations*; National Agricultural Library: Salinas, CA, USA, 2008.
30. Bragoszewska, E.; Mainka, A.; Pastuszka, J.S. Concentration and Size Distribution of Culturable Bacteria in Ambient Air during Spring and Winter in Gliwice: A Typical Urban Area. *Atmosphere* **2017**, *8*, 239. [CrossRef]
31. Li, M.; Qi, J.; Zhang, H.; Huang, S.; Li, L.; Gao, D. Concentration and Size Distribution of Bioaerosols in an Outdoor Environment in the Qingdao Coastal Region. *Sci. Total Environ.* **2011**, *409*, 3812–3819. [CrossRef]
32. Prohaska, M.C. *Climates of Central and South America.*; Swerdtfeger, W., Ed.; Elsevier Scientific Pub. Co.: Amsterdam, the Netherlands, 1976.
33. Kottek, M.; Grieser, J.; Beck, C.; Rudolf, B.; Rubel, F. World Map of the Köppen-Geiger Climate Classification Updated. *Meteorol. Z.* **2006**, *15*, 259–263. [CrossRef]
34. Lässig, J.L.; Cogliati, M.G.; Bastanski, M.A.; Palese, C. Wind Characteristics in Neuquen, North Patagonia, Argentina. *J. Wind Eng. Ind. Aerodyn.* **1999**, *79*, 183–199. [CrossRef]
35. Gassmann, M.I.; Mazzeo, N.A. Air Pollution Potential: Regional Study in Argentina. *Environ. Manag.* **2000**, *25*, 375–382. [CrossRef]
36. Pasquill, F. The Estimation of the Dispersion of Windborne Material. *Meteorol. Mag.* **1961**, *90*, 33–49.
37. Li, Y.; Lu, R.; Li, W.; Xie, Z.; Song, Y. Concentrations and Size Distributions of Viable Bioaerosols under Various Weather Conditions in a Typical Semi-Arid City of Northwest China. *J. Aerosol Sci.* **2017**, *106*, 83–92. [CrossRef]
38. Paez, P.A.; Cogliati, M.G.; Pianciola, L.A.; Mut, P.N.; Ciaramaría, E.; Caputo, M.; Tesán, I. Dispersion of PM10 Bioaerosols Associated with Intensive Cattle Breeding in Argentinian Patagonia Provinces. In Proceedings of the III Jornadas Internacionales y V Nacionales de Ambiente, Buenos Aires, Argentina, 12–14 May 2021; pp. 3–6.
39. EPA. Screening Procedures for Estimating the Air Quality Impact of Stationary Sources, revised (Technical Report). 1992. Available online: https://www.osti.gov/biblio/6018929 (accessed on 19 October 2022).
40. USEPA. *Detecting and Mitigating the Environmental Impact of Fecal Pathogens Originating from Confined Animal Feeding Operations: Review*; Bibliogov: Cincinnati, OH, USA, 2005.
41. Stein, A.F.; Draxler, R.R.; Rolph, G.D.; Stunder, B.J.B.; Cohen, M.D.; Ngan, F. NOAA's HYSPLIT Atmospheric Transport and Dispersion Modeling System. *Bull. Am. Meteorol. Soc.* **2015**, *96*, 2059–2077. [CrossRef]
42. Zhong, X.; Qi, J.; Li, H.; Dong, L.; Gao, D. Seasonal Distribution of Microbial Activity in Bioaerosols in the Outdoor Environment of the Qingdao Coastal Region. *Atmos. Environ.* **2016**, *140*, 506–513. [CrossRef]
43. Ingraham, A.G.; Marr, J.L. *Escherichia coli and Salmonella: Cellular and Molecular Biology*, 2nd ed.; Microbiology, A.S., Ed.; American Society for Microbiology: Washington, DC, USA, 1996.
44. Alghamdi, M.A.; Shamy, M.; Redal, M.A.; Khoder, M.; Awad, A.H.; Elserougy, S. Microorganisms Associated Particulate Matter: A Preliminary Study. *Sci. Total Environ.* **2014**, *479–480*, 109–116. [CrossRef]

atmosphere MDPI

Article

Bayesian Inverse Modelling for Probabilistic Multi-Nuclide Source Term Estimation Using Observations of Air Concentration and Gamma Dose Rate

Kasper Skjold Tølløse [1,2,*] and Jens Havskov Sørensen [1]

1 Danish Meteorological Institute, DK-2100 Copenhagen, Denmark
2 Niels Bohr Institute, University of Copenhagen, DK-2100 Copenhagen, Denmark
* Correspondence: ktoe@dmi.dk

Abstract: In case of a release of hazardous radioactive matter to the atmosphere from e.g., a nuclear power plant accident, atmospheric dispersion models are used to predict the spatial distribution of radioactive particles and gasses. However, at the early stages of an accident, only limited information about the release may be available. Thus, there is a need for source term estimation methods suitable for operational use shortly after an accident. We have developed a Bayesian inverse method for estimating the multi-nuclide source term describing a radioactive release from a nuclear power plant. The method provides a probabilistic source term estimate based on the early available observations of air concentration and gamma dose rate by monitoring systems. The method is intended for operational use in case of a nuclear accident, where no reliable source term estimate exists. We demonstrate how the probabilistic formulation can be used to provide estimates of the released amounts of each radionuclide as well as estimates of future gamma dose rates. The method is applied to an artificial case of a radioactive release from the Loviisa nuclear power plant in southern Finland, considering the most important dose-contributing nuclides. The case demonstrates that only limited air concentration measurement data may be available shortly after the release, and that to a large degree one will have to rely on gamma dose rate observations from a frequently reporting denser monitoring network. Further, we demonstrate that information about the core inventory of the nuclear power plant can be used to constrain the release rates of certain radionuclides, thereby decreasing the number of free parameters of the source term.

Keywords: source characterization; atmospheric dispersion modelling; inverse modelling; Bayesian inference

Citation: Tølløse, K.S.; Sørensen, J.H. Bayesian Inverse Modelling for Probabilistic Multi-Nuclide Source Term Estimation Using Observations of Air Concentration and Gamma Dose Rate. *Atmosphere* **2022**, *13*, 1877. https://doi.org/10.3390/atmos13111877

Academic Editor: Daniel Viúdez-Moreiras

Received: 14 October 2022
Accepted: 7 November 2022
Published: 10 November 2022

1. Introduction

In case of a nuclear accident, radioactive particles and gasses may be released to the atmosphere. Consequently, an important part of emergency preparedness is to run simulations with atmospheric dispersion models, thereby predicting the atmospheric distribution as well as deposition of radioactive particles and gasses on the surface of the Earth. However, such models are subject to a number of uncertainties, the most important being the uncertainties of the meteorological predictions, inaccurate physics parameterizations in the dispersion model, and uncertainties of the estimated source term. Immediately after an accident in a nuclear power plant, only limited information about the release may be available. Thus, at the early stages of the accident, the dominating source of uncertainty is most likely the source term. If this is the case, inverse modelling can be used to obtain a source term estimate, which in turn can be used for running the atmospheric dispersion model. The aim of this study is to develop an inverse method for source term estimation, which is suited for operational use for emergency preparedness at the early stages of an accident, i.e., providing a source term estimate based on the limited data available shortly after the accident.

In the early phase of a nuclear power plant accident, a limited number of air concentration observations will be available, and these will typically have a low spatial and temporal resolution, e.g., the filters in such measurement stations may be changed every 24 h or even less frequently. In addition, there may exist gamma dose rate observations at much higher resolution, both spatially and temporally. However, since such measurements are the sum of contributions from all the different radionuclides, it is not clear a priori if they are useful for source term estimation.

Previous studies have used inverse methods for source term estimation. Lately, the still unaccounted for release of Ru-106 in the fall of 2017, was subject to several studies, e.g., [1–4]. However, since the release location has still not been confirmed, the main focus of these studies is localization of the source. The Fukushima Daiichi nuclear disaster in 2011, on the other hand, demonstrated that in-plant monitoring systems may not be working during a severe accident. Thus, different inverse methods have been applied in order to assess the source term. Some studies have estimated the release of certain radionuclides based solely on air concentration measurements [5,6], other include surface deposition measurements [7,8], while other again also include gamma dose rates [9]. Saunier et al. [9] demonstrate that information about ratios between the amounts of certain radionuclides can be used to further constrain the release rates. They use a variational approach to assess the source term, thereby providing a deterministic estimate. However, by using different Bayesian approaches, Liu et al. [6] show that significant uncertainties are associated with the estimated source term, indicating that probabilistic methods are better suited for this type of problem.

Most previous studies in this field aim at estimating the source term associated with accidents a long time after they occurred. However, for emergency preparedness, it is also important to be able to estimate source terms during the early stages, where especially air concentration measurement data are limited. This was addressed by Saunier et al. [9], who further developed their method to be applicable in real-time in case of an accident [10]. Our method is inspired by Saunier et al. [9,10], but instead we use a Bayesian inference method to be able to realistically account for uncertainties of the estimated source term, similar to Liu et al. [6].

The method is applied to an idealized artificial release case from the Finnish Loviisa nuclear power plant. A set of simulated air concentration measurements and gamma dose rate measurements have been created as described in Section 2.1. The same meteorological data and dispersion model have been used for data creation and for the source term estimation. Thus, the study demonstrates the uncertainties of the estimated source term arising only from the information loss due to the limited measurement capabilities. Due to the idealized nature of our study, our results apply to a situation, where model errors are negligible. In reality, meteorological uncertainties and model errors will further increase the uncertainty of the estimated source term.

Section 2 describes the data and methodology; Section 2.1 describes the synthetic measurement data set, Sections 2.2 and 2.3 describe the meteorological data and the dispersion model used, while Sections 2.4–2.7 describe the methodology. Next, the results are presented and discussed in Section 3. Finally, Section 4 presents a summary and the conclusions of the study.

2. Materials and Methods

2.1. *Artificial Loviisa Release Case*

For the artificial release from the Loviisa nuclear power plant in south Finland, the selected source term describes a core melt event without functioning mitigation systems. The initial event is a total loss of all power systems without battery back-up. The filtered containment venting system is assumed disconnected, and instead comprises an exhaust pathway from the reactor containment. It is postulated that the exhaust pathway was open at the time of melt-through of the reactor vessel. The released activity was corrected for decay and ingrowth for the time period between the emergency shutdown of the nuclear reactor (SCRAM) and the time of the release starting three hours later. It is assumed that there was no significant heat release associated with the accident, and therefore all material is released from a fixed height of 27 m above ground.

The time evolution is given in one-hour time steps starting at the onset of the accident (time of the SCRAM) and the following 12 h, intended to represent the first part of the release to undergo subsequent detection by the gamma monitoring stations and capture by the air filter stations. The source term was developed for the research project SOurce CHAracterizatiOn accounting for meTeorologIcal unCertainties (SOCHAOTIC), for further details, see [11].

Figure 1 shows the gamma dose field at the end of the simulation, 63 h after the release starts, as well as the locations of gamma dose rate stations and filter stations. The source term is given in Section 3.

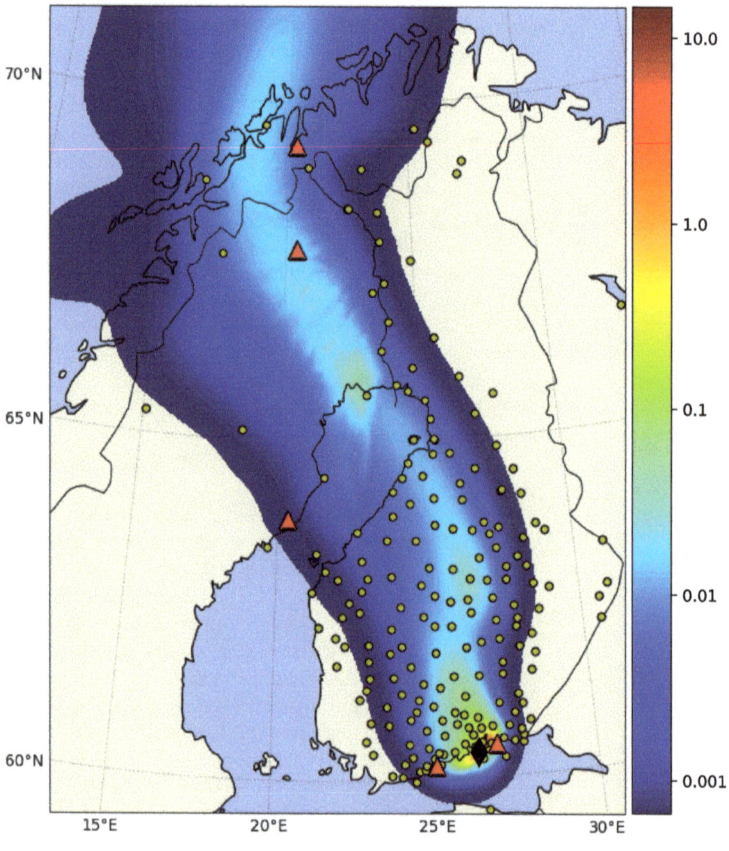

Figure 1. Total gamma dose in units of mSv at 63 h after the release start. Areas only influenced by background radiation are left uncolored. The black diamond shows the release location, the yellow circles show the locations of the gamma stations, and the red triangles show the locations of the filter stations.

2.1.1. Simulated Gamma and Filter Station Measurements

The total dose rate at the gamma monitoring stations is the sum of the contributions from cloud and ground since the stations are not shielded from activity deposited on the ground. Over time, the contamination of the station itself will also contribute to the measurements.

A set of 11 nuclides was selected to represent the most important nuclides for human doses: Kr-88*, Xe-133*, Xe-135*, Xe-135m*, Cs-134#, Cs-137, I-131#, I-132*#, I-133#, I-135# and Te-132. The list consists of the expected top five for the gamma monitoring stations (denoted by ∗), and top five for the air filter stations (denoted by #), expected to represent more than 90% of the dose rate contribution in the first 12 h of the postulated event. Moreover, two nuclides from the top ten list, Cs-137 and Te-132, were included since they represent key nuclides as seen from historical releases. For further details, see [11].

The artificial scenario consisting of simulated filter station and gamma station measurements was derived by predicting the atmospheric dispersion of radionuclides from a 9-hour release at the Loviisa nuclear power plant starting at 08:00 UTC on 22 September 2021. The DERMA atmospheric dispersion model was applied to the release scenario described above and using Harmonie data, cf. Sections 2.2 and 2.3, thereby providing average concentration values at existing filter stations, and gamma dose rates at gamma stations by using the ARGOS gamma dose model [12,13]. The filter concentration values are computed as 24 h averages from 08:00 UTC to 08:00 UTC the next day. Further, the filter measurement stations are assumed to have a detection limit of 0.1 $\mathrm{mBq\,m^{-3}}$. For the gamma dose rates, we have assumed a background radiation of 0.1 $\mathrm{\mu Sv\,h^{-1}}$, which has been added to all modelled dose rates.

2.2. Meteorological Data

The simulations have been carried out using meteorological data derived by the non-hydrostatic convection-permitting limited-area numerical weather prediction model Harmonie [14]. The horizontal grid resolution is approximately 2.5 km, and the vertical dimension is resolved by 65 levels with a terrain-influenced hybrid coordinate. The lowest model level is about 12 m above ground, and the highest at approximately 10 hPa. The model is configured with three-hourly data assimilation cycling. For the Loviisa case, the model simulation starts on 22 September 2021, at 00:00 UTC and runs until 24 September 2021, at 23:00 UTC.

2.3. Dispersion Modelling

The atmospheric dispersion is modelled by using the Danish Emergency Response Model of the Atmosphere (DERMA) [15,16]. DERMA is used operationally for a number of Danish emergency preparedness purposes [17–21] including nuclear [13]. The three-dimensional model is of Lagrangian type making use of a hybrid stochastic particle-puff diffusion description [15,16]. The model uses aerosol size dependent dry and wet deposition parameterizations as described by [22].

DERMA is interfaced with the nuclear decision-support system ARGOS (Accident Reporting and Guidance Operational System) [12,13], where the integration is accomplished through automatic online exchange of data between ARGOS and the DMI High Performance Computing (HPC) facility. The dose calculation modules are incorporated in ARGOS.

2.4. Problem Description

The temporal release profiles of the radionuclides considered are estimated by using observations of both air concentration and gamma dose rate combined with a series of forward runs by the dispersion model DERMA. We assume an overall start time t_0 and end time t_n of the release. We then separate the total release period into n time bins of duration Δt_{bin} and for each of these assume a unit release of each of the included radionuclides. The releases are assumed to be point releases at ground level. As described in Section 2.1.1, we assume that only a selection of all released radionuclides contributes significantly to the gamma dose rates, while other radionuclides will be ignored. Let C^{o}_{ik} be the k'th observed average concentration of the i'th radionuclide, measured over a specified time period at a specified filter station. Similarly, let $\Gamma^{\mathrm{o}}_{\kappa}$ be the κ'th observed gamma dose rate measured at a specified time and gamma station.

The atmospheric dispersion model DERMA is run forward in time for each of the unit releases, and for the j'th release of the i'th radionuclide the average activity concentrations $\bar{c}_{ijk}$ are calculated, where the k-index corresponds to the location and time of an existing filter measurement. Further, instantaneous activity concentrations $c_{ij\kappa}$ and deposition values $d_{ij\kappa}$ are calculated, where the κ-index corresponds to the location and time of an existing gamma dose rate observation. By using the gamma dose model as described in Section 2.3, the contributions to the gamma dose rates $\gamma_{ij\kappa} = \gamma_{ij\kappa}(c_{ij\kappa}, d_{ij\kappa})$ are calculated. For a given set of non-negative coefficients, λ_{ij}, the predicted average concentrations and gamma dose rates corresponding to existing measurements are calculated:

$$\begin{aligned} C^{\mathrm{m}}_{ik} &= \sum_j \lambda_{ij}\bar{c}_{ijk} \\ \Gamma^{\mathrm{m}}_{\kappa} &= \sum_i \sum_j \lambda_{ij}\gamma_{ij\kappa}. \end{aligned} \tag{1}$$

2.5. Bayesian Inversion and Sampling Method

Given a set of observations, $(\mathbf{C}^{\mathrm{o}}, \boldsymbol{\Gamma}^{\mathrm{o}})$, the coefficients, $\boldsymbol{\lambda}$ can be determined by applying Bayes' theorem:

$$P(\boldsymbol{\lambda}, \boldsymbol{\theta}|\mathbf{C}^{\mathrm{o}}, \boldsymbol{\Gamma}^{\mathrm{o}}, I) = \frac{P(\boldsymbol{\lambda}, \boldsymbol{\theta}|I)\, P(\mathbf{C}^{\mathrm{o}}, \boldsymbol{\Gamma}^{\mathrm{o}}|\boldsymbol{\lambda}, \boldsymbol{\theta}, I)}{P(\mathbf{C}^{\mathrm{o}}, \boldsymbol{\Gamma}^{\mathrm{o}}|I)}, \tag{2}$$

where $\boldsymbol{\theta}$ denotes any so-called nuisance parameters, i.e., unknown parameters, which are not of direct interest. One way to account for these is to treat them just like the parameters of interest and consider $P(\boldsymbol{\lambda}, \boldsymbol{\theta}|\mathbf{C}^{\mathrm{o}}, \boldsymbol{\Gamma}^{\mathrm{o}}, I)$, which is the posterior probability distribution for the combined set of parameters $(\boldsymbol{\lambda}, \boldsymbol{\theta})$. $P(\boldsymbol{\lambda}, \boldsymbol{\theta}|I)$ is then the prior probability distribution for $(\boldsymbol{\lambda}, \boldsymbol{\theta})$, $P(\mathbf{C}^{\mathrm{o}}, \boldsymbol{\Gamma}^{\mathrm{o}}|\boldsymbol{\lambda}, \boldsymbol{\theta}, I)$ is the likelihood, and $P(\mathbf{C}^{\mathrm{o}}, \boldsymbol{\Gamma}^{\mathrm{o}}|I)$ is the evidence; a statistical constant independent of $(\boldsymbol{\lambda}, \boldsymbol{\theta})$. I is any background information that may be available, e.g., amount of material present in the core at the time of the accident.

To evaluate Equation (2), the quantities $P(\boldsymbol{\lambda}, \boldsymbol{\theta}|I)$ and $P(\mathbf{C}^{\mathrm{o}}, \boldsymbol{\Gamma}^{\mathrm{o}}|\boldsymbol{\lambda}, \boldsymbol{\theta}, I)$ must be estimated for a selection of realizations of $(\boldsymbol{\lambda}, \boldsymbol{\theta})$, and the resulting posterior probability distribution $P(\boldsymbol{\lambda}, \boldsymbol{\theta}|\mathbf{C}^{\mathrm{o}}, \boldsymbol{\Gamma}^{\mathrm{o}}, I)$ can then be estimated by normalizing the distribution. The posterior probability distribution for $\boldsymbol{\lambda}$ can then be determined by marginalizing:

$$P(\boldsymbol{\lambda}|\mathbf{C}^{\mathrm{o}}, \boldsymbol{\Gamma}^{\mathrm{o}}, I) = \int_{\boldsymbol{\theta}} P(\boldsymbol{\lambda}, \boldsymbol{\theta}|\mathbf{C}^{\mathrm{o}}, \boldsymbol{\Gamma}^{\mathrm{o}}, I)\, \mathrm{d}\boldsymbol{\theta}. \tag{3}$$

To get a good estimate of the probability distribution, the relevant parts of the parameter space must be sampled. One option is to use random-walk based Markov Chain Monte Carlo (MCMC) methods, such as Metropolis-Hastings or Gibbs [23,24]. However, these methods generally require a large number of iterations, because the random-walk based model proposals do not sample the parameter space of the posterior probability distribution in the most efficient way. Further, parameters such as the step size of the random-walk typically need to be tuned to the specific case. Instead, we use the Hamiltonian Monte Carlo (HMC) method No U-Turn Sampling (NUTS) [25], implemented in the Python library PyMC3 [26]. HMC methods generally have an advantage over random-walk based MCMC methods, because the model proposals are not generated by a random-walk but instead based on estimated gradients of the posterior distribution. Thus, much fewer iterations are typically needed to sufficiently sample the probability distribution. However, the efficiency of HMC algorithms strongly depends on the step size parameter. The NUTS algorithm uses adaptive step sizing such that the step size does not need to be set by the user. Further, as the name suggests, the algorithm is constructed such that trajectories in the parameter space avoid making "U-turns", i.e., retracing their own steps. Thus, it should produce more independent samples in fewer iterations. When the aim is to use Bayesian inverse modelling operationally, the NUTS algorithm is ideal, since very little parameter tuning is necessary [25]. In addition, when using the PyMC3 implementation [26], Gelman-Rubin convergence diagnostics [27] are automatically calculated, when sampling with two or more chains. This makes it easy to control that the sampler has converged. For further details on the NUTS algorithm, see [25].

2.6. Prior Probability Distributions

Defining useful prior probability distributions for the release rates is challenging, since the magnitude of the release is unknown. To allow for variation over several orders of magnitude while ensuring non-negative values, we use log-normal prior distributions. Assuming a normal distributed variable $x \sim \mathcal{N}(\mu, \sigma)$, then the variable $z = e^x \sim \text{Lognormal}(\mu, \sigma)$ is log-normal distributed with parameters μ and σ. Thus, these denote the mean and standard deviation of x and not of the log-normal distributed variable z. The prior probability distribution for the coefficients λ_{ij} can be written as:

$$P(\lambda_{ij}|I) = \text{Lognormal}(\mu_i, \sigma_i), \tag{4}$$

where μ_i and σ_i are parameters to be determined for the specific radionuclide. Given that total amount of the i'th radionuclide in the core inventory is S_i in units of Bq, the upper limit for λ_{ij} is $S_i/\Delta t_{\text{bin}}$, where Δt_{bin} is the duration in seconds of each assumed unit release. To allow for release rates approaching the upper limit with reasonable probability, we set $\mu_i + 2\sigma_i = \log(S_i/\Delta t_{\text{bin}})$, where log() denotes the natural logarithm. The lower limit must be small compared the "typical" release rate, μ_i. Since the typical release rate is unknown, we assume $\mu_i = \log(fS_i/\Delta t_{\text{bin}})$, where f is some (small) fraction. Assuming a sufficiently low value for f will result in a conservative prior distribution, which allows for a broader range than necessary. In this study, we use $f = 10^{-3}$, which means that $\mu_i \pm 2\sigma_i$ includes six orders of magnitude for each release rate. Thus, the mean and standard deviations for the prior probability distributions are given as:

$$\mu_i = \log(10^{-3} S_i/\Delta t_{\text{bin}}) \quad \text{and} \quad \sigma_i = \frac{1}{2}\log(10^3). \tag{5}$$

Further, we can use information about the core inventory to reduce the parameter space by imposing correlations between release rates of certain radionuclides, inspired by the method by Saunier et al. [9,10]. For example, two different isotopes of the same element will largely behave similarly during a release. Thus, if the half-lives of two such isotopes are long compared to the duration of the release and if there is no significant ingrowth from other processes, the ratio of the release rates between two isotopes can be assumed constant and equal to the ratio of the amounts in the core inventory. For example, ^{134}Cs and ^{137}Cs have half-lives of approximately 2 and 30 years, respectively, and thus, the ratio of their activity concentrations in the core inventory can be considered constant during the release. Accordingly, based on the amounts of the two isotopes in the core, we can assume the ratio of their release rates to be constant.

For other isotope pairs, it is necessary to take into account the difference in half-lives in order to set realistic constraints on the release rates. In this case, knowing the amount of the two isotopes at the time of SCRAM gives one limit for the isotopic ratios, while estimating the activity concentration n hours later will provide an estimate of the other limit, assuming no significant ingrowth. An example is the isotope pair ^{131}I and ^{133}I, which has half-lives of approximately 8 days and 20.8 h, respectively. Let the release rates of these isotopes be $q_{^{131}\text{I}}$ and $q_{^{133}\text{I}}$, respectively. Based on their activity concentrations in the core at the time of the accident, we have $\frac{q_{^{133}\text{I}}}{q_{^{131}\text{I}}} < 2.1$. Assuming that the duration of the main release is less than 24 h, we can determine the other limit. Due to radioactive decay during these 24 h, the amount of ^{133}I is decreased by a factor of 0.45, while we assume that the amount of ^{131}I is unchanged due to its relatively long half-life. Thus, a lower limit can be determined $\frac{q_{^{133}\text{I}}}{q_{^{131}\text{I}}} > 0.9$. Following this approach, we determine the following constraints:

$$\frac{q_{^{134}\text{Cs}}}{q_{^{137}\text{Cs}}} = 1.4, \;\; 0.001 < \frac{q_{^{132}\text{I}}}{q_{^{131}\text{I}}} < 1.5, \;\; 0.9 < \frac{q_{^{133}\text{I}}}{q_{^{131}\text{I}}} < 2.1 \;\text{ and }\; 0.15 < \frac{q_{^{135}\text{I}}}{q_{^{131}\text{I}}} < 2.0. \tag{6}$$

For ^{134}Cs and ^{137}Cs, this effectively means that only one release rate needs to be determined instead of two, and that the combined set of measurements of the two isotopes can be used. For the other isotope pairs, the constraints allow us to define log-normal distributions with upper and lower bounds, which depend on the release rate of one of the other nuclides. Let λ_{mj} and λ_{nj} be the coefficients for two release rates, which are related by the flexible constraints $r_{\text{lower}} < \lambda_{nj}/\lambda_{mj} < r_{\text{upper}}$. Then, the prior probability distribution for λ_{mj} will be defined as in Equation (4), while the prior probability distribution for λ_{nj} can be written as:

$$P(\lambda_{nj}|I, \text{constraints}) \propto \begin{cases} P(\lambda_{nj}|I) & r_{\text{lower}} < \frac{\lambda_{nj}}{\lambda_{mj}} < r_{\text{upper}} \\ 0 & \text{otherwise} \end{cases}. \quad (7)$$

It might be possible to impose further constraints, i.e., across the type of element, such that the release rates of the iodine isotopes can also be related to the release rates of the caesium isotopes, Te-132 and the noble gasses. However, the underlying assumptions in this case are less trivial.

2.7. Likelihood and Uncertainty Quantification

The likelihood is the probability of observing the set of measurements $(\mathbf{C}^{o}, \mathbf{\Gamma}^{o})$, given a proposed source term, $\boldsymbol{\lambda}$. The likelihood is evaluated by assuming a probability distribution for the residuals $C^{o}_{ik} - C^{m}_{ik}(\lambda_{ij})$ and $\Gamma^{o}_{\kappa} - \Gamma^{m}_{\kappa}(\lambda_{ij})$. In this study, we use a log-normal likelihood, which is less sensitive to outliers than the Gaussian distribution and automatically gives a higher weight to measurements/predictions of low values. This makes it useful when dealing with measurement values over several orders of magnitude [6]. One practical challenge when dealing with log-normal distributions is that only positive values are mathematically allowed, while the physical quantity may in principle be zero. For the gamma dose rates, this is not an issue, since we add background radiation to the modelled measurements, thereby ensuring that values are always positive. For the air concentration measurements, on the other hand, modelled predictions may be zero, while the measured predictions may be below the detection limit. Assume that for a given measurement, C^{o}_{ik}, the detection limit is ϵ_{ik}. To avoid zero-values, we use these altered observations and model predictions $\widetilde{C^{o}_{ik}} = \max(\epsilon_{ik}, C^{o}_{ik})$ and $\widetilde{C^{m}_{ik}} = \max(\epsilon_{ik}, C^{m}_{ik})$. These altered forms have the additional benefit that they provide a theoretically sound way of using non-detections, since these will only contribute to the likelihood, when the modelled concentration is above the detection limit. Thus, there is no risk of falsely interpreting a low value as a zero. The likelihood is given as:

$$P\left(\widetilde{\mathbf{C}^{o}}, \mathbf{\Gamma}^{o}|\boldsymbol{\lambda}, I\right) = \prod_{k}\prod_{i} \text{Lognormal}\left(\widetilde{C^{m}_{ik}}, \sigma_{f}\right) \prod_{\kappa} \text{Lognormal}\left(\Gamma^{m}_{\kappa}, \sigma_{g}\right), \quad (8)$$

where C^{m}_{ik} and Γ^{m}_{κ} are as defined in Equation (1). σ_{f} and σ_{g} are related to the uncertainty of the measurements as well as the unknown model errors. In this study, both are negligible as discussed in Section 1. However, in order to make the method as general as possible, the uncertainty parameters are treated as nuisance parameters, i.e., they are kept as free parameters and sampled by the Monte Carlo algorithm. In practice, a wide uniform distribution has been used as prior distribution for the nuisance parameters $\sigma_{f}, \sigma_{g} \sim U(0, 10)$, which allows for a broad range of shapes of log-normal distributions.

3. Results and Discussions

As described in Section 2.5, the results are obtained by using the NUTS algorithm [25], which is implemented in the PyMC3 python library [26]. The algorithm is constructed in such a way that almost no parameter tuning is necessary. To ensure convergence, the target acceptance rate was increased from the default 0.8 to 0.99. Aside from this, everything was kept at PyMC3's default values; two simultaneously running chains, each with 1000 tuning steps and 1000 draws from the target distribution. This provides a total of 2000 realizations of the posterior probability distribution. For further details on the NUTS parameters, see [25,26].

In our analysis, we include 10 of the 11 radionuclides described in Section 2.1.1, excluding Xe-135m based on the rationale that its short half-life of approximately 15 min makes it unimportant on longer temporal, and thus also spatial scales. This means that there is not enough information in the measurement data to sufficiently constrain the release rate of Xe-135m. The other three noble gasses are included, although there are no filter measurements to help constrain their release rates. However, as long as their half-lives are sufficiently different, we expect the gamma dose rate patterns to differ enough to be able to distinguish between their contributions. The prior probability distributions for the release rates of Kr-88, Xe-133, Xe-135, Cs-137, I-131 and Te-132 were defined as log-normal distributions, Equation (4) with mean and standard deviations given by Equation (5). The release rate for Cs-134 was defined as a deterministic variable, equal to the release rate for Cs-137 multiplied by the fixed ratio 1.4. Finally, the prior distributions for the release rates of I-132, I-133 and I-135 were defined as bound log-normal distributions Equation (7), where the bounds are given by the flexible constraints, Equation (6).

We assume that the time of the emergency shutdown of the nuclear reactor (SCRAM), 22 September, 05:00 UTC, is known. We therefore consider this as the first possible time of release. We then consider the release during the following 24 h by assuming twelve 2-h constant releases, i.e., $\Delta t_{\text{bin}} = 7200$ s. The source term estimation is based on the simulated measurements described in Section 2.1.1, but only measurements until 23 September, 08:00 UTC are used for the source term estimation, leaving the remaining measurements for validation of model predictions based on the estimated source term. Thus, for all particles, only two 24-h filter measurements from each of the five filter stations are available, i.e., ten filter measurements per particle. However, first, all measurements without any information are discarded; if a given measurement is not influenced by any of the time-binned unit releases, it is removed from the data set. After this automatic removal of data, only one filter measurement per particle from each of the two filter stations in southern Finland are left. Thus, even when using the additional constraints described in Section 2.6, the amount of filter measurement data is very limited.

The gamma dose rates, on the other hand, are measured every hour at 214 different locations, see Figure 1. Thus, from 22 September, 05:00 UTC to 23 September, 08:00 UTC, a total of 5778 measurements. After the automatic removal of data without information, 1918 measurements are left.

Given the high dimensionality of the parameter space, it is not possible to visualize all elements of the actual posterior distribution. Instead the individual release rates are shown in Figure 2. The plots show the median release rates as well as the 10th and 90th percentiles based on marginal distributions for each 2-h release period. Further, Figure 3 shows histograms of the marginal distributions of time integrated releases for all radionuclides. The only release rate, which is well determined for most time bins is that of Xe-133. This makes sense, since it is the only relatively long-lived noble gas; the half-life is approximately five days, while Xe-135 and Kr-88 have half-lives of roughly nine and three hours, respectively. Further, since the noble gasses do not deposit, the gamma dose rate pattern of Xe-133 will also be easy to distinguish from those of the long-lived particles. For the particles, the estimated release rates clearly indicate the effects of the constraints in Equation (6); the release rates of the four iodine isotopes, which are all "tied together", are better estimated than those of both the caesium isotopes and of Te-132. Since the release rates of the two caesium isotopes are forced to differ only by a factor, we also expect these to be better estimated than the release rate of Te-132. While it is not easy to see that this is the case, it is clear from Figure 3 that the released amounts of the two caesium isotopes are better estimated than Te-132.

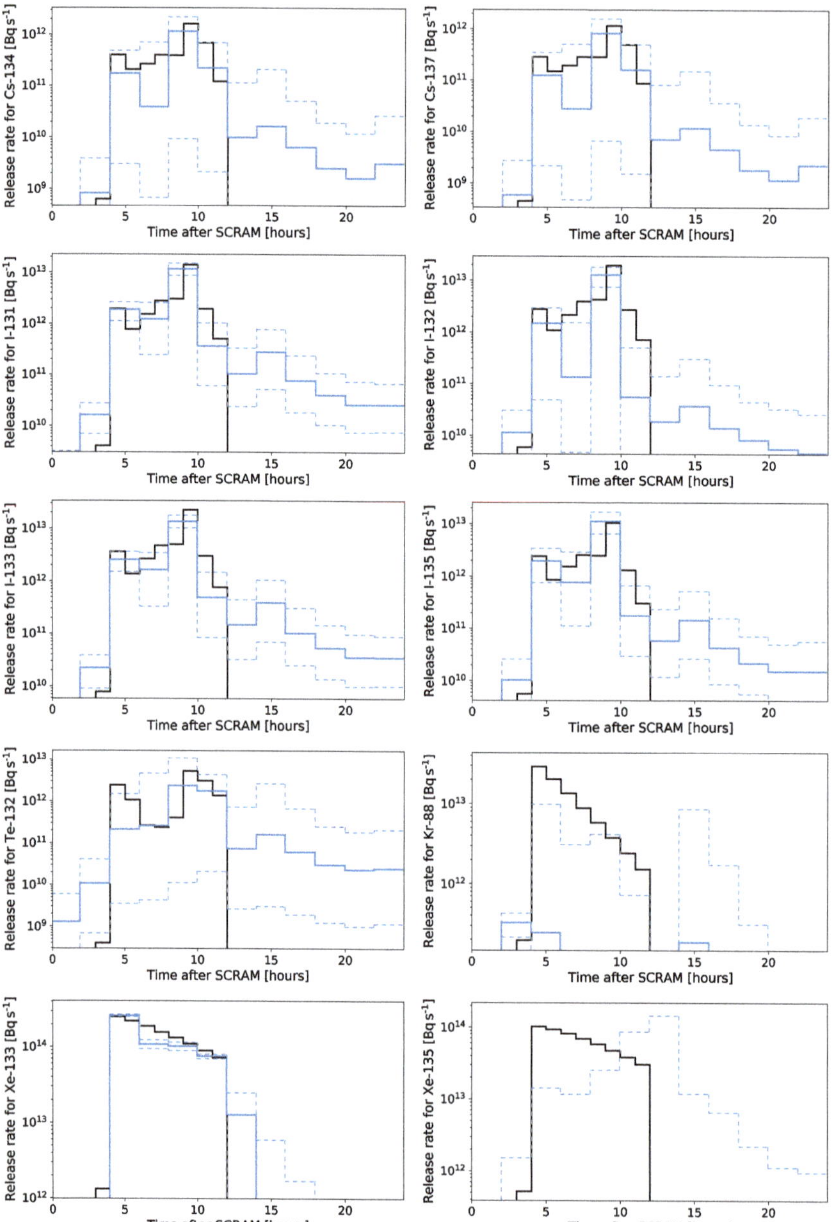

Figure 2. Release rates for each radionuclide in each 2-h time bin. The solid blue lines show the medians of the marginal distributions, while the dashed blue lines show the 10th and 90th percentiles. For comparison, the solid black lines show the true release profile. To focus on the release rates of high magnitude, we have set the minimum value on the y-axis to 10% of the lowest true release rate. Thus, predicted release rates below this limit are not shown in the plot, e.g., the predicted release rate of Xe-135 only shows the 90th percentile, while both the 10th percentile and the median are below the axis limit.

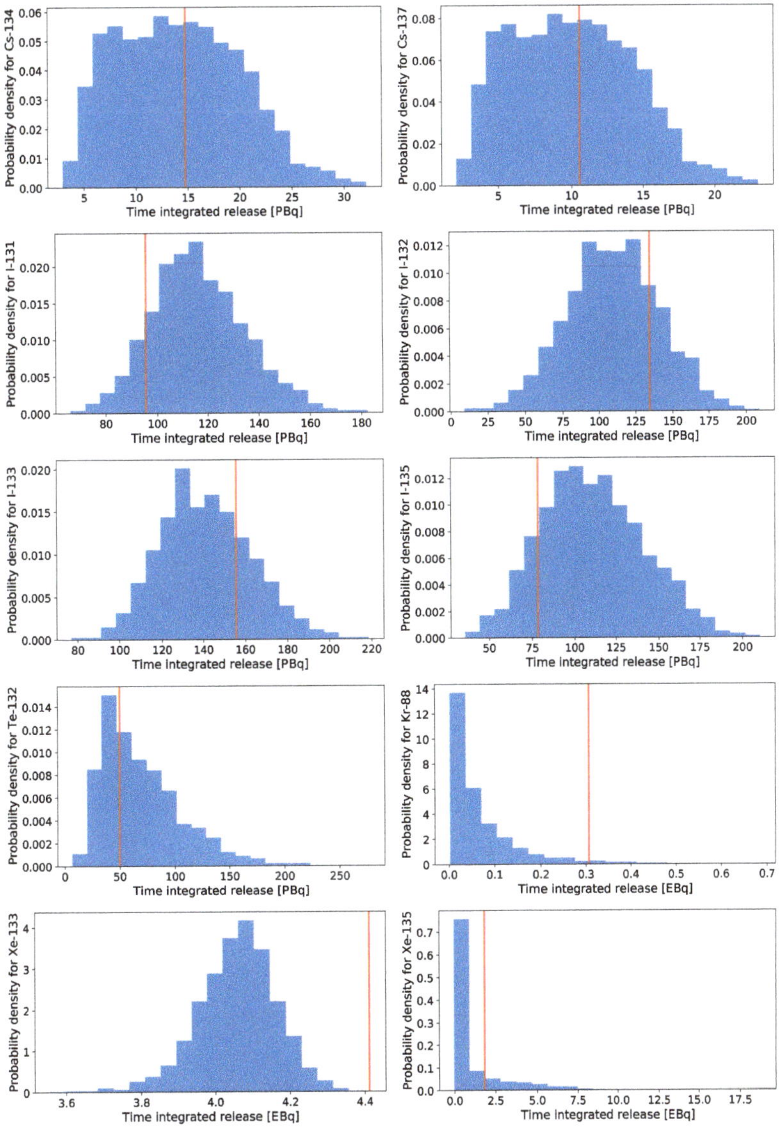

Figure 3. Probability density for each radionuclide as a function of time-integrated release. The vertical red lines show the actual released amounts.

The histograms in Figure 3 show that for some radionuclides, the amounts are quite well constrained, e.g., the release of I-131, which varies from roughly 70 PBq to 180 PBq, and Xe-133, which varies from roughly 3.6 EBq to 4.4 EBq. The latter, however, only barely include the true released amount in the probability distribution. For the remaining radionuclides, the released amounts are not very accurately estimated, especially not for Kr-88 and Xe-135. Given the limited amount of measurement data, this result is not surprising. Further, it is important to note that the log-normal prior distribution ensures release rates of positive values. Hence, the estimated release will necessarily have the same duration as the considered release period, 24 h in this case. However, we see from Figure 2 that most release rates drop significantly in magnitude after 12 h from SCRAM.

From Figures 2 and 3, it may seem that the source term is not sufficiently constrained by the data. Clearly, release rates for some nuclides are poorly estimated, e.g., Kr-88 and Xe-135, and it may therefore be tempting to exclude these from the source term. However, we found that when excluding these, the estimated release rates of the remaining nuclides are less accurate. Thus, it seems that the release of some of the other nuclides compensate for their lacking contribution. On the other hand, it is important to note that including Kr-88 and Xe-135 in the source term does not seem to compromise the release rates of the remaining nuclides. Thus, when it is not known a priori which nuclides constitute the best possible source term, the safer choice seems to be to include more nuclides than necessary. Further, the marginal distributions are obtained by integrating over the remaining parameters of the multi-nuclide source term, and therefore all correlations between parameters are ignored. As demonstrated below, though the marginal distributions of individual releases might be uncertain, the gamma dose rate patterns of different realizations of the multi-nuclide source term vary significantly less.

Figure 4 shows predicted air concentrations and gamma dose rates as function of observations. The upper plots show filter measurements, and the lower plots show gamma dose rates. The left plots show measurements before 23 September, 08:00 UTC, i.e., the measurements that are used for the source term estimation. The right plots show measurements after 23 September, 08:00 UTC and therefore show a prediction of future values based on the estimated source term. The percentiles are estimated by first calculating the concentrations and gamma dose rates from all source terms in the posterior distribution and then finding the percentiles in the calculated values. The plots with the gamma dose rates show a randomly selected subset of 300 observations, since more data in the plot makes it impossible to distinguish the different data points. The figure shows that the average activity concentrations at the filter stations are generally estimated to match the observations within the uncertainties, although some allow for a wide variation. On the other hand, the predicted gamma dose rates fit very well with the observed even for the predicted values. Considering the fact that a total of 1918 gamma measurements and only 2 filter measurements for each nuclide are used for the inversion, it is not surprising that the gamma dose rates are more accurately estimated.

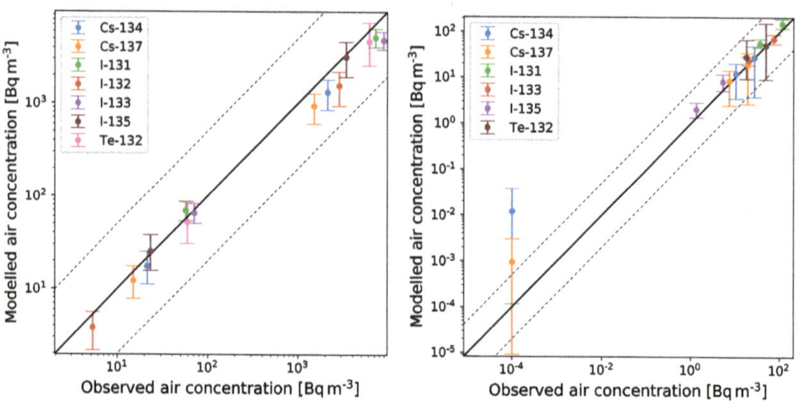

(a) (b)

Figure 4. *Cont.*

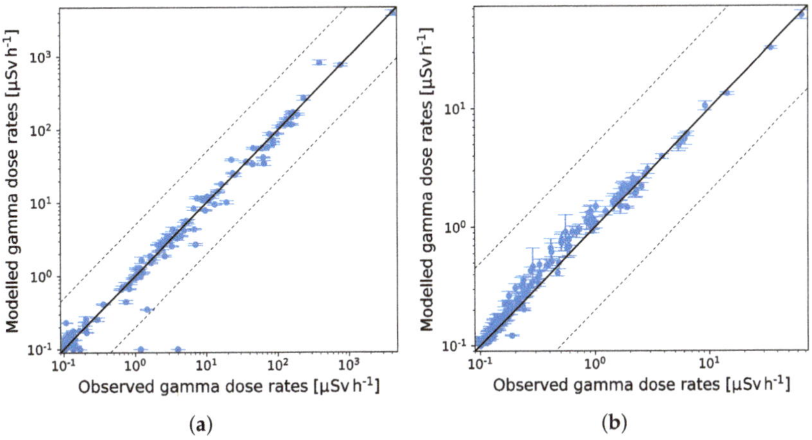

Figure 4. Model predictions with uncertainties (median and 10th and 90th percentile) on the *y*-axis, and observations on the *x*-axis. The solid black lines indicate a perfect correlation, while the dashed black lines indicate a factor of 5 between model and observation. (**a**) shows the filter measurements until 23 September, 08:00 UTC, i.e., the measurements that are used for the source term estimation, whereas (**b**) shows the filter measurements after 23 September, 08:00 UTC, i.e., predicted future air concentrations. (**c**) similarly shows the gamma dose rates until 23 September, 08:00 UTC, and (**d**) shows the gamma dose rates after 23 September, 08:00 UTC.

Figure 5 shows the predicted gamma dose rates at the locations of six selected gamma stations, viz. the six stations that measured the highest values. The plots show that there is good agreement between modelled observed gamma dose rates and that even the time evolution is captured very well.

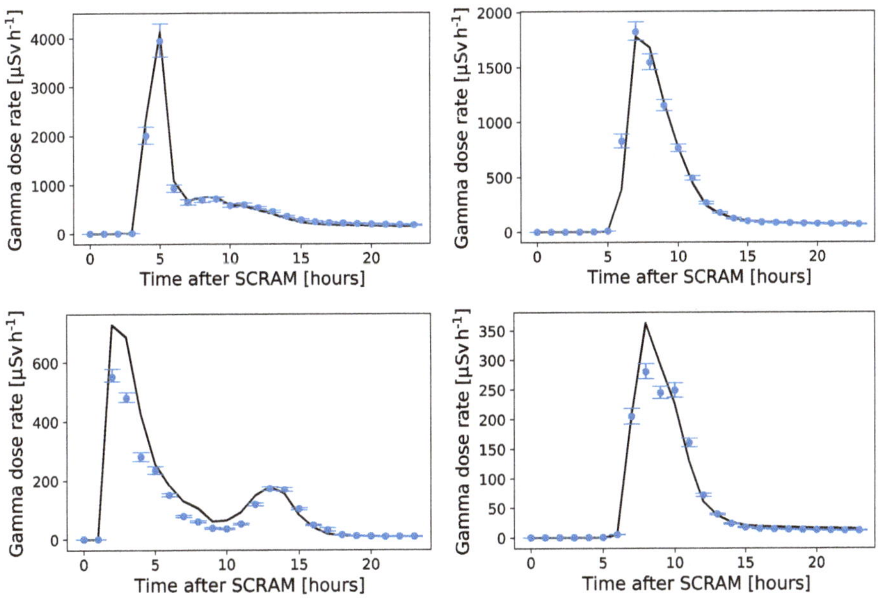

Figure 5. *Cont.*

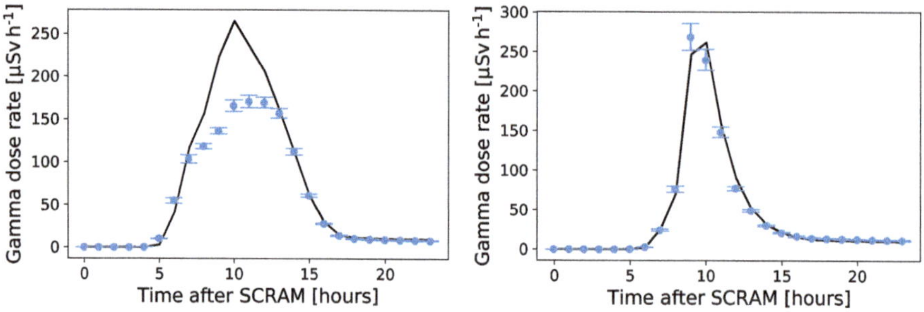

Figure 5. Gamma dose rates at locations of gamma stations during the first 24 h after the accident. Model predictions with uncertainties (median and 10th and 90th percentile) are shown by the blue dots and error bars, while the true gamma dose rates are shown by the black solid line. The selected gamma stations are all close to release locations, viz. the six stations that measured the highest values during the first 24 h.

Finally, Figure 6 shows the probability distributions of the two uncertainty parameters σ_f and σ_g; both parameter distributions indicate relatively narrow log-normal distributions, which is expected given that model errors are negligible.

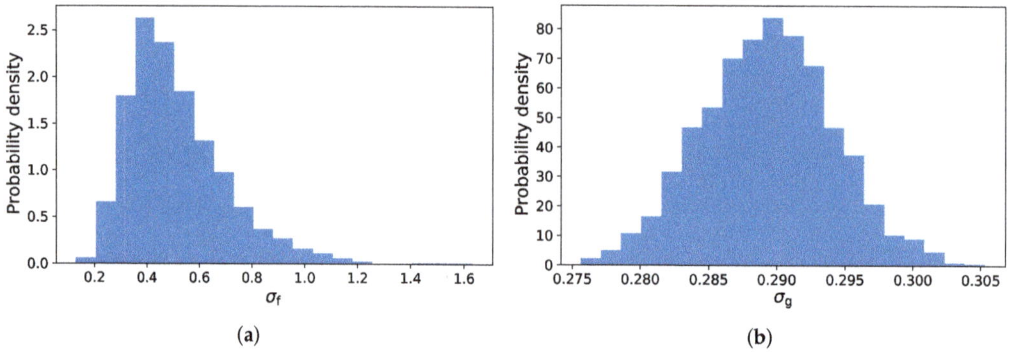

Figure 6. Marginal probability distributions of the uncertainty parameters, (**a**) σ_f and (**b**) σ_g.

3.1. Including All Data

For comparison, we show the estimated source term when including all measurements. Figure 7 shows the release rates and probability densities of released amounts for three selected nuclides, Cs-134, I-131 and Xe-133. Interestingly, the release rates are all better defined than the previous result, i.e., the distributions are narrower. However, the release rate estimates are not necessarily more accurate. On the other hand, comparison with Figure 2 shows that the use of later measurements allows for a better estimate of the duration, as all release rates are very low after 16 h from the SCRAM.

As discussed previously, there are not many filter measurements available, and therefore the gamma dose rates are dominant; thus, the estimated source term is more likely to match the gamma dose rates than the filter measurements. This is apparent from Figure 8, which shows the modelled air concentrations and gamma dose rates as function of observations, similar to Figure 4. There is a very good agreement for gamma dose rates, while for air concentrations, the discrepancy is somewhat larger.

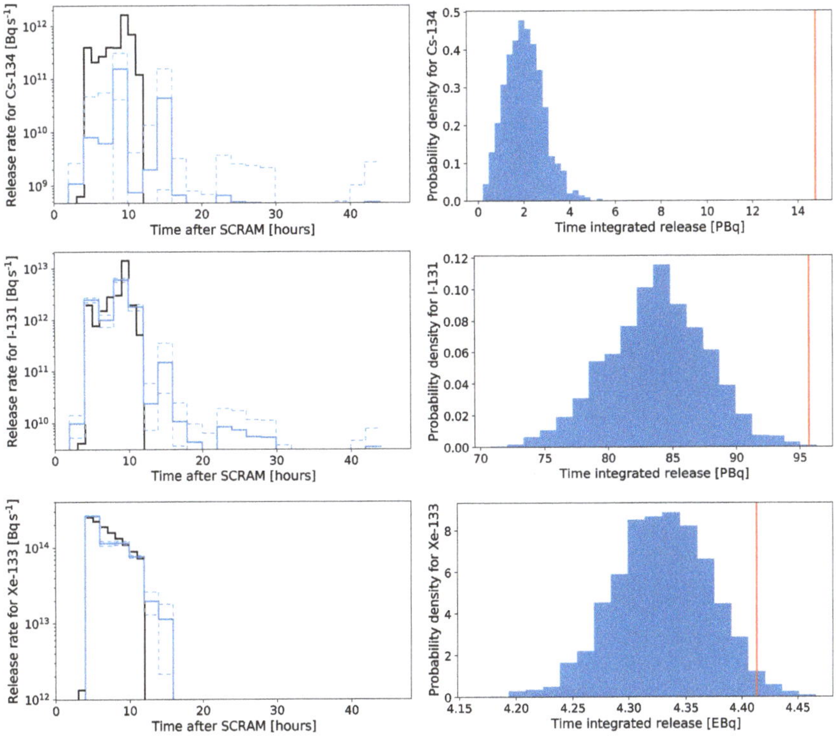

Figure 7. Release rates and probability densities for selected nuclides. For further description of the plots, see captions of Figures 2 and 3.

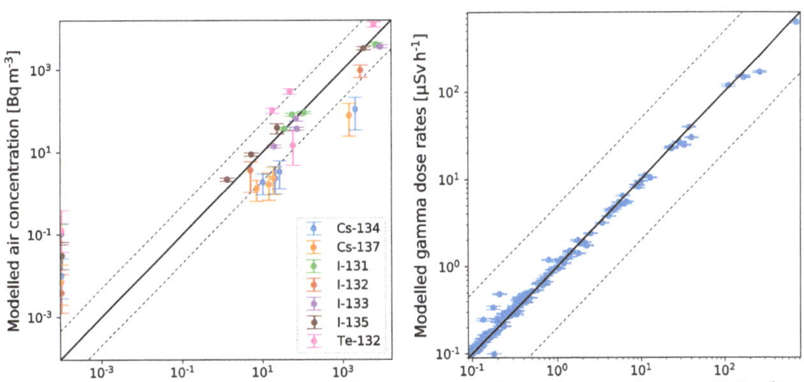

(**a**) (**b**)

Figure 8. Model predictions with uncertainties (median and 10th and 90th percentile) on the y-axis, and observations on the x-axis. The solid black lines indicate a perfect correlation, while the dashed black lines indicate a factor of 5 between model and observation. (**a**) shows the filter measurements, whereas (**b**) shows the gamma dose rates.

3.2. *Efficiency*

Regarding efficiency, we only have rough estimates of the computation time. However, we see that the time depend strongly on the amount of data included. The computation time for the first result, using data from only the first 24 h, was approximately half an hour. When including all data, the computation time was approximately 3.5 h. These estimates are the wall times of the runs of the NUTS algorithm, when running the algorithm in parallel on two CPUs on a standard modern laptop. In addition, some time is of course required for running the dispersion model and restructuring the data.

When operationalized, the code should be adapted to run on an HPC facility to further decrease computation time. In addition, the total set of gamma dose rate observations constitute 8953 measurements from a relatively dense network sampling at every hour. We suspect that there is a lot of redundant information in this data set, so instead using a subsample of this data set might be sufficient and would reduce computation time significantly.

4. Summary and Conclusions

We have developed a Bayesian inverse method for probabilistic source term estimation to be used for accidental nuclear releases to the atmosphere. The source term probability distribution is sampled using the Hamiltonian Monte Carlo algorithm NUTS, which is robust and needs only limited parameter tuning. In theory, this makes it directly applicable to other cases without making significant changes to the method.

The method is applied to a synthetic data set derived by running an atmospheric dispersion model for a realistic accident at a nuclear power plant. The data set consists of air concentration measurements at existing filter stations as well as gamma dose rates at gamma stations. We have shown that even with a limited set of air concentration measurements, realistic source term estimation is possible based on early observations of gamma dose rates. Further, the results indicate that additional constraints on the release rates based on information on the nuclear reactor core inventory can be used to improve the accuracy of the predictions. The estimated released amounts of most individual radionuclides are described by relatively wide probability distributions. However, the gamma dose rates predicted using the probabilistic source term correspond well with observations.

Of course, when applied to a real-world case, we expect that model errors will reduce the accuracy of the predictions to some extent. However, if the models used are unbiased, we anticipate that the predicted gamma dose rates will still be more accurately estimated than the release rates of the individual radionuclides. Further, to make the method as generally applicable as possible, we treat the uncertainty parameters as nuisance parameters. Hence, no assumptions about the magnitude of the uncertainties are made; the only assumption is that the residuals are log-normal distributed.

In conclusion, we have developed a method that performs well applied to the simulated release case, and the results indicate that even with limited measurement data available, it is possible to construct a probabilistic source term that provides accurate predictions of gamma dose rates and reasonable estimates of the released amounts of most of the radionuclides considered. Due to the few assumptions made and the robust theoretical foundation, we expect the method to generalize well. However, in order to fully examine the performance of the method, future application to real-world cases is necessary.

Author Contributions: Conceptualization, K.S.T. and J.H.S.; methodology, K.S.T.; software, K.S.T.; validation, K.S.T.; formal analysis, K.S.T.; investigation, K.S.T.; writing—original draft preparation, K.S.T.; writing—review and editing, K.S.T. and J.H.S.; visualization, K.S.T.; supervision J.H.S.; funding acquisition, K.S.T. and J.H.S. All authors have read and agreed to the published version of the manuscript.

Funding: This research was funded by the Innovation Fund Denmark grant number 0196-00017B, by the Danish Meteorological Institute, and by the Nordic Nuclear Safety Research (NKS) through the

SOCHAOTIC (SOurce CHAracterizatiOn accounting for meTeorologIcal unCertainties) project, grant number AFT/B(22)1.

Institutional Review Board Statement: Not applicable

Informed Consent Statement: Not applicable

Conflicts of Interest: The authors declare no conflict of interest.

Abbreviations

The following abbreviations are used in this manuscript:

DERMA	Danish Emergency Response Model of the Atmosphere
MCMC	Markov Chain Monte Carlo
HMC	Hamiltonian Monte Carlo
NUTS	No U-Turn Sampling
SOCHOTIC	SOurce CHAracterizatiOn accounting for meTeorologIcal unCertainties

References

1. Sørensen, J.H. Method for source localization proposed and applied to the October 2017 case of atmospheric dispersion of Ru-106. *J. Environ. Radioact.* **2018**, *189*, 221–226. [CrossRef] [PubMed]
2. Saunier, O.; Didier, D.; Mathieu, A.; Masson, O.; Le Brazidec, J.D. Atmospheric modeling and source reconstruction of radioactive ruthenium from an undeclared major release in 2017. *Proc. Natl. Acad. Sci. USA* **2019**, *116*, 24991–25000. [CrossRef] [PubMed]
3. Le Brazidec, J.D.; Bocquet, M.; Saunier, O.; Roustan, Y. MCMC methods applied to the reconstruction of the autumn 2017 Ruthenium-106 atmospheric contamination source. *Atmos. Environ. X* **2020**, *6*, 100071. [CrossRef]
4. Tølløse, K.S.; Kaas, E.; Sørensen, J.H. Probabilistic Inverse Method for Source Localization Applied to ETEX and the 2017 Case of Ru-106 including Analyses of Sensitivity to Measurement Data. *Atmosphere* **2021**, *12*, 1567. [CrossRef]
5. Winiarek, V.; Bocquet, M.; Saunier, O.; Mathieu, A. Estimation of errors in the inverse modeling of accidental release of atmospheric pollutant: Application to the reconstruction of the cesium-137 and iodine-131 source terms from the Fukushima Daiichi power plant. *J. Geophys. Res. Atmos.* **2012**, *117*, D05122. [CrossRef]
6. Liu, Y.; Haussaire, J.M.; Bocquet, M.; Roustan, Y.; Saunier, O.; Mathieu, A. Uncertainty quantification of pollutant source retrieval: Comparison of Bayesian methods with application to the Chernobyl and Fukushima Daiichi accidental releases of radionuclides. *Q. J. R. Meteorol. Soc.* **2017**, *143*, 2886–2901. [CrossRef]
7. Stohl, A.; Seibert, P.; Wotawa, G.; Arnold, D.; Burkhart, J.F.; Eckhardt, S.; Tapia, C.; Vargas, A.; Yasunari, T.J. Xenon-133 and caesium-137 releases into the atmosphere from the Fukushima Dai-ichi nuclear power plant: Determination of the source term, atmospheric dispersion, and deposition. *Atmos. Chem. Phys.* **2012**, *12*, 2313–2343. [CrossRef]
8. Terada, H.; Nagai, H.; Tsuduki, K.; Furuno, A.; Kadowaki, M.; Kakefuda, T. Refinement of source term and atmospheric dispersion simulations of radionuclides during the Fukushima Daiichi Nuclear Power Station accident. *J. Environ. Radioact.* **2020**, *213*, 106104. [CrossRef]
9. Saunier, O.; Mathieu, A.; Didier, D.; Tombette, M.; Quélo, D.; Winiarek, V.; Bocquet, M. An inverse modeling method to assess the source term of the Fukushima Nuclear Power Plant accident using gamma dose rate observations. *Atmos. Chem. Phys.* **2013**, *13*, 11403–11421. [CrossRef]
10. Saunier, O.; Korsakissok, I.; Didier, D.; Doursout, T.; Mathieu, A. Real-time use of inverse modeling techniques to assess the atmospheric accidental release of a nuclear power plant. *Radioprotection* **2020**, *55*, 107–115. [CrossRef]
11. Sørensen, J.H.; Feddersen, H.; Tølløse, K.S.; Kouznetsov, R.; Sofiev, M.; Uppstu, A.; Klein, H.; Ulimoen, M.; Robertson, L.; Pehrsson, J.; et al. *NKS-458: SOurce CHAracterizatiOn Accounting for meTeorologIcal unCertainties (SOCHAOTIC)—First-Year Report*; Technical Report; Nordic Nuclear Safety Research (NKS): Roskilde, Denmark, 2021; ISBN 978-87-7893-551-9.
12. Hoe, S.; Sørensen, J.H.; Thykier-Nielsen, S. The nuclear decision support system ARGOS NT and early warning systems in some countries around the Baltic sea. In Proceedings of the 7th Topical Meeting on Emergency Preparedness and Response, Santa Fe, NM, USA, 14–17 September 1999.
13. Hoe, S.; Müller, H.; Gering, F.; Thykier-Nielsen, S.; Sørensen, J.H. ARGOS 2001 a decision support system for nuclear emergencies. In Proceedings of the Radiation Protection and Shielding Division Topical Meeting. Santa Fe, NM, USA, 14–17 April 2002.
14. Bengtsson, L.; Andrae, U.; Aspelien, T.; Batrak, Y.; Calvo, J.; de Rooy, W.; Gleeson, E.; Hansen-Sass, B.; Homleid, M.; Hortal, M.; et al. The HARMONIE–AROME Model Configuration in the ALADIN–HIRLAM NWP System. *Mon. Weather. Rev.* **2017**, *145*, 1919–1935. [CrossRef]
15. Sørensen, J.H. Sensitivity of the DERMA long-range Gaussian dispersion model to meteorological input and diffusion parameters. *Atmos. Environ.* **1998**, *32*, 4195–4206. [CrossRef]
16. Sørensen, J.H.; Baklanov, A.; Hoe, S. The Danish emergency response model of the atmosphere (DERMA). *J. Environ. Radioact.* **2007**, *96*, 122–129. [CrossRef] [PubMed]

17. Sørensen, J.H.; Mackay, D.K.J.; Jensen, C.Ø.; Donaldson, A.I. An integrated model to predict the atmospheric spread of foot-and-mouth disease virus. *Epidemiol. Infect.* **2000**, *124*, 577–590. [CrossRef]
18. Sørensen, J.H.; Jensen, C.Ø.; Mikkelsen, T.; Mackay, D.; Donaldson, A.I. Modelling the atmospheric spread of foot-and-mouth disease virus for emergency preparedness. *Phys. Chem. Earth* **2001**, *26*, 93–97. [CrossRef]
19. Mikkelsen, T.; Alexandersen, S.; Astrup, P.; Champion, H.J.; Donaldson, A.I.; Dunkerley, F.N.; Gloster, J.; Sørensen, J.H.; Thykier-Nielsen, S. Investigation of airborne foot-and-mouth disease virus transmission during low-wind conditions in the early phase of the UK 2001 epidemic. *Atmos. Chem. Phys* **2003**, *3*, 2101–2110. [CrossRef]
20. Gloster, J.; Jones, A.; Redington, A.; Burgin, L.; Sørensen, J.H.; Turner, R. International approach to atmospheric disease dispersion modelling. *Vet. Rec.* **2010**, *166*, 369. [CrossRef]
21. Gloster, J.; Jones, A.; Redington, A.; Burgin, L.; Sørensen, J.H.; Turner, R.; Hullinger, P.; Dillon, M.; Astrup, P.; Garner, G.; et al. Airborne spread of foot-and-mouth disease – model intercomparison. *Vet. J.* **2010**, *183*, 278–286. [CrossRef]
22. Baklanov, A.; Sørensen, J.H. Parameterisation of radionuclide deposition in atmospheric dispersion models. *Phys. Chem. Earth* **2001**, *26*, 787–799. [CrossRef]
23. Hastings, W.K. *Monte Carlo Sampling Methods Using Markov Chains and Their Applications;* Oxford University Press: Oxford, UK, 1970.
24. Casella, G.; George, E.I. Explaining the Gibbs sampler. *Am. Stat.* **1992**, *46*, 167–174.
25. Hoffman, M.D.; Gelman, A. The No-U-Turn sampler: Adaptively setting path lengths in Hamiltonian Monte Carlo. *J. Mach. Learn. Res.* **2014**, *15*, 1593–1623.
26. Salvatier, J.; Wiecki, T.V.; Fonnesbeck, C. Probabilistic programming in Python using PyMC3. *PeerJ Comput. Sci.* **2016**, *2*, e55. [CrossRef]
27. Gelman, A.; Rubin, D.B. Inference from iterative simulation using multiple sequences. *Stat. Sci.* **1992**, *7* , 457–472. [CrossRef]

atmosphere

MDPI

Article

Ensemble of Below-Cloud Scavenging Models for Assessing the Uncertainty Characteristics in Wet Raindrop Deposition Modeling

Alexey Kiselev *, **Alexander Osadchiy, Anton Shvedov and Vladimir Semenov**

Nuclear Safety Institute of the Russian Academy of Sciences (IBRAE RAN), 115191 Moscow, Russia
* Correspondence: aak@ibrae.ac.ru

Abstract: This work is devoted to the development of an ensemble of below-cloud scavenging models of pollutant aerosol transport into the atmosphere. Among other factors contributing to the uncertainty of the forecasts of the dispersion and deposition of technogenic gas-aerosol releases in the atmosphere, precipitation scavenging is one of the least studied and, in case of precipitation, can be the dominant mechanism for aerosol deposition. To form the ensemble of below-cloud scavenging models, appropriate experimental data, raindrop-aerosol capture models, raindrop terminal velocity parameterizations, and raindrop size distributions were chosen. The pool of models was prepared and then evaluated to adequately describe the experimental data using statistical analysis. Rank diagrams were used to analyze the adequacy of meteorological ensembles; together with the ensemble distribution construction, they allowed selecting the groups of models with such properties as to produce unbiased estimates and dispersion corresponding to the dispersion of the experimental data. The model calculations of the concentration fraction deposited due to below-cloud scavenging were performed using a log-normal distribution with characteristics corresponding to those observed during the accidents at the Chernobyl NPP and Fukushima-1 NPP. The results were compared with those obtained using the models of the NAME and FLEXPART codes. The results of this work can be used to improve the current approaches applied for modelling the distribution of pollutants in the atmosphere in the case of emergency, enhancing the reliability of forecasts by taking into account uncertainties in the results. The formed multi-model ensemble will be included in the decision support system used in responding to releases of radioactive substances into the atmosphere.

Keywords: ensemble forecast; below-cloud scavenging; raindrop; contamination; aerosols; pollutants

Citation: Kiselev, A.; Osadchiy, A.; Shvedov, A.; Semenov, V. Ensemble of Below-Cloud Scavenging Models for Assessing the Uncertainty Characteristics in Wet Raindrop Deposition Modeling. *Atmosphere* **2023**, *14*, 398. https://doi.org/10.3390/atmos14020398

Academic Editor: Daniel Viúdez-Moreiras

Received: 27 January 2023
Revised: 14 February 2023
Accepted: 15 February 2023
Published: 18 February 2023

1. Introduction

Atmospheric transport modeling is one of the most important tasks of emergency preparedness and response [1,2] because atmospheric transport is the fastest way to deliver pollutants from the source to humans. Atmospheric transport modeling is performed using diagnostic or predictive weather forecast models, which generally determine the quality of the modeling. Depending on the problem, the forecast can be considered on different time scales, from "nowcasting" [3,4] for early detection of warning situations, to medium-range forecasting [5,6], which can be used for assessments of possible hazards to human health due to pollutant emissions, and finally long-term forecasts [7,8] for long-term safety analysis issues. Using models of atmospheric transport, the parameters of pollution of environmental objects are modeled. Important parameters such as surface deposition density and time-integrated concentration are used for assessment of the population exposure and hence the population safety [9]. The accuracy (mean and dispersion) of such estimates is studied in terms of processes that affect the result: turbulent and advective transport [10,11], dry deposition [12–14], chemical transformations [15,16], and precipitation scavenging [17,18]. At present, we know the accuracy of the models describing the turbulent transport for the landscapes, where the Monin–Obukhov similarity theory [19] is valid; dry deposition for areas with typified landscape conditions [20]; and wet deposition for precipitations that are relatively homogeneous over composition and territory [21]. In practice, when modeling the distribution of atmospheric releases during the accident at the Fukushima-1 NPP, the differences between the actual data and forecasts differed by orders of magnitude or more due to inaccuracy of knowledge about the source term [22] as well as to the inaccuracy of physical processes modeling [23]. An ensemble forecast [24] was developed to solve such problems, taking into account all uncertainties (inaccuracy of the initial data and inaccuracy of modeling processes). It was previously used for modeling weather conditions [25], to obtain the weather forecast for impurity dispersion problems. In [26], the first ensemble forecasts were successfully applied to the ETEX experiment and then to the accident forecasting system [6,27–29]. It is essential for the ensemble approach to take into account all types of uncertainties. This can be achieved by using a priori data on the distribution function of given parameters [30] or, for example, using a multi-model ensemble [24].

In this paper, we present the results of assessment of the uncertainties caused by the below-cloud scavenging of aerosol particles by raindrops, using the multi-model ensemble method. The main result is a well-defined structure of the distribution function. This is critical if the ensemble is used for dose estimates with assurance levels (e.g., probability of exceeding a threshold) [31].

When forming an ensemble of models, the following processes should be taken into account in the below-cloud scavenging models of aerosols by precipitation: microphysical processes of the interaction of aerosol particles with droplets using Brownian diffusion, interception, impaction, thermo- and diffusiophoresis, and viscous (aerodynamic) drag, as well as the processes of coalescence and breaking of raindrops when falling in the atmosphere [18]. One of the determining factors, similar to the modelling of dry deposition [12], is the particle size distribution of aerosols. We will evaluate the quality of the ensemble on the basis of comparison with experimental data and field measurements by metrics used in meteorological models for describing the quality of reproduction of calculated data (in comparison with the observation data).

Wet deposition is commonly divided into in-cloud scavenging and below-cloud scavenging [17]. The below-cloud scavenging of aerosol particles by raindrops is studied in this paper. There are two types of below-cloud scavenging models: integral models that are based on the dependence of the scavenging coefficient Λ on the intensity of precipitation $\Lambda = aI^b$ (where a and b are empirical constants, depending on the type of precipitation) [32], and differential models that take into account the dependence of the scavenging constant on the diameter of the aerosol particle d [33].

To select the models for a multi-model ensemble, we analyzed the experimental data from studies of below-cloud scavenging, as well as a set of models of below-cloud scavenging that consider microphysical processes of interaction (Brownian diffusion, interception, impaction, thermo- and diffusiophoresis), models describing the terminal droplet fall velocity, and the polydispersity of the aerosol and the raindrop system.

The prepared set of models was tested against the experimental data; the estimates of the *FB*, *Pearson*, *FAC5*, and *FAC10* criteria were obtained. Then, the models were selected using the rank histograms to form a multi-model ensemble that provides an unbiased estimate. We also demonstrated that the ensemble dispersion covers the scatter of the experimental data if the ensemble is approximated by a normal distribution function. Additionally, this ensemble was applied for the log-normal distribution of aerosol particles with diameters with the characteristics of Chernobyl aerosols observed in the first days after the accident. Comparative calculations were carried out using the integral models in order to assess the fraction of the washed-out concentration. They showed a significant (more than an order of magnitude) increase in the modeling error when there is a deviation from the average dimensional characteristics of aerosols.

The formed multi-model ensemble could be used to describe the quantitative characteristics of the forecast uncertainty associated with the lack of knowledge about the environment parameters and the parameters of the aerosol interaction with hydrometeors in the atmosphere, providing a well-defined structure of the distribution function of the results. This is necessary for the development of codes predicting the consequences for the population by improving the quality of the forecast.

2. Materials and Methods

2.1. Below-Cloud Scavenging Models

Below-cloud scavenging of aerosol particles in atmospheric transport models is described by the below-cloud scavenging coefficient, Λ (1/s):

$$\frac{dC(d)}{dt} = -\Lambda(d) \times C(d) \quad (1)$$

where $C(d)$ is the concentration of aerosol particles with diameter d in the volume under study, and t is the interaction time of precipitation and aerosol. Determining Λ seems to be the main difficulty in modeling the interaction between aerosols and precipitation.

In case of the polydisperse system of raindrops, the below-cloud scavenging coefficient Λ is determined by the following expression:

$$\Lambda(d) = \int_0^\infty E(D,d)V(D)\frac{\pi D^2}{4}N(D)dD \quad (2)$$

where $N(D)$ is the size distribution function of raindrops, m^{-4}, $E(D)$ is the collision efficiency of raindrops, $V(D)$ is the raindrop terminal velocity, m/s, and D is the raindrop diameter, m.

2.1.1. Collision Efficiency E

The collision efficiency $E(D, d)$ is, by definition, the ratio of the actual collision cross section to the geometric cross section of the raindrop. The value $E = 1$ means that all particles in the geometric volume swept by the falling raindrop will be collected by the raindrop. Usually the values of $E << 1$ [34].

There are a number of different models for evaluating the collision efficiency. The following models for calculating E were used in this paper to form a multi-model ensemble: the Slinn model [35] (hereinafter the SL83 model); the extended model SL83+ [36], which takes into account the contributions of thermo- and diffusiophoresis; the Pripachkin model [37] (hereinafter PR19).

Figure 1 shows a schematic representation of the contributions of processes that affect the wet deposition for various aerosol particle diameters, built according to the SL83+ model, as well as the contribution of thermo- and diffusiophoresis at a fixed raindrop diameter $D = 1$ mm and a difference between particle and raindrop absolute temperatures of $T_a - T_s = 3$ °C. It can be seen that the Brownian diffusion has a greater effect for small particles (with a diameter of less than 0.01 μm), while impaction affects large particles (with a diameter of more than 2 μm). Aerosols in the range of 0.01–2 μm (the so-called Greenfield Gap range) are least scavenged. The contribution of thermophoresis decreases with particle diameter increase. Diffusiophoresis does not depend on the aerosol particle diameter.

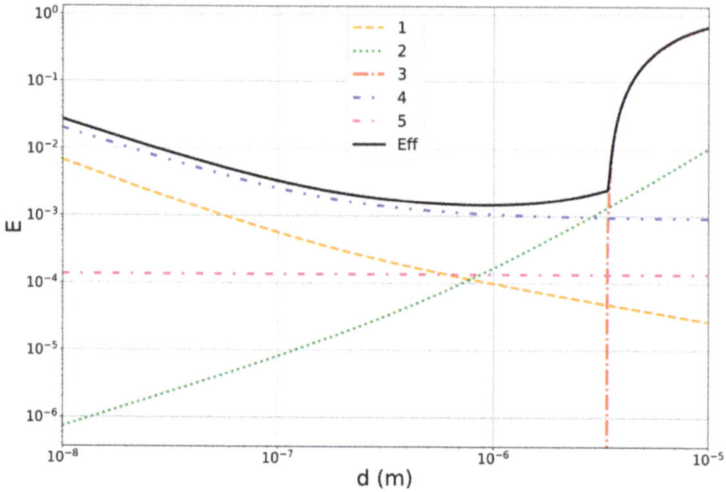

Figure 1. Schematic representation of the contributions of processes. 1—Brownian diffusion, 2—interception, 3—impaction, 4—thermophoresis, 5—diffusiophoresis, Eff—total collision efficiency.

Parameterization of collision efficiency of an aerosol particle by a raindrop in the SL_83 model

The SL83 model takes into account the influence of Brownian diffusion processes, impaction, and interception. The collision efficiency E is expressed as [30]:

$$E(d,D) = E_{Br} + E_{int} + E_{ine} \tag{3}$$

where E_{Br}, E_{int}, and E_{ine} are the efficiencies of diffusion, interception and inertial impaction, respectively.

$$E_{Br} = \frac{4}{Re_D Sc}\left(1 + 0.4 Re_D^{\frac{1}{2}} Sc^{\frac{1}{3}} + 0.16 Re_D^{\frac{1}{2}} Sc^{\frac{1}{2}}\right) \tag{4}$$

$$E_{int} = 4\frac{d}{D}\left[\frac{\mu_a}{\mu_w} + \frac{d}{D}\left(1 + 2Re_D^{\frac{1}{2}}\right)\right] \tag{5}$$

$$E_{ine} = \left(\frac{St - St_*}{St - St_* + \frac{2}{3}}\right)^{3/2} \tag{6}$$

where D and d are the diameters of a raindrop and an aerosol particle, correspondingly, m; $St = \frac{2\tau(V(D)-v(d))}{D}$ is the Stokes number; $V(D)$ and $v(d)$ are the terminal velocities of raindrops and particles, correspondingly, $(v(d) << V(D))$; $\tau = \frac{(\rho_p-\rho_a)d^2 C_c}{18\mu_a}$ is the relaxation time of particles; μ_a, μ_w are the dynamic viscosity of air and water, correspondingly, kg/(m·s); $St_* = \frac{1.2+1/12\ln(1+Re_D)}{1+\ln(1+Re_D)}$ is the critical Stokes number calculated for a raindrop; $Re_D = \frac{D \cdot V(D)}{2\nu}$ is the Reynolds number calculated for a raindrop with a diameter D; $Sc = \frac{\mu_a}{\rho_a \cdot D_{diff}}$ is the Schmidt number calculated for a raindrop; ρ_a, ρ_p are the air density and particle density, kg/m^3; $C_c = 1 + \frac{2\lambda_a}{d}\left(1.257 + 0.4e^{-0.55\frac{d}{\lambda_a}}\right)$ is the Cunningham correction for aerosol particle glide; $\lambda_a = 0.0651 \cdot 10^{-6}$ is the mean free path of air molecules, m; $D_{diff} = \frac{C_c k_b T_a}{3\pi\mu d}$ is the diffusion coefficient of aerosol particles in air, m^2/s; and $k_b = 1.38 \cdot 10^{-23}$ is the Boltzmann constant, J/K. For particles of mass density ρ_p different from 1.0, g/cm^3, the inertial contribution to the collision efficiency should be scaled by $\left(\frac{\rho_w}{\rho_p}\right)^{1/2}$ [35].

Parameterization of the collision efficiency in the SL_83+ model

The model is based on the SL83 model, which takes into account the models of thermo- and diffusiophoresis processes. The aerosol particle collision efficiency E is as follows:

$$E(d,D) = E_{Br} + E_{int} + E_{ine} + E_{th} + E_{dph} \tag{7}$$

According to [36], the contributions of thermophoresis E_{th} and diffusiophoresis E_{dph} to the efficiency E can be expressed as follows (other terms of the sum are similar to the SL83 model):

$$E_{th}(d,D) = \frac{4\alpha_{th}\left(2 + 0.6{Re_D}^{\frac{1}{2}} Pr^{\frac{1}{3}}\right)(T_a - T_s)}{V(D)\, D} \tag{8}$$

$$E_{dph}(d,D) = \frac{4\beta_{dph}\left(2 + 0.6{Re_D}^{\frac{1}{2}} Sc_w^{\frac{1}{3}}\right)\left(\frac{p_s^0}{T_s} - \frac{p_a^0 RH}{T_a}\right)}{V(D)\, D} \tag{9}$$

Coefficients α_{th} and β_{dph} are defined by the following expressions, correspondingly:

$$\alpha_{th} = \frac{2C_c\left(k_a + \frac{5\lambda_a}{D} k_p\right) k_a}{5P\left(1 + \frac{6\lambda_a}{D}\right)\left(2k_a + k_p + \frac{10\lambda_a}{D} k_p\right)} \tag{10}$$

$$\beta_{dph} = \frac{T_a D_{diffwater}}{P}\sqrt{\frac{M_w}{M_a}} \tag{11}$$

where $Pr = \frac{c_p \mu_a}{k_a}$ is the Prandtl number; $Sc_w = \frac{\mu_a}{\rho_a D_{diffwater}}$ is the Schmidt number for water vapor in air; $D_{diffwater} = 0.25 \cdot 10^{-4}$ [38] is the water vapor diffusion coefficient in the air, m^2/s; RH is the relative air humidity; P is the normal atmospheric pressure, Pa; T_a is the absolute air temperature, K; T_s is the absolute temperature of the raindrop surface, K (in [18,36,39], the values of T_s and T_a were chosen to obtain $T_a - T_s = 3$ °C); p_s^0 p_a^0 is the pressure of water vapor at temperatures T_a T_s, Pa; k_a, k_p is the thermal conductivity of air and particles, W/m·K; $M_w = 18.0$ and $M_a = 28.97$ are the molecular masses of water and air, a.m.u.

2.1.2. Terminal (Falling) Velocity V

The terminal raindrop velocity (raindrop fall velocity) is the terminal vertical component of the raindrop velocity, which is achieved by balancing drag force of the medium with the raindrop gravity.

Table 1 lists the parametrizations of the terminal raindrop velocity, m/s, from [18].

Table 1. Formulas for calculating the terminal raindrop velocity, m/s.

Kessler 1969	$V(D) = 13(D \cdot 100)^{0.5}$
Atlas 1977	$V(D) = 1.7 \cdot 10^5 (D \cdot 100)^{0.67}$
Willis 1984	$V(D) = 48.54 \cdot (D \cdot 100) \cdot e^{-1.95(D \cdot 100)}$
Best 1950	$V(D) = 9.58 \cdot (1 - e\left(-\frac{D \cdot 100}{0.171}\right)^{1.147}\Big)$

Figure 2 shows the dependence of the calculated terminal velocity on the raindrop diameter for the selected models. One can see that around the raindrop diameters of ~0.1 mm and ~10.0 mm, the terminal velocity values differ by several times for the selected parametrizations, which in turn will affect the results of calculations, if the size distribution of raindrops is taken into account.

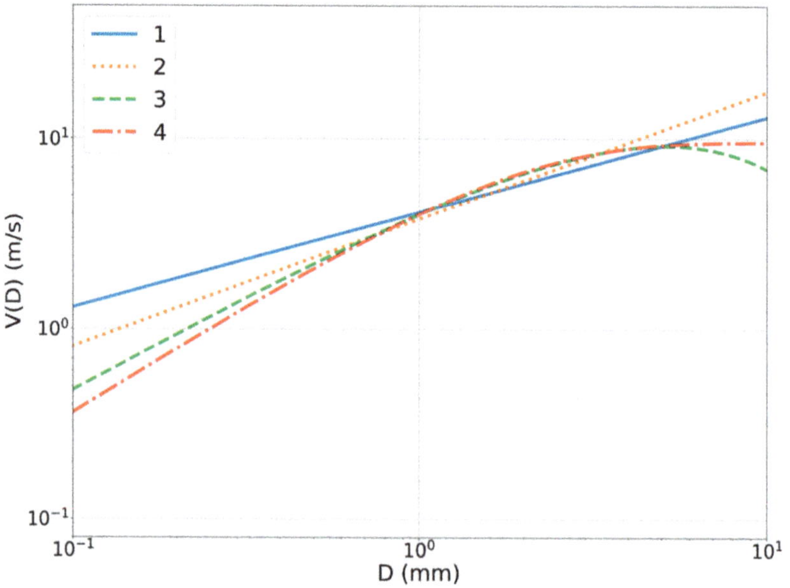

Figure 2. Dependence of the calculated terminal velocity on the diameter of the raindrop for the selected models. 1—Kessler 1969, 2—Atlas 1977, 3—Willis 1984, 4—Best 1950.

2.1.3. Raindrop Size Distribution N(D)

The following raindrop size distributions were chosen for the current study:

Raindrop size distribution ND_MP48

Exponential distribution [40]:

$$N(D) = N_{0e}e^{-\beta_e D} \tag{12}$$

where N_{0e} = $0.08 \cdot 10^8$ m^{-4} is the entanglement parameter (at D = 0); D is the raindrop diameter, m; $\beta_e = 4100 \times I^{-0.21}$ m^{-1} is the slope parameter; I is the rain intensity, mm/hour.

Raindrop size distribution ND_FL86

Lognormal distribution [41]:

$$N(D) = \frac{N_{total}}{\sqrt{2\pi}D\ \log(\sigma_D)} \exp\left(-\frac{\left(\log\left(D/D_g\right)\right)^2}{2log^2(\sigma_D)}\right) \tag{13}$$

where D is the raindrop diameter, mm; N_{total} is the total raindrop density, m^{-3}; D_g is the raindrop median diameter, mm; σ_D is the standard geometric deviation.

$$\sigma_D^* = 1.43 - 3.10 \times 10^{-4}I \tag{14}$$

$$N_{total} = 172\ I^{0.22} \tag{15}$$

$$D_g = 0.75\ I^{0.21} \tag{16}$$

where (*) is used for the variables, for which the approximation is based on the intensity of precipitation I; σ^* is only applicable for $I > 5$ mm/h.

2.1.4. Models for Comparison of Concentration Loss for the Integral Spectrum of Aerosol Particles

Pripachkin microphysical model PR19

In this model, the collision efficiency takes into account the processes of Brownian diffusion, impaction, and the interception of an aerosol particle by a raindrop and does not take into account the thermo-diffusiophoresis and the electrostatic interaction. The raindrops are represented as a system of spherical obstacles that form a homogeneous filtering medium for capturing the aerosol particles [37]. The below-cloud scavenging coefficient Λ is determined by the following expression:

$$\Lambda(d) = \frac{3}{2}E(d, D)I\frac{1}{D} \tag{17}$$

The aerosol collision efficiency *E(d, D)* is related to the capture coefficient η, as follows:

$$E(d, D) = 1 - e^{-\gamma\eta} \tag{18}$$

where $\gamma = \frac{4\alpha H}{\pi D}$ is a coefficient depending on the macroscopic parameters of the medium (~1.0), $\alpha = \frac{\pi D^3 n(D)}{6}$ is the partial density of raindrops (varies from 10^{-5} to 10^{-10} for raindrops with a size of 0.1–6 mm, respectively), *n(D)* is the volume concentration of raindrops, m^{-3}, H is the vertical size of the reference volume (from 10^2 to 10^3 m), $\eta = \eta_{st} + \eta_D + \eta_{DK}$ are the coefficients of inertial, diffusion capture, and capture due to entanglement, respectively.

$$\eta_{st} = \frac{St - St_*}{St} \tag{19}$$

$$\eta_D = \frac{4 \times \left(1 + 0.4 Re_D^{\frac{1}{2}} Sc^{\frac{1}{3}}\right)}{Re_D Sc} \tag{20}$$

$$\eta_{DK} = 4r\left[\frac{\mu_a}{\mu_w} + r\left(1 + 2Re_D^{\frac{1}{2}}\right)\right] \tag{21}$$

where *St* is the Stokes number calculated for a raindrop, St_* is the critical Stokes number, Sc is the Schmidt number, Re_D is the Reynolds number calculated for a raindrop of diameter *D* (St, St_*, Sc, Re_D μ_a, μ_w; see model SL83), and *r* = *d/D* is the entanglement parameter.

Parameterized model FLEXPART v10.3

This model was used in the transport code FLEXPART (version v10.3 and higher) to calculate the below-cloud scavenging coefficient [33]; a detailed description of the model is presented in [42].

$$\log\left(\frac{\Lambda}{\Lambda_0}\right) = a + bD_p^{-4} + cD_p^{-3} + dD_p^{-2} + eD_p^{-1} + f\left(\frac{I}{I_0}\right)^{0.5} \tag{22}$$

where $\Lambda_0 = 1.0$ 1/s, $I_0 = 1.0$ mm/h, $D_p = \log\frac{d}{d_0}$, $d_0 = 1.0$ m, $a = 274.36$, $b = 332839.6$, $c = 226656$, $d = 58005.9$, $e = 6588.38$, and $f = 0.24498$.

Integral models of the codes NAME and FLEXPART

These models were used in the transport code NAME [32] and FLEXPART (versions below v10.3; a detailed description is presented in [33]) to calculate the below-cloud scavenging coefficient in the form of dependence of the type $\Lambda = aI^b$.

$$\text{NAME} \quad \Lambda = 8.4 \times 10^{-5}\, I^{0.79} \tag{23}$$

$$\text{FLEXPART} \quad \Lambda = 0.33 \times 10^{-4} I^{0.42} + 1.0 \times 10^{-4} I^{0.58} \tag{24}$$

2.2. Statistics

The following values were calculated as statistical metrics for selecting the ensemble members: *FB*—fractional bias, *R*—Pearson's correlation coefficient.

$$FB = \frac{\overline{x_o} - \overline{x_p}}{0.5 \cdot \left(\overline{x_o} + \overline{x_p}\right)} \tag{25}$$

$$R = \frac{\sum_{i=1}^{n}(x_{pi} - \overline{x_p})(x_{oi} - \overline{x_o})}{\sqrt{\sum_{i=1}^{n}(x_{pi} - \overline{x_p})^2}\sqrt{\sum_{i=1}^{n}(x_{oi} - \overline{x_o})^2}} \quad (26)$$

$$\text{FAC5} = \text{Proportion of data satisfying } 0.2 \leq \frac{x_p}{x_o} \leq 5 \quad (27)$$

$$\text{FAC10} = \text{Proportion of data satisfying } 0.1 \leq \frac{x_p}{x_o} \leq 10 \quad (28)$$

where x_p are predicted values, and x_o are experimental values.

2.3. Ensemble Verification

The rank histograms (Talagrand diagrams) [43] characterize bias of ensemble estimate and allow demonstrating the reliability of the ensemble forecast in comparison with a sample of experimental data. The interpretation of the rank histogram is based on the assumption that all ensemble members, as well as observation data, have the same distribution. If the ensemble units are chosen in such a way that the observations equally "fit" between the ensemble members, then the rank histogram is flat. The asymmetric type of the histogram indicates a bias of the mean value of the forecasts. If the rank histogram is symmetrical and U-shaped, this means that the ensemble distribution does not cover the observations. If the histogram is Λ-shaped, the distribution is scattered. When the experimental value is less than the ensemble calculations, bin 1 is assigned to it; if the experimental value is in second place in ascending order, bin 2 is assigned to it, etc.

2.4. Experimental Data

The experimental data were taken from the studies of below-cloud scavenging of aerosols by raindrops [35,44–47]. Unfortunately, at the moment, there is a lack of experimental data on below-cloud scavenging, and in most studies the data are presented in the form of an integral approximation, both in terms of precipitation intensity and disperse composition of aerosols. The data given in the study conducted by Sparmacher were obtained under controlled conditions and are the best in terms of their "validity". Table 2 presents the experimental values of the below-cloud scavenging coefficient Λ, as well as the conditions under which they were obtained.

Table 2. Experimental conditions and experimental values of the below-cloud scavenging coefficient Λ from various studies.

N	Source	Precipitation Intensity, mm/h	Density of Aerosol Particles, g/cm³	Coefficient Λ, 1/s
1	Sparmacher $0.1 \leq d_p \leq 1.0$ µm	2.0–5.0	1.0	$(1.4–4.9) \times 10^{-7}$
2		5.0–12.0		$(0.6–2.0) \times 10^{-6}$
3	Sparmacher $1.0 \leq d_p \leq 10.0$ µm	2.0–5.0	1.0	$(0.2–2.6) \times 10^{-6}$
4		5.0–12.0		$(4.2–6.2) \times 10^{-6}$
5	Slinn $1.0 \leq d_p \leq 10.0$ µm	2.0–5.0	3.0 *	$(1.6–8.6) \times 10^{-4}$
6		5.0–12.0		$(1.0–2.1) \times 10^{-3}$
7	Baklanov $1.0 \leq d_p \leq 10.0$ µm	5.0	3.0 *	$(0.1–15.0) \times 10^{-4}$
8	Blanco-Alegre $0.1 \leq d_p \leq 1.0$ µm	1.0–3.0	1.0 **	$(0.2–8.0) \times 10^{-5}$
9	Zikova $0.1 \leq d_p \leq 1.0$ µm	2.0–5.0 ***	1.0 **	$(1.7–5.5) \times 10^{-5}$

* the values of density of aerosol particles were not explicitly indicated in the papers and were chosen according to [48]; ** the values of density of aerosol particles also were not indicated in these papers and were taken equal to 1.0 g/cm³ due to the fact that in the scavenging models, the particle density contributes only to large particles (>1.0 µm); *** the values of precipitation intensity were taken equal to the generally accepted moderate rains intensity [49].

The experimental data from experiments 8 and 9 also could be described using the ensemble approach, but the conditions under which such experiments were carried out led to dominance of the in-cloud scavenging process [18]. In this paper, we decided to focus on below-cloud scavenging models only, but we cannot ignore the recent experimental data, so we performed assessments by combining in-cloud and below-cloud scavenging models to ensure that these experimental data can be also covered. For this purpose, models (Jylha's model [50], Environ's model [51], Pudykiewicz's model [52])were taken and applied under experimental conditions.

$$\text{Jylha 1991} \quad \Lambda = 3.4 \times 10^{-5} I^{0.55} \tag{29}$$

$$\text{Environ 2018} \quad \Lambda = 4.2 \times 10^{-4} I^{0.79} \tag{30}$$

$$\text{Pudykiewicz 1989} \quad \Lambda = 3.5 \times 10^{-5} \frac{RH - RH_0}{RH_c - RH_0}, \ RH > 80\% \tag{31}$$

where Λ is an in-cloud scavenging coefficient, 1/s; I is rain intensity, mm/h; RH is relative humidity, %; RH_0 is a threshold value of the relative humidity above which the subgrid scale condensation occurs; RH_c is a relative humidity for the saturation state. As can be seen in Figure 3, the in-cloud scavenging process determines the scavenging coefficient and deposition rate. Thus, we will focus on the experimental data 1–7 from Table 2, and in further work we will consider the cleaning process in the cloud as a separate process, the models of which, also using the ensemble preparation method studied in this work, will be tested and built in the same way.

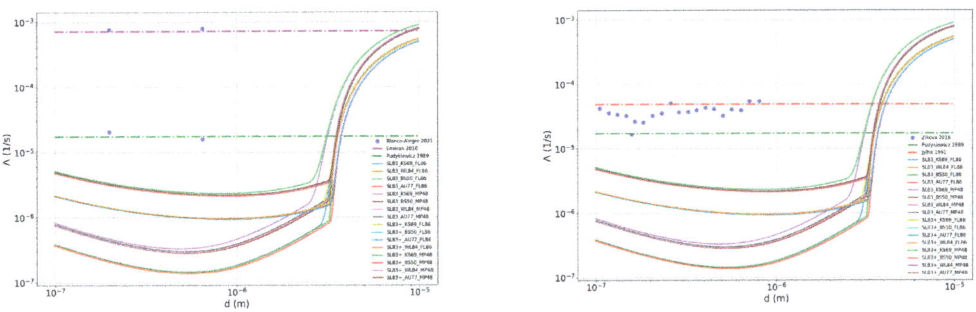

(**a**) (**b**)

Figure 3. Dependence of the calculated values of scavenging coefficient on the diameter of scavenged particles calculated for 16 models of below-cloud scavenging and for 3 models of in-cloud scavenging; the values from experiments (**a**) No. 8 and (**b**) No. 9.

3. Results and Discussion

3.1. Set of Below-Cloud Scavenging Models

Based on the collision efficiency models, the terminal raindrop velocity, and the size distribution of raindrops, a set of models (Table 3) of below-cloud scavenging was formed, which was then tested against the experimental data.

Table 3. Set of below-cloud scavenging models.

N	Collision Efficiency	Distribution of Raindrops	Terminal Raindrop Velocity	N	Collision Efficiency	Distribution of Raindrops	Terminal Raindrop Velocity
1	SL83	MP48	KS69	9	SL83+	MP48	KS69
2	SL83	FL86	KS69	10	SL83+	FL86	KS69
3	SL83	MP48	AU77	11	SL83+	MP48	AU77
4	SL83	FL86	AU77	12	SL83+	FL86	AU77
5	SL83	MP48	WL84	13	SL83+	MP48	WL84
6	SL83	FL86	WL84	14	SL83+	FL86	WL84
7	SL83	MP48	BS50	15	SL83+	MP48	BS50
8	SL83	FL86	BS50	16	SL83+	FL86	BS50

The curves in Figure 4 represent all 16 combinations of below-cloud scavenging models calculated using the parameters of experiment Nos. 1 and 2 (Table 2). Integrals for models were calculated by the numerical Simpson's 1/3 rule.

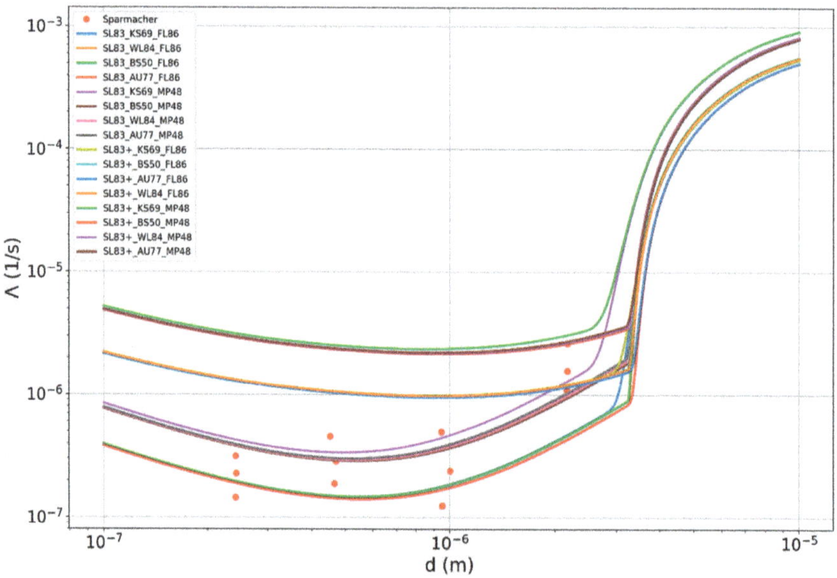

Figure 4. Dependence of the calculated values of below-cloud scavenging coefficient on the diameter of scavenged particles calculated for all 16 models; the values from experiments Nos. 1 and 2 were used as parameters.

3.2. Partitioning of the Study Area According to Diameter Ranges

Due to the fact that the values of below-cloud scavenging coefficient Λ differ by several orders of magnitude (from 10^{-7} to 10^{-3}) depending on the diameter of aerosol particles, the study area was divided into two groups according to the characteristic sizes of aerosol particles: the fine aerosol group: 0.1–1.0 µm; and the coarse group: 1.0–10.0 µm. Figure 5 show the histograms of normalized systematic errors for both groups of diameters. The histogram in Figure 5a shows that for the fine aerosol group, all combinations of models that include the SL83 collision efficiency model underestimate the values of the coefficient Λ, while the models that include the SL83+ model (that takes into account the thermo- and diffusiophoresis in addition to SL83), on the contrary, overestimate them. This is probably related with the limited experimental data in the fine aerosol group. In the coarse group (Figure 5b), the models have both positive and negative FB values, depending on the transition between exponential and lognormal raindrop distribution systems.

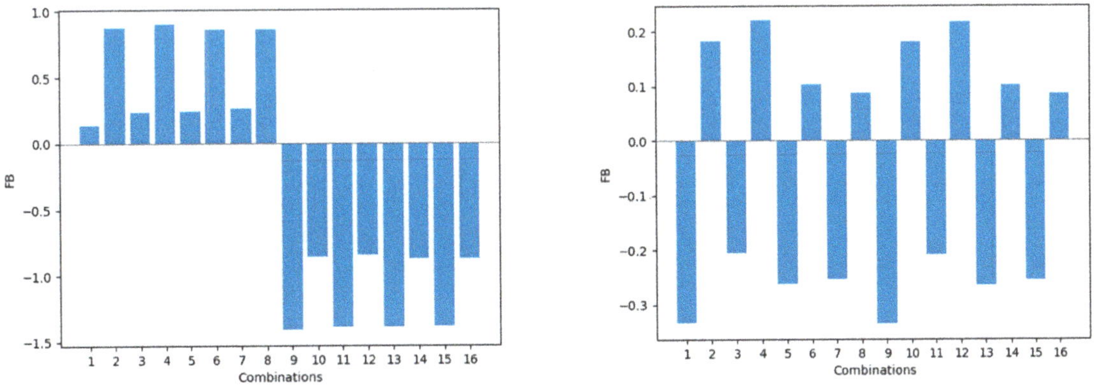

(a) (b)

Figure 5. Histograms of FB values for below-cloud scavenging models numbered 1–16: (**a**) fine aerosol group, (**b**) coarse aerosol group.

Since the choice of parameterization of terminal raindrop fall velocity does not strongly affect the scavenging coefficient, it was decided for ensemble constructing to use the models with different descriptions of collision efficiency, as well as different systems of raindrop distribution, to give an unbiased estimate in total.

3.3. Results of Calculations of Statistical Metrics

The values of statistical metrics for the constructed set of below-cloud scavenging models are given in Table 4. Pearson correlation coefficient values are about 0.6 for the fine group and about 0.5 for the coarse group; FAC5 values range from 0.28–1.0 for the fine group and are about 0.5 for the coarse group; FAC10 values range from 0.64–1.0 for the fine group and are approximately 0.7 for the coarse group. The results indicate that the used set of disperse models is valid in terms of reproducing experimental data both for fine and coarse aerosol fractions. Further construction of a multi-model ensemble involves the enumeration of possible combinations from the selected set that satisfies the criteria of unbiased estimates and compliance of dispersion of the ensemble with dispersion of experimental data. This is done using the rank histograms.

Table 4. A set of below-cloud scavenging models and the values of statistical metrics compared with experimental data (combinations selected for ensemble models are in bold).

N	Collision Efficiency	Distribution of Raindrops	Terminal Raindrop Velocity	FB (Fine)	FB (Coarse)	Pearson (Fine)	Pearson (Coarse)	FAC5 (Fine)	FAC5 (Coarse)	FAC10 (Fine)	FAC10 (Coarse)
1	**SL83**	**MP48**	**KS69**	**0.14**	**−0.33**	**0.68**	**0.50**	**1.00**	**0.63**	**1.00**	**0.70**
2	SL83	FL86	KS69	0.88	0.18	0.61	0.50	0.93	0.59	1.00	0.74
3	SL83	MP48	AU77	0.23	−0.21	0.64	0.51	1.00	0.59	1.00	0.74
4	**SL83**	**FL86**	**AU77**	**0.90**	**0.22**	**0.61**	**0.51**	**0.93**	**0.59**	**1.00**	**0.70**
5	SL83	MP48	WL84	0.24	−0.26	0.65	0.51	1.00	0.63	1.00	0.74
6	SL83	FL86	WL84	0.86	0.10	0.63	0.52	0.93	0.59	1.00	0.74
7	SL83	MP48	BS50	0.26	−0.25	0.64	0.52	1.00	0.63	1.00	0.74
8	SL83	FL86	BS50	0.86	0.09	0.63	0.52	0.93	0.59	1.00	0.74
9	SL83+	MP48	KS69	−1.41	−0.34	0.54	0.50	0.29	0.63	0.64	0.74
10	SL83+	FL86	KS69	−0.86	0.18	0.55	0.50	0.71	0.67	1.00	0.74
11	SL83+	MP48	AU77	−1.39	−0.21	0.54	0.51	0.29	0.59	0.64	0.78
12	SL83+	FL86	AU77	−0.85	0.22	0.55	0.51	0.71	0.67	1.00	0.70
13	**SL83+**	**MP48**	**WL84**	**−1.39**	**−0.26**	**0.54**	**0.51**	**0.29**	**0.63**	**0.64**	**0.78**
14	**SL83+**	**FL86**	**WL84**	**−0.87**	**0.10**	**0.56**	**0.52**	**0.71**	**0.67**	**1.00**	**0.74**
15	SL83+	MP48	BS50	−1.38	−0.26	0.54	0.52	0.29	0.63	0.64	0.78
16	SL83+	FL86	BS50	−0.88	0.08	0.56	0.52	0.71	0.67	1.00	0.74

3.4. Construction of Rank Histograms

The applicability of a set of models as an ensemble can be demonstrated by rank histogram. For this task, combinations of models with different collision efficiency descriptions (SL83 and SL83+) as well as different raindrop distribution systems (MP48 and FL86) were taken. A total of 225 combinations, collected from a set of models from Table 3, were analyzed. Rank histograms were used to analyze how the models describe the statistical dispersion of observed values for all experimental data from Table 2. Given the small amount of experimental data, a four-member ensemble was chosen. A larger set, as the calculations showed, led to a strong nonuniformity in the histograms for individual members. Therefore, histograms were constructed for each combination (for each measurement in each experiment, taking into account the conditions, the values of below-cloud scavenging coefficients Λ were calculated for the selected models, they were ranked, and the rank corresponding to the experimental value was determined; the number of experimental values that fell into the corresponding ranks is represented as the column height in the rank histogram).

Examples of the rank histograms for combinations of models are shown in Figure 6a–f. The remaining combinations show a similar behavior in terms of distribution; therefore, they are not presented in this paper. We can conclude that the ensemble with models 1, 4, 13, and 14 in Figure 6f is the best. It can be seen that all experimental values are evenly distributed among the ensemble members, which indicates that the estimate of the multi-model ensemble is unbiased. A similar behavior was observed in the analysis of meteorological ensemble in [16], where many more experimental points are available; nevertheless, there are also bins with behavior that is not ideal with respect to the overall distribution. The constructed ensemble has an unbiased estimate. If we consider the ensemble by all measurements, then the average characteristics are FB~−0.3, Pearson~0.6, FAC5~0.7, and FAC10~0.9 for the fine group, and FB~0.1, Pearson~0.5, FAC5~0.6, and FAC10~0.7 for the coarse group, which is acceptable considering a wide range of Λ. We shall consider further the results of calculations for this ensemble based on experimental data separately for each experiment, integrating the spectrum of aerosol particles over diameters for the postulated form of size distribution of aerosol particles.

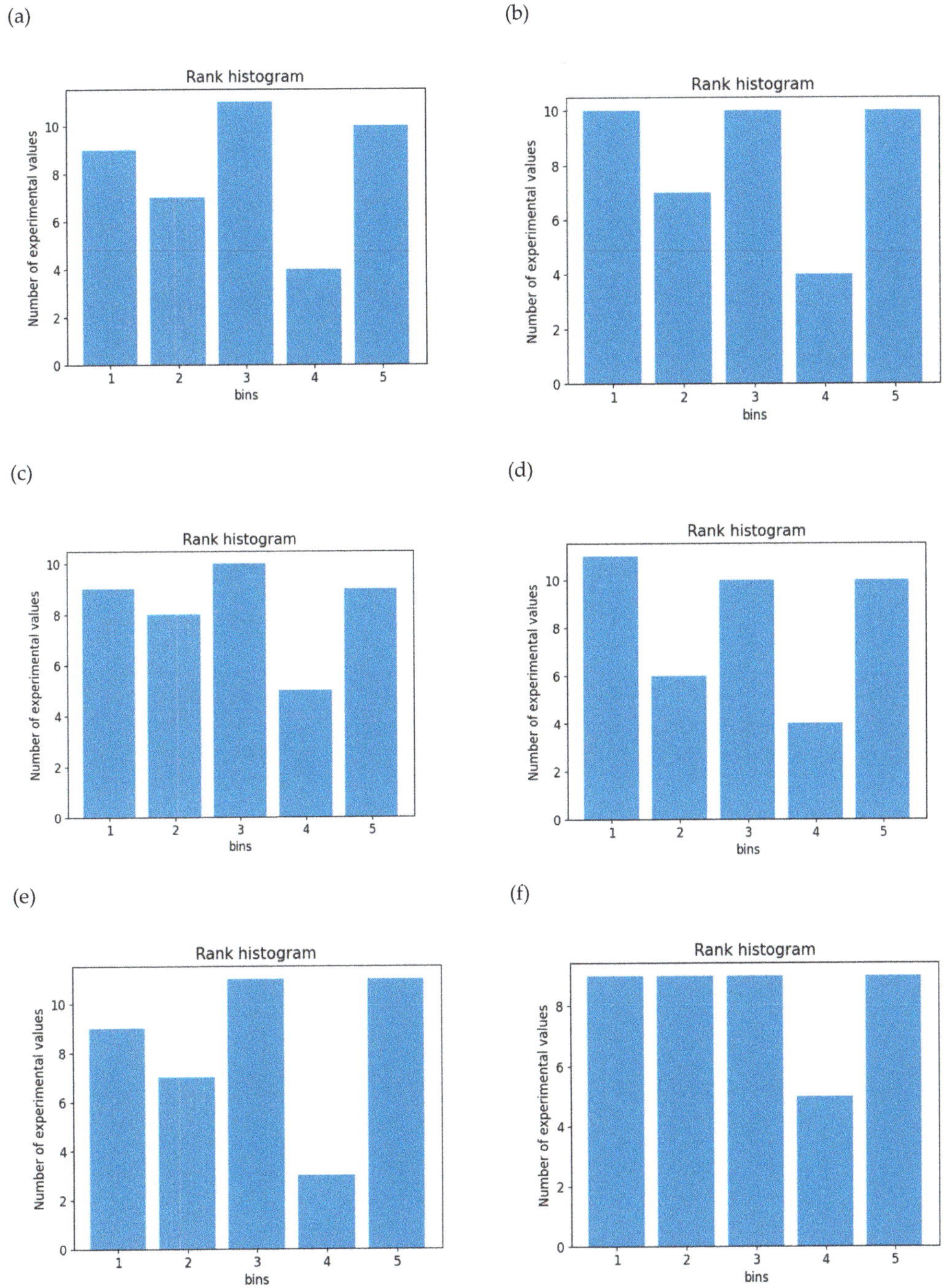

Figure 6. Rank histogram of a multi-model ensemble, built on combinations of models: 2, 3, 12, 13 (**a**); 2, 5, 13, 16 (**b**): 3, 6, 9, 12 (**c**); 5, 8, 11, 14 (**d**); 4, 7, 10, 15 (**e**); 1, 4, 13, 14 (**f**). Ranges for bins are different for each ensemble depending on the measurable quantities.

3.5. Comparison of the Results of Model Calculations for the Ensemble with Experimental Data

To estimate the statistical dispersion of the ensemble with respect to the experimental data, we used an approach based on the approximation of the results of ensemble calculation by the normal distribution function. For this purpose:

1. Groups of diameters were selected in the experimental data, for which the scavenging value was within the error of the experimental data. This was used to separate the coarse areas from the fine areas;
2. Model calculations of the below-cloud scavenging coefficient Λ were carried out for the average diameter of an aerosol particle $\overline{d}$ corresponding to the experiment in the considered group of diameters;
3. Then, using the obtained values, the average value $\mu = \frac{\sum_{i=1}^{4} \Lambda(\overline{d})_i}{4}$ and standard deviation $\sigma = \sqrt{\frac{\sum_{i=1}^{4} \Lambda(\overline{d})_i^2}{4} - \left(\frac{\sum_{i=1}^{4} \Lambda(\overline{d})_i}{4}\right)^2}$ was calculated;
4. The fractions of the experimental values that fall into the ranges $\mu \pm \sigma$, $\mu \pm 2\sigma$, and $\mu \pm 3\sigma$ were analyzed.

Figure 7a–g shows the results of the model calculations for the ensemble (curves e1, e4, e13, and e14) in comparison with the experimental data from Table 2. Figure 8a–g shows the resulting distribution densities of the ensemble (curve ens) together with the experimental results from Table 2. One can see from the figures that the ensemble distributions cover 100% of experimental data Nos. 1, 2 within σ (Figure 8a,b), 100% of experimental data No. 3 within 2σ (Figure 8c), and 80–100% of experimental data Nos. 4–7 within 3σ (Figure 8d–g).

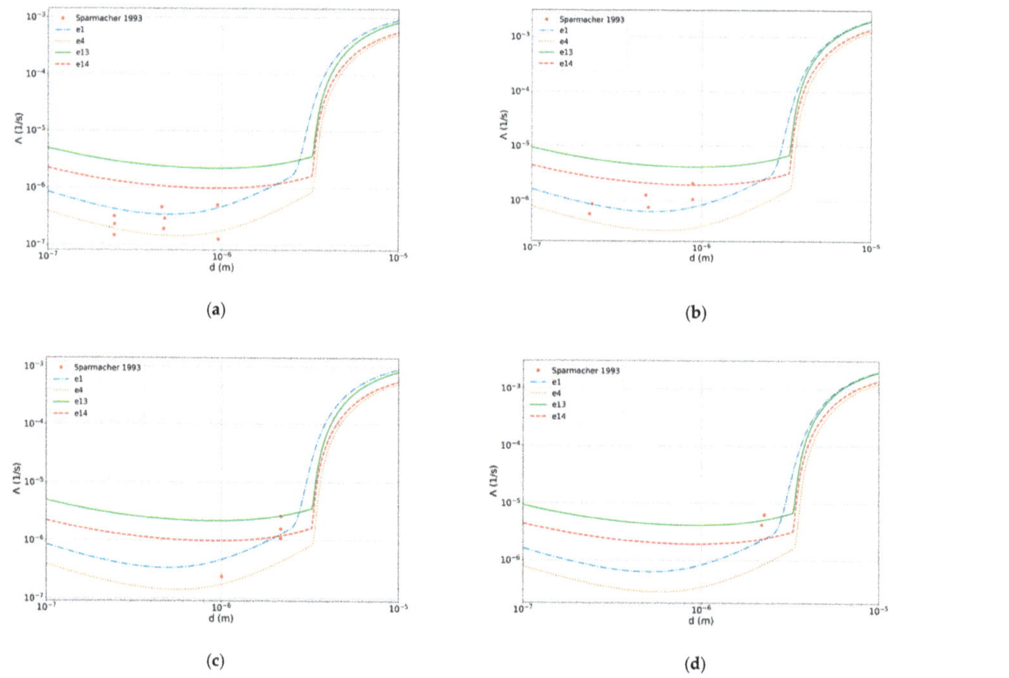

Figure 7. *Cont.*

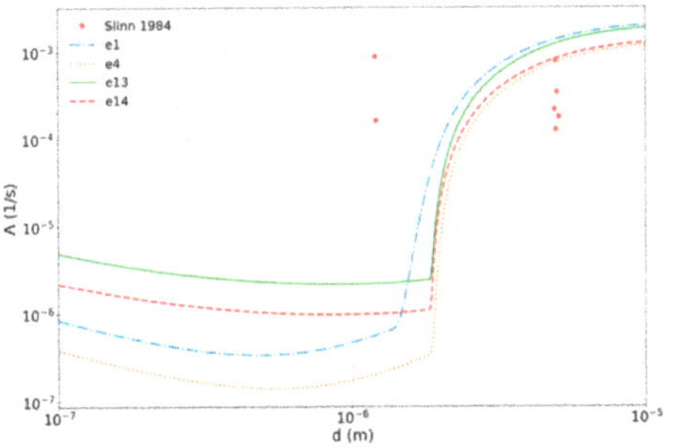

(**e**)

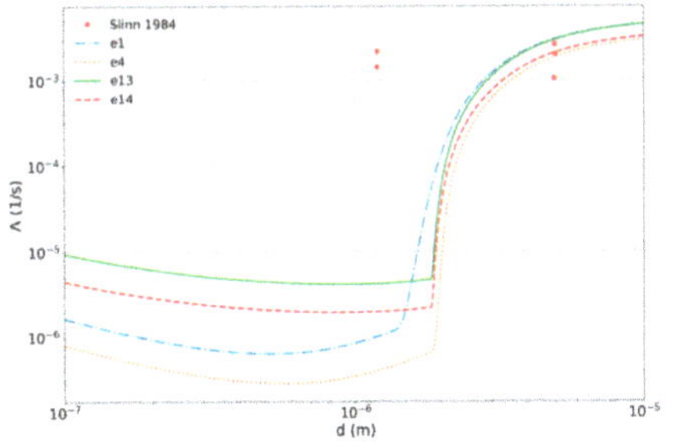

(**f**)

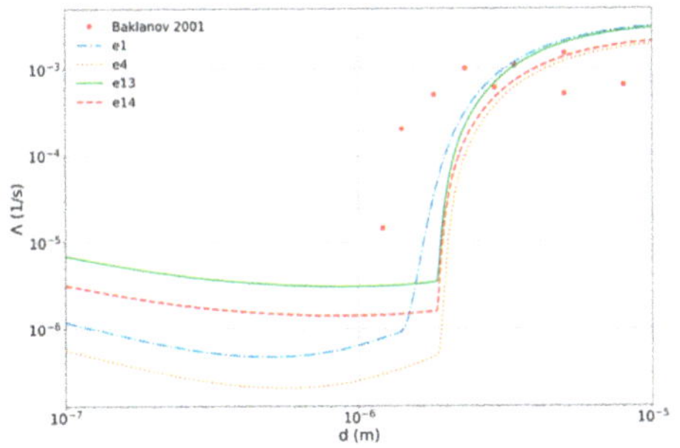

(**g**)

Figure 7. Dependence of the calculated values of the below-cloud scavenging coefficients Λ, 1/s, on the diameter of the aerosol particle d, m: (**a**) experiment No. 1, (**b**) experiment No. 2, (**c**) experiment No. 3, (**d**) experiment No. 4; (**e**) experiment No. 5, (**f**) experiment No. 6, (**g**) experiment No. 7.

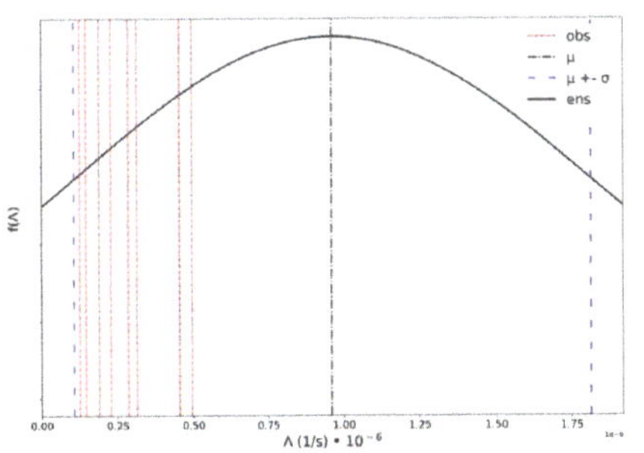

(**a**)

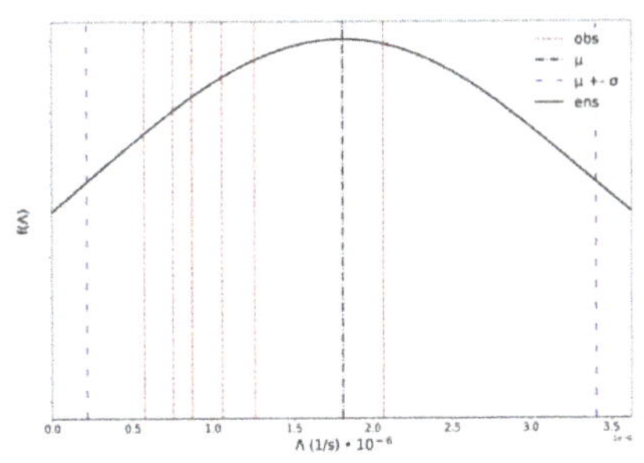

(**b**)

Figure 8. *Cont.*

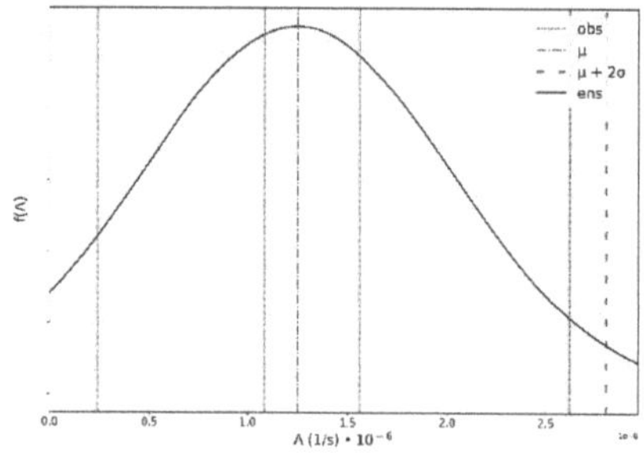

(c)

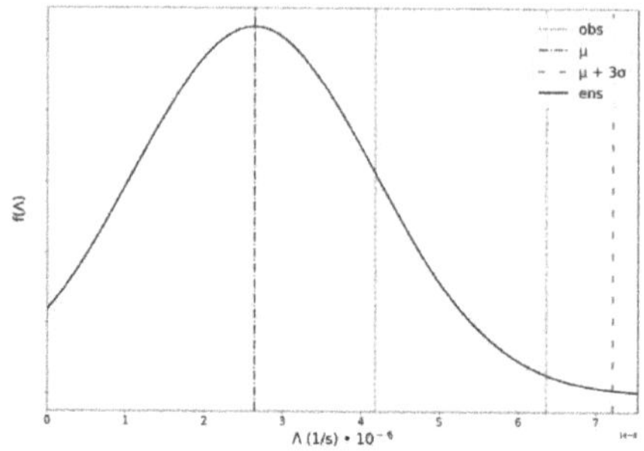

(d)

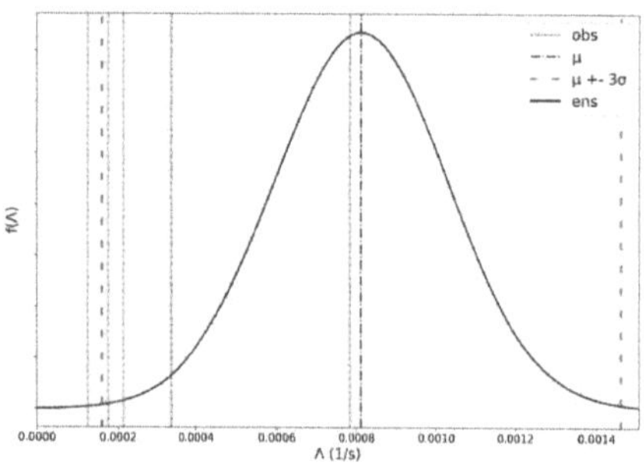

(e)

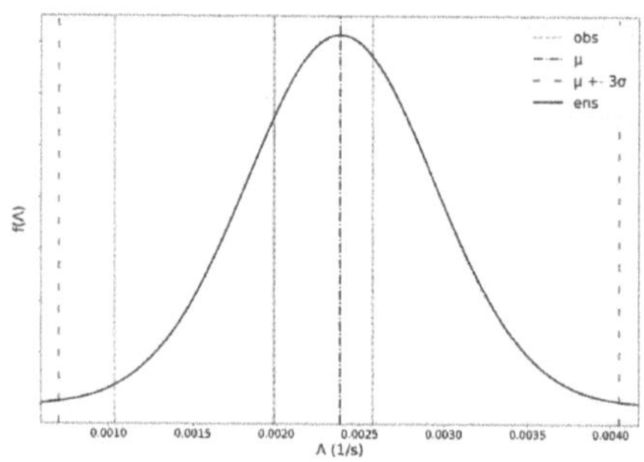

(f)

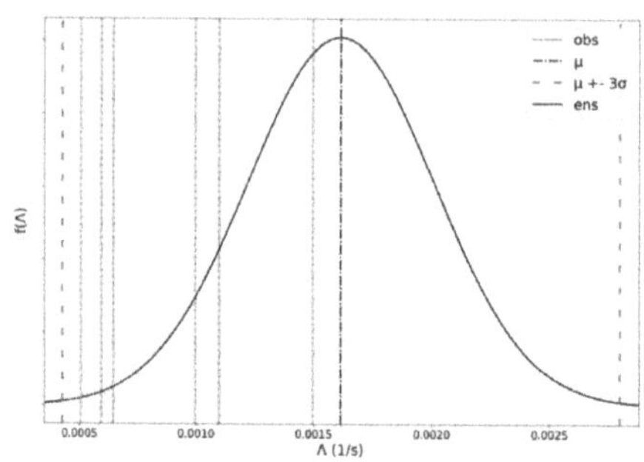

(g)

Figure 8. Approximated distribution density of the ensemble f(Λ)—ens curve, and experimental values Λ, 1/s—obs corresponding to the experiment number from Table 2: (**a**) experiment No. 1, (**b**) experiment No. 2, (**c**) experiment No. 3, (**d**) experiment No. 4; (**e**) experiment No. 5, (**f**) experiment No. 6, (**g**) experiment No. 7.

3.6. Calculation of Polydisperse Aerosol Scavenging from a Control Volume

In this section, we analyze how the choice of below-cloud scavenging model affects the results of estimates of changes in the concentration of aerosol particles in the test volume due to scavenging. The models described in Section 2.1.4, and the used multi-model ensemble were studied for this analysis.

First, the lognormal aerosol size distribution with different CMAD (count median aerodynamic diameter) and σ = 2.5 was considered; the scavenging time was varied between 0 to 3 h. This test was conducted to compare results integrated over droplet and particle size distributions in comparison with previous studies. The FLEXPART integral model was not tested here due to limitation of 500 nm in maximum particle size. The CMAD of 1 μm corresponds to an MMAD (mass median aerodynamic diameter) of about 12.4 μm (for 4000 kg/m^3 density), and a major part of the mass is concentrated in the range exceeding the model applicability.

As can be seen from Figure 9 for CMAD = 1 μm for precipitation intensities of 2.5 and 10 mm/h, the results of calculations using ensemble models are close to the Pripachkin model. For lower intensities, the results are in good agreement with the FLEXPART integral model. With the increase of precipitation, the intensity difference becomes significant and reaches the values of 0.2–0.5; the results obtained using ensemble models are higher than those calculated by the Pripachkin model. The other models considered significantly overestimate fraction of mass concentration in comparison with the ensemble models (underestimation of deposited values). The same behavior of mass concentration in the control volume could be found in studies [18,53].

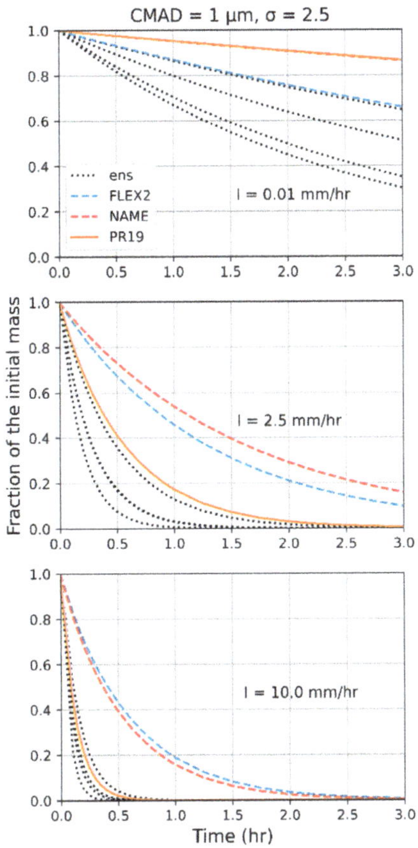

Figure 9. Time evolution of the normalized bulk aerosol particle mass concentration deposited due to below-cloud scavenging for different aerosol particle number distribution CMAD = 1 μm, σ = 2.5. ens—ensemble members, FLEX2—FLEXPART integral model, NAME—NAME integral model, PR19—Pripachkin dispersion model.

The next set of results is devoted to the analysis of the characteristic distributions of pollutant emissions into the atmosphere. The lognormal distribution with different MMAD and $\sigma = 2.5$ was considered as the aerosol size distribution; the scavenging time was varied from 0 up to 3 h. Since the actual diameter is used in the below-cloud scavenging models for Λ calculations, the aerodynamic diameter of aerosol spectrum was recalculated according to $d = d_a / \sqrt{\frac{\rho_a}{\rho_0}}$, where d_a is the aerodynamic diameter and $\rho_0 = 1000\ \text{kg/m}^3$. Two cases were studied. In the first case, the dependence of concentration fraction in control volume on the timing of precipitations was studied. Values of 1.0 μm [54] and 8 μm [55] were taken as the MMAD (of the lognormal aerosol distributions representative of accidental releases). It was seen (Figure 10) that for low intensities, the results of calculation of concentration loss are in good agreement with each other. For an MMAD of 1 μm, the FLEXPART parameterized dispersion model gives values within the ensemble distribution, and the FLEXPART curve is close to ensemble mean. For MMADs of 1 μm and 8 μm, the integral models give underestimated and overestimated results of concentration loss in the control volume. In all cases, the ensemble covers at least one of the integral models, which corresponds to ensemble forming objectives.

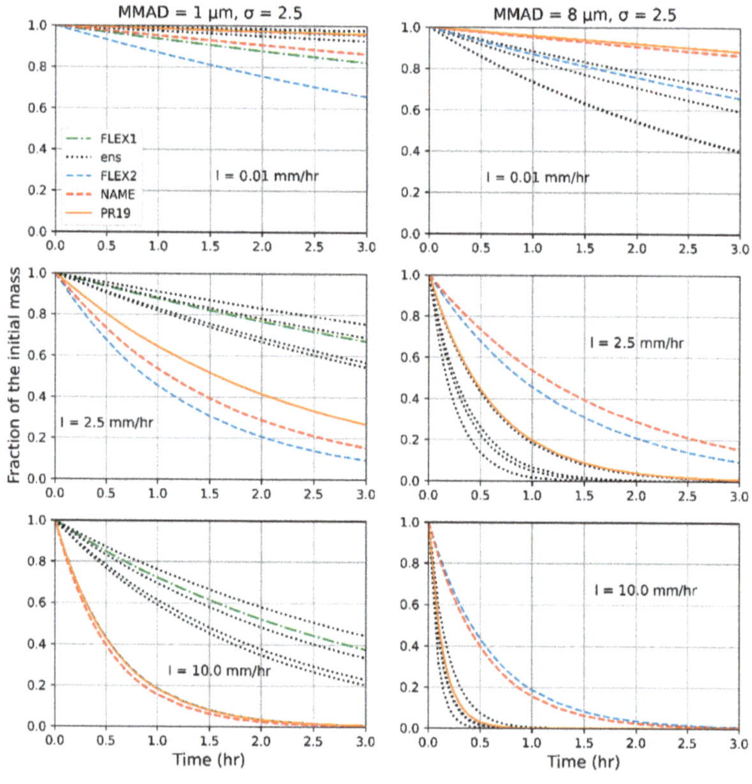

Figure 10. Time evolution of the aerosol mass concentration deposited due to below-cloud scavenging for different aerosol mass distributions (*left column*: MMAD = 1 μm, σ = 2.5; *right column*: MMAD = 8 μm, σ = 2.5). ens—ensemble memebers, FLEX1—FLEXPART parameterized dispersion model, FLEX2—FLEXPART integral mode, NAME—NAME integral model, PR19—Pripachkin dispersion model.

In the second case, the dependence of the fraction of scavenging concentration on the MMAD of the lognormal aerosol distribution was studied; the precipitation intensity I was taken equal to 0.01, 2.5, and 10 mm/h, and the duration of precipitations was set to 1 h.

Figure 11 shows that the concentration loss increases with increasing MMAD for models that take into account the dispersed structure of aerosols. At a value of MMAD ~10^{-5} m, the proportion of the washed out concentration tends toward 100% at high values of precipitation intensity.

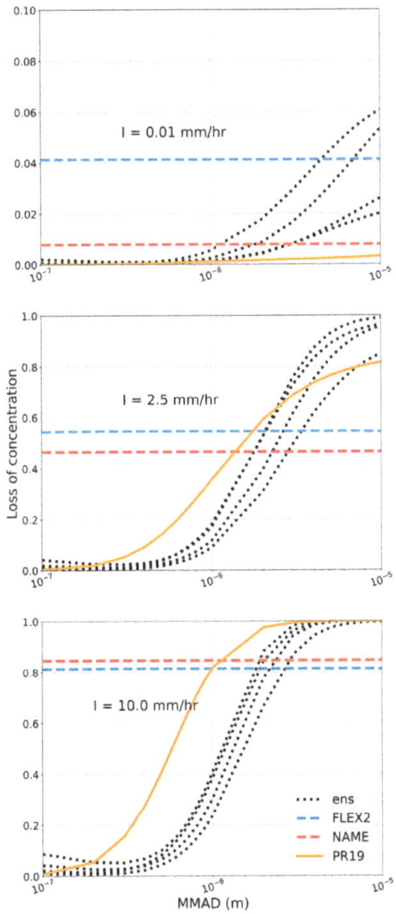

Figure 11. Dependence of fraction of aerosol concentration deposited due to below-cloud scavenging(%) from MMAD of lognormal aerosol distribution for different intensities (m), ens—ensemble memebers, FLEX2—FLEXPART integral model, NAME—NAME integral model, PR19—Pripachkin dispersion model.

For lower values of intensity, the ensemble gives reasonable dispersion up to 160% from the results averaged over ensemble members. The spread of the results decreases with the MMAD's increase: for 0.01 mm/h, from 240% to 100%; for 2.5 mm/h, from 200% to 14%; for 10 mm/h, from 200% to 0.2%. The curves intersect the FLEXPART integral model, the NAME integral model, and the Pripachkin model in the MMAD range between 10^{-6} and 10^{-5}. In general, there is an overestimation of washed out concentration when using integral models compared with the results based on the models that resolve the spectrum of aerosol particles for aerosols with MMAD of 2 μm or less, and an underestimation for coarse ones. In addition, after integration over the spectrum of aerosol particles, a decrease in dispersion of the ensemble is seen. However, it should be noted that the source of release can also be practically monodisperse (for example, a filtered release [56] or the release of molecular forms of radionuclides [57]).

4. Conclusions

The paper presents the results of a multi-model ensemble construction to determine the below-cloud scavenging coefficient in atmospheric transport models. Combinations of models describing the efficiency of aerosol capture by raindrops, the terminal raindrop fall velocity, and the size distribution of raindrops were considered as members of the ensemble. The following criteria for constructing a multi-model ensemble were used: the unbiased estimate, characterized by a flat form of a rank histogram, as well as ensemble dispersion of no more than 3σ of the normal distribution covering the statistical spread of experimental data. As a result of the study, it was found that the following combinations of models satisfy these requirements: the Slinn 83 collision efficiency model with the Kessler 1969 raindrop velocity model and exponential raindrop size distribution, as well as with the Atlas 1977 model and lognormal distribution. In addition, the modified Slinn 83 model, which takes into account the processes of thermo- and diffusiophoresis, was included in the constructed ensemble in combination with the Willis 1984 model using both exponential and lognormal raindrop size distributions. The following range of metric values as a result of comparison with experimental data is found: FB (−1.4–0.9), FAC10 (0.64–1.0), Pearson (~0.6) for the fine fraction and FB (−0.3–0.2), FAC10 (~0.7), and Pearson (~0.5) for the coarse fraction.

Based on experimental data in two ranges of particle diameters, using the ensemble approximation by a normal distribution, the FB metric, and Talagrand rank diagrams, it was found that the constructed ensemble provides an unbiased estimate. Perhaps, when using an ensemble in an atmospheric dispersion model, it is worth using this ensemble to reproduce the statistical characteristics of the normal distribution and then use the Monte Carlo method in the complex ensemble forecast. The results were compared with integrated models (not taking into account the size spectrum of aerosol). It was found that the ensemble forecast is physically adequate; the integral models tend to overestimate the concentration decrease (increase in predicted deposition) in the range of fine fractions (up to 1–5 μm) and underestimate the results for coarse fractions. The spread of the ensemble after integration over the spectra of raindrops and aerosol particles decreased to 250%, which decreases with MMAD increase to 100% for low intensities and from 200% to 0.2% for high precipitation rates. Nevertheless, taking into account the different characteristics of aerosols released into the atmosphere due to dispersion and dust suppression, the authors believe it correct to include the advanced approaches that take into account the size characteristics of raindrops and aerosols in the atmospheric transport model. In the future, the authors consider it expedient to develop this approach in describing in-cloud scavenging, which in some cases can also affect estimates of wet deposition.

Author Contributions: Conceptualization, A.K.; methodology, A.K.; validation, A.K., A.O. and V.S.; formal analysis, A.K., A.S. and V.S.; investigation, A.K., A.O. and A.S.; writing—original draft preparation, A.K., A.O., A.S. and V.S.; writing—review and editing, A.K., A.O., A.S. and V.S.; visualization, A.O.; supervision, A.K. All authors have read and agreed to the published version of the manuscript.

Funding: The APC was funded by Nuclear Safety Institute of the Russian Academy of Sciences (IBRAE RAN).

Informed Consent Statement: Not applicable.

Data Availability Statement: Data sharing not applicable.

Conflicts of Interest: The authors declare no conflict of interest.

Nomenclature

μ, μ_w	Dynamic viscosity of air and water, kg/m·s
$C(d)$	Concentration of aerosol particles with diameter d, 1/m^3
Cc	Cunningham correction for aerosol particle glide
D	Diameter of a raindrop, m
d	Diameter of a particle, m
$Ddiff$	Diffusion coefficient of aerosol particles in air, m^2/s
$Ddiff\ water$	Water vapor diffusion coefficient in the air, m^2/s
$E(D)$	Aerosol particle collision efficiency
E_{Br}	Efficiency of Brownian diffusion
E_{ine}	Efficiency of inertial impaction
E_{int}	Efficiency of interception
E_{dph}	Efficiency of diffusiophoresis
E_{th}	Efficiency of thermophoresis
H	Vertical size of the reference volume (from 10^2 to 10^3 m)
I	Rain intensity, mm/hour
ka, kp	Thermal conductivity of air, particles, W/m·K
kb	Boltzmann constant, J/K
Mw, Ma	Molecular masses of water and air, a.m.u.
$N(D)$	Raindrop size distribution, m^{-4}
$n(d)$	Volume concentration of raindrops, m^{-3}
P	Normal atmospheric pressure, Pa
p_s^0 p_a^0	Pressure of water vapor at temperatures T_a T_s, Pa
Pr	Prandtl number
r	Entanglement parameter
ReD	Reynolds number calculated for a raindrop with a diameter D
Sc	Schmidt number
Scw	Schmidt number for water vapor in air
St	Stokes number
St^*	Critical Stokes number
t	Interaction time of precipitation and aerosol, s
Ta, Ts	Absolute air temperature and absolute temperature of the raindrop surface, K
$V(D)$	Raindrop terminal velocity, m/s
$v(d)$	Particle velocity, m/s
α	Partial density of raindrops (varies from 10^{-5} to 10^{-10} for raindrops with a size of 0.1–6 mm, respectively)
γ	Coefficient depending on the macroscopic parameters of the medium (~1.0)
η	Coefficients of inertial diffusion capture and capture due to entanglement
Λ	Below-cloud scavenging coefficient, 1/s
λa	Rean free path of air molecules, m
ν	Rinematic viscosity, m^2/s
ρ_a, ρ_p	Rir density, particle density, kg/m^3
τ	Relaxation time of particles, s

References

1. International Atomic Energy Agency. *Dispersion of Radioactive Material in Air and Water and Consideration of Population Distribution in Site Evaluation for Nuclear Power Plants*; Safety Standards Series No. NS-G-3.2; International Atomic Energy Agency: Vienna, Austria, 2002.
2. International Atomic Energy Agency. *Operations Manual for IAEA Assessment and Prognosis During a Nuclear or Radiological Emergency*; Emergency Preparedness and Response; IAEA: Vienna, Austria, 2020.
3. Wang, Y.; Schmid, F.; Harou, A. *Guidelines for Nowcasting Techniques*; World Meteorological Organization: Geneva, Switzerland, 2017; ISBN 978-92-63-11198-2.
4. Chen, X.; Wang, M.; Wang, S.; Chen, Y.; Wang, R.; Zhao, C.; Hu, X. Weather Radar Nowcasting for Extreme Precipitation Prediction Based on the Temporal and Spatial Generative Adversarial Network. *Atmosphere* **2022**, *13*, 1291. [CrossRef]
5. Leadbetter, S.J.; Jones, A.R.; Hort, M.C. Assessing the value meteorological ensembles add to dispersion modelling using hypothetical releases. *Atmos. Chem. Phys.* **2022**, *22*, 577–596. [CrossRef]
6. Bakin, R.I.; Gubenko, I.; Dolganov, K.; Ignatov, R.; Ilichev, E.; Kiselev, A.; Krasnoperov, S.; Konyaev, P.; Rubinshtein, K.; Tomashchik, D. Application of ensemble method to predict radiation doses from a radioactive release during hypothetical severe accidents at Russian NPP. *J. Nucl. Sci. Technol.* **2020**, *58*, 635–650. [CrossRef]
7. Elahi, E.; Khalid, Z.; Tauni, M.Z.; Zhang, H.; Lirong, X. Extreme weather events risk to crop-production and the adaptation of innovative management strategies to mitigate the risk: A retrospective survey of rural Punjab, Pakistan. *Technovation* **2022**, *117*, 102255. [CrossRef]
8. Abid, M.; Scheffran, J.; Schneider, U.A.; Elahi, E. Farmer Perceptions of Climate Change, Observed Trends and Adaptation of Agriculture in Pakistan. *Environ. Manag.* **2019**, *63*, 110–123. [CrossRef] [PubMed]
9. *COST ES1006*; Best Practice Guidelines. COST Action ES1006. University of Hamburg: Hamburg, Germany, 2015.
10. Hegarty, J.; Draxler, R.R.; Stein, A.F.; Brioude, J.; Mountain, M.; Eluszkiewicz, J.; Nehrkorn, T.; Ngan, F.; Andrews, A. Evaluation of Lagrangian Particle Dispersion Models with Measurements from Controlled Tracer Releases. *J. Appl. Meteorol. Climatol.* **2013**, *52*, 2623–2637. [CrossRef]
11. Olesen, H.R. *User's Guide to the Model Validation Kit*; Research Notes from NERI no. 226; National Environmental Research Institute: Roskilde, Denmark, 2005; p. 72. Available online: http://research-notes.dmu.dk (accessed on 26 January 2023).
12. Khan, T.R.; Perlinger, J.A. Evaluation of five dry particle deposition parameterizations for incorporation into atmospheric transport models. *Geosci. Model Dev.* **2017**, *10*, 3861–3888. [CrossRef]
13. Saylor, R.D.; Baker, B.D.; Lee, P.; Tong, D.; Li, P.; Hicks, B.B. The particle dry deposition component of total deposition from air quality models: Right, wrong or uncertain? *Tellus B Chem. Phys. Meteorol.* **2019**, *71*, 1550324. [CrossRef]
14. Wesely, M.L.; Hicks, B.B. A Review of The Current Status of Knowledge on Dry Deposition. *Atmos. Environ.* **2000**, *34*, 2261–2282. [CrossRef]
15. Mun, C.; Cantrel, L.; Madic, C. A Literature Review on Ruthenium Behaviour in Nuclear Power Plant Severe Accidents. *Nucl. Technol.* **2007**, *156*, 332–346. [CrossRef]
16. Paesler-Sauer, J. *Comparative Calculations and Validation Studies with Atmospheric Dispersion Models*; Kernforschungszentrum Karlsruhe: Karlsruhe, Germany, 1986.
17. Sportisse, B. A review of parameterizations for modelling dry deposition and scavenging of radionuclides. *Atmos. Environ.* **2007**, *41*, 2683–2698. [CrossRef]
18. Wang, X.; Zhang, L.; Moran, M.D. Uncertainty assessment of current size-resolved parameterizations for below-cloud particle scavenging by rain. *Atmos. Chem. Phys.* **2010**, *10*, 5685–5705. [CrossRef]
19. Monin, A.S.; Obukhov, A.M. The main regularities of turbulent mixing in the surface layer of the atmosphere. *Proc. Geophys. Inst. Acad. Sci. USSR* **1954**, *151*, 163–187. (In Russian)
20. Piskunov, V.N. Parameterization of Aerosol Dry Deposition Velocities onto Smooth and Rough Surfaces. *J. Aerosol Sci.* **2009**, *40*, 664–679. [CrossRef]
21. Fang, S.; Zhuang, S.; Goto, D.; Hu, X.; Sheng, L.; Huang, S. Coupled modeling of in- and below-cloud wet deposition for atmospheric 137Cs transport following the Fukushima Daiichi accident using WRF-Chem: A self-consistent evaluation of 25 scheme combinations. *Environ. Int.* **2022**, *158*, 106882. [CrossRef]
22. Ulimoen, M.; Berge, E.; Klein, H.; Salbu, B.; Lind, O.C. Comparing model skills for deterministic versus ensemble dispersion modelling: The Fukushima Daiichi NPP accident as a case study. *Sci. Total Environ.* **2022**, *806*, 150128. [CrossRef]
23. Leadbetter, S.J.; Andronopoulos, S.; Bedwell, P.; Chevalier-Jabet, K.; Geertsema, G.; Gering, F.; Hamburger, T.; Jones, A.; Klein, H.; Korsakissok, I.; et al. Ranking uncertainties in atmospheric dispersion modelling following the accidental release of radioactive material. *Radioprotection* **2020**, *55*, S51–S55. [CrossRef]
24. Galmarinia, S.; Bianconi, R.; Klug, W.; Mikkelsen, T.; Addis, R.; Andronopoulos, S.; Astrup, P.; Baklanov, A.; Bartniki, J.; Bartzis, J.C.; et al. Ensemble dispersion forecasting—Part I: Concept, approach and indicators. *Atmos. Environ.* **2004**, *38*, 4607–4617. [CrossRef]
25. Sorensen, J.H.; Amstrup, B.; Feddersen, H.; Bartnicki, J.; Klein, H.; Simonsen, M.; Lauritzen, B.; Cordt Hoe, S.; Israelson, C.; Lindgren, J. *Fukushima Accident: UNcertainty of Atmospheric Dispersion Modelling (FAUNA)*; NKS Secretariat: Roskilde, Denmark, 2016; p. 43; ISBN 978-87-7893-421-5.

26. Straume, A.G.; Koffi, E.N.; Nodop, K. Dispersion Modeling Using Ensemble Forecasts Compared to ETEX Measurements. *J. Appl. Meteorol.* **1998**, *37*, 1444–1456. [CrossRef]
27. Perillat, R.; Korsakissok, I.; Mallet, V.; Mathieu, A.; Sekiyama, T.; Kajino, M.; Adachi, K.; Igarashi, Y.; Didier, D. Using meteorological ensembles for atmospheric dispersion modelling of the Fukushima nuclear accident. In Proceedings of the 18th International Conference on Harmonisation within Atmospheric Dispersion Modelling for Regulatory Purposes, Bologna, Italy, 9–12 October 2017.
28. Kajino, M.; Sekiyama, T.T.; Mathieu, A.; Korsakissok, I.; Périllat, R.; Quélo, D.; Quérel, A.; Saunier, O.; Adachi, K.; Girard, S.; et al. Lessons learned from atmospheric modeling studies after the Fukushima nuclear accident: Ensemble simulations, data assimilation, elemental process modeling, and inverse modeling. *Geochem. J.* **2018**, *52*, 85–101. [CrossRef]
29. Kajino, M.; Sekiyama, T.T.; Igarashi, Y.; Katata, G.; Sawada, M.; Adachi, K.; Zaizen, Y.; Tsuruta, H.; Nakajima, T. Deposition and dispersion of radio-cesium released due to the Fukushima nuclear accident: Sensitivity to meteorological models and physical modules. *J. Geophys. Res. Atmos.* **2019**, *124*, 1823–1845. [CrossRef]
30. Yegnan, A.; Williamson, D.G.; Graettinger, A.J. Uncertainty analysis in air dispersion modeling. *Environ. Model. Softw.* **2002**, *17*, 639–649. [CrossRef]
31. Mandal, S.; Madhipatla, K.K.; Guttikunda, S.; Kloog, I.; Prabhakaran, D.; Schwartz, J.D. GeoHealth Hub India Team, Ensemble averaging based assessment of spatiotemporal variations in ambient PM2.5 concentrations over Delhi, India, during 2010–2016. *Atmos. Environ.* **2020**, *224*, 117309. [CrossRef] [PubMed]
32. Webster, H.; Thomson, D. The NAME wet deposition scheme Forecasting Research. *Tech. Rep.* **2014**, *584*, 41.
33. Pisso, I.; Sollum, E.; Grythe, H.; Kristiansen, N.I.; Cassiani, M.; Eckhardt, S.; Arnold, D.; Morton, D.; Thompson, R.L.; Zwaaftink, C.D.G.; et al. The Lagrangian particle dispersion model FLEXPART version 10.4. *Geosci. Model Dev.* **2019**, *12*, 4955–4997. [CrossRef]
34. Seinfeld, J.H.; Pandis, S.N. *Atmospheric Chemistry and Physics: From Air Pollution to Climate Change*, 2nd ed.; John Wiley & Sons, Inc.: Hoboken, NJ, USA, 2006; pp. 900–929.
35. Slinn, W.G.N. Precipitation scavenging. In *Atmospheric Sciences and Power Production—1979, Chap. 11*; Division of Biomedical Environmental Research, U.S. Department of Energy: Washington, DC, USA, 1983.
36. Andronache, C.; Grönholm, T.; Laakso, L.; Phillips, V.; Venäläinen, A. Scavenging of ultrafine particles by rainfall at a boreal site: Observations and model estimations. *Atmos. Chem. Phys.* **2006**, *6*, 4739–4754. [CrossRef]
37. Pripachkin, D.A.; Budyka, A.K. Influence of the parameters of aerosol particles on their scavenging from the atmosphere by raindrops, Izvestiya RAN. *Phys. Atmos. Ocean.* **2020**, *56*, 203–209. (In Russian)
38. Montgomery, R.B. Viscosity And Thermal Conductivity Of Air And Diffusivity Of Water Vapor In Air. *J. Atmos. Sci.* **1947**, *4*, 193–196. [CrossRef]
39. Bae, S.Y.; Jung, C.H.; Kim, Y.P. Relative contributions of individual phoretic effect in the below-cloud scavenging process. *J. Aerosol Sci.* **2009**, *40*, 621–632. [CrossRef]
40. Marshall, J.S.; Palmer, W.M. The distribution of raindrop with size. *J. Meteorol.* **1948**, *5*, 165–166. [CrossRef]
41. Feingold, G.; Levin, Z. The lognormal fit to raindrop spectra from frontal convective clouds in Israel. *J. Clim. Appl. Meteorol.* **1986**, *25*, 1346–1363. [CrossRef]
42. Laakso, L. Ultrafine particle scavenging coefficients calculated from 6 years field measurements. *Atmos. Environ.* **2003**, *37*, 3605–3613. [CrossRef]
43. Anctil, F.; Ramos, M.H. *Verification Metrics for Hydrological Ensemble Forecasts*; Springer: Berlin/Heidelberg, Germany, 2017. [CrossRef]
44. Sparmacher, H.; Fülber, K.; Bonka, H. Below-cloud scavenging of aerosol particles: Particle-bound radionuclides—Experimental. *Atmos. Environ.* **1993**, *27A*, 605–618. [CrossRef]
45. Baklanov, B.; Sørensen, J.H. Parameterization of radionuclides deposition in atmospheric long-range transport modelling. *Phys. Chem. Earth B* **2001**, *26*, 787–799. [CrossRef]
46. Blanco-Alegre, C.; Calvo, A.; Castro, A.; Oduber, F.; Alonso-Blanco, E.; Fraile, R. Scavenging of submicron aerosol particles in a suburban atmosphere: The raindrop size factor. *Environ. Pollut.* **2021**, *285*, 117371. [CrossRef]
47. Zikova, N.; Zdimal, V. Precipitation scavenging of aerosol particles at a rural site in the Czech Republic. *Tellus B Chem. Phys. Meteorol.* **2016**, *68*, 27343. [CrossRef]
48. Gonfiotti, B.; Paci, S. Stand-alone containment analysis of Phébus FPT tests with ASTEC and MELCOR codes: The FPT-2 test. *Heliyon* **2018**, *4*, e00553. [CrossRef]
49. Met Office: National Meteorological Library and Archive. *Fact Sheet No. 3—Water in the Atmosphere*; Met Office: Devon, UK, 2007.
50. Jylhä, K. Relationship between the Scavenging Coefficient for Pollutants in Precipitation and the Radar Reflectivity Factor. Part I: Derivation. *J. Appl. Meteorol.* **1999**, *38*, 1421–1434. [CrossRef]
51. *User's Guide Comprehensive Air Quality Model with Extensions Version 7.10*; Ramboll Environment and Health: Copenhagen, Denmark, 2020.
52. Pudykiewicz, J. Simulation of the Chernobyl dispersion with a 3-D hemispheric tracer model. *Tellus* **1989**, *41B*, 391–412. [CrossRef]
53. Jones, C.; Hill, A.; Hemmings, J.; Lemaitre, P.; Quérel, A.; Ryder, C.L.; Woodward, S. Below-cloud scavenging of aerosol by rain: A review of numerical modelling approaches and sensitivity simulations with mineral dust in the Met Office's Unified Model. *Atmos. Chem. Phys.* **2022**, *22*, 11381–11407. [CrossRef]

54. Ogorodnikov, B.I.; Budyka, A.K.; Pavlyuchenko, N.I. Observation of radioactive aerosol emissions from the sarcophagus at the chernobyl nuclear power plant. *At. Energy* **2004**, *96*, 202–213. [CrossRef]
55. Dorrian, M.; Bailey, M. Particle size distributions of radioactive aerosols measured in workplaces. *Radiat. Prot. Dosim.* **1995**, *60*, 119–133. [CrossRef]
56. Sugiyama, G.; Gowardhan, A.; Simpson, M.; Nasstrom, J. *Deposition Velocity Methods for DOE Site Safety Analyses*; Lawrence Livermore National Lab.: Livermore, CA, USA, 2014.
57. Telly Bah, O.; Hebert, D.; Connan, O.; Solier, L.; Laguionie, P.; Bourlès, D.; Maro, D. Measurement and modelling of gaseous elemental iodine (I_2) dry deposition velocity on grass in the environment. *J. Environ. Radioact.* **2020**, *219*, 106253. [CrossRef] [PubMed]

atmosphere
MDPI

Article

Assessment of Land Surface Schemes from the WRF-Chem for Atmospheric Modeling in the Andean Region of Ecuador

Rene Parra

Instituto de Simulación Computacional (ISC-USFQ), Colegio de Ciencias e Ingenierías, Universidad San Francisco de Quito USFQ, Quito 170901, Ecuador; rrparra@usfq.edu.ec

Abstract: Surface interactions occur near the land–atmosphere interface, thus affecting the temperature, convection, boundary layer, and stability of the atmosphere. A proper representation of surface interactions is a crucial component for numerical atmospheric and air quality modeling. We assessed four land surface schemes—1. 5-layer thermal diffusion scheme (1 5-Layer); 2. unified Noah land surface model (2 Noah); 3. rapid update cycle (3 RUC) land surface model; and 4. Pleim–Xiu land surface model (4 Pleim–Xiu)—from the Weather Research and Forecasting with Chemistry (WRF-Chem V3.2) model for the purposes of atmospheric modeling in Cuenca, which is a region with a complex topography and land use configuration and which is located in the Southern Andean region, in Ecuador. For this purpose, we modeled the meteorological and air quality variables during September 2014. It was found that the meteorological and short-term air quality variables were better modeled through the 2 Noah scheme. Long-term (mean monthly) air quality variables were better modeled by the 1 5-Layer and 3 RUC options. On average, the 2 Noah scheme was better at modeling meteorology and air quality. In addition, we assessed the 2 Noah scheme combined with the urban canopy model (UCM) (5 Noah UCM), which was developed as an option to represent the urban effects at a subgrid-scale. Results indicated that the performance of the 5 Noah UCM scheme was not better at modeling than the 2 Noah scheme alone. Moreover, the 5 Noah UCM scheme notably decreased the modeling performance for carbon monoxide and fine particulate matter. These results complement previous assessments of other schemes, allowing us to recommend a basic configuration of parameters for atmospheric modeling in the Andean region of Ecuador.

Keywords: Cuenca; parameterization; modeling performance; land surface model; lsm

Citation: Parra, R. Assessment of Land Surface Schemes from the WRF-Chem for Atmospheric Modeling in the Andean Region of Ecuador. *Atmosphere* **2023**, *14*, 508. https://doi.org/10.3390/atmos14030508

Academic Editor: Daniel Viúdez-Moreiras

Received: 31 December 2022
Revised: 26 February 2023
Accepted: 1 March 2023
Published: 6 March 2023

1. Introduction

Surface processes occur near the land surface–atmosphere interface [1]. The movement of heat and water through this interface affects components, such as temperature, convection, boundary layer height, and stability, thus conditioning the performance of atmospheric and air quality modeling [2–5]. As solar radiation reaches the surface, a part of it is reflected, but much is absorbed. In addition, the atmosphere emits infrared radiation, thus warming the surface [6]. Energy from these two sources is stored in the soil or transferred to the atmosphere through sensible and heat fluxes. These interactions define the temperature and humidity near the ground surface and are modeled by the land surface schemes, which provide heat and moisture fluxes over the surface points. The different land surface models have various degrees of sophistication. Studies have found a high degree of variability in their results [7] and reported that surface temperature is a sensitive variable to land surface schemes [5,8]. Therefore, dedicated assessments of these land surface models are required to identify the option with the best performance for atmospheric modeling.

For modeling land surface fluxes, such as in the Weather Research and Forecasting with Chemistry model (WRF-Chem V3.2), information from the surface layer scheme, radiation, and precipitation from the microphysics and convective components, are combined with land variables and properties in order to provide an account of the heat and moisture fluxes [9].

The 5-layer thermal diffusion scheme (1 5-Layer); is a simple land surface model that is based on the MM5 five-layer soil temperature model [10]. This scheme works with five layers; below them, the temperature is fixed at a deep-layer average. The soil moisture is set with land use and season-dependent constant value, without vegetation effects. This scheme is the simplest in structure and physics when compared to the following land surface options.

The unified Noah land surface model (2 Noah) [11] is a four-layer soil temperature and moisture model with canopy moisture and snow cover prediction, as well as including the root zone, evapotranspiration, soil drainage, and runoff variables in its modeling. It also considers the influence of vegetation categories and soil texture. This scheme provides an account of the sensible and latent heat fluxes with respect to the boundary layer and has an improved urban treatment.

As default, the rapid update cycle (3 RUC) land surface scheme [12,13] works with six soil levels, solving heat diffusion and moisture transfer equations with higher resolution at the top of the soil. It includes vegetation effects and canopy water, using a layered approach to solve the energy and moisture budgets. The layer spans the ground surface, including half of the first atmospheric layer and half of the top soil layer.

The Pleim–Xiu land surface scheme (4 Pleim–Xiu) [14,15] works with a two-layer soil temperature and moisture model. This scheme considers three components for moisture: evapotranspiration, soil evaporation, and evaporation from wet canopies. The soil parameters are derived from the fractional coverages of land use categories and soil texture types.

Table 1 summarizes the main features and configuration of soil layers of these four schemes. They are coded in WRF-Chem V3.2 for modeling the land surface processes.

Table 1. Land surface scheme options in the WRF-Chem V3.2, based on [9].

Scheme	Vegetation Processes	Soil Variables and Features	Number of Soil Layers	Thickness of Soil Layers (Top to Bottom, cm)	Total Thickness (cm)
1 5-Layer	No	Temperature. No moisture or frozen soil	5	1, 2, 4, 8, 16	31
2 Noah	Yes	Temperature, water+ice, water. Moisture and frozen soil	4	10, 30, 60, 100	200
3 RUC	Yes	Temperature, ice, water+ice. Moistures and frozen soil	6 (default)	0, 5, 20, 40, 160, 300	525
4 Pleim–Xiu	Yes	Temperature, moisture	2	1, 99	100

An urban canopy model (UCM) can complement a land surface scheme to consider the influence of urban areas at the subgrid-scale [1]. The UCM component parameterizes the effects of street canyons, shadowing from buildings, a wind profile in the canopy layer, and a heat model from roof and road surfaces.

In 2021, in Cuenca–a city located in the Southern of the country (Figure 1)–$PM_{2.5}$ yearly mean concentrations between 5.7 to 9.8 $\mu g\ m^{-3}$ were measured, which were larger than the current World Health Organization (WHO) guideline (5.0 $\mu g\ m^{-3}$) [16,17]. Moreover, during forty days from 2021, the $PM_{2.5}$ 24 h mean levels in the historic center were higher than the current WHO guideline (15 $\mu g\ m^{-3}$). In addition, the O_3 levels can reach concentrations that are more elevated than the corresponding WHO guideline (100 $\mu g\ m^{-3}$, maximum 8 h daily mean) [18], typically during September, which is when high levels of solar radiation reach the Ecuadorian region at the surface, promoting photochemical O_3 production [19].

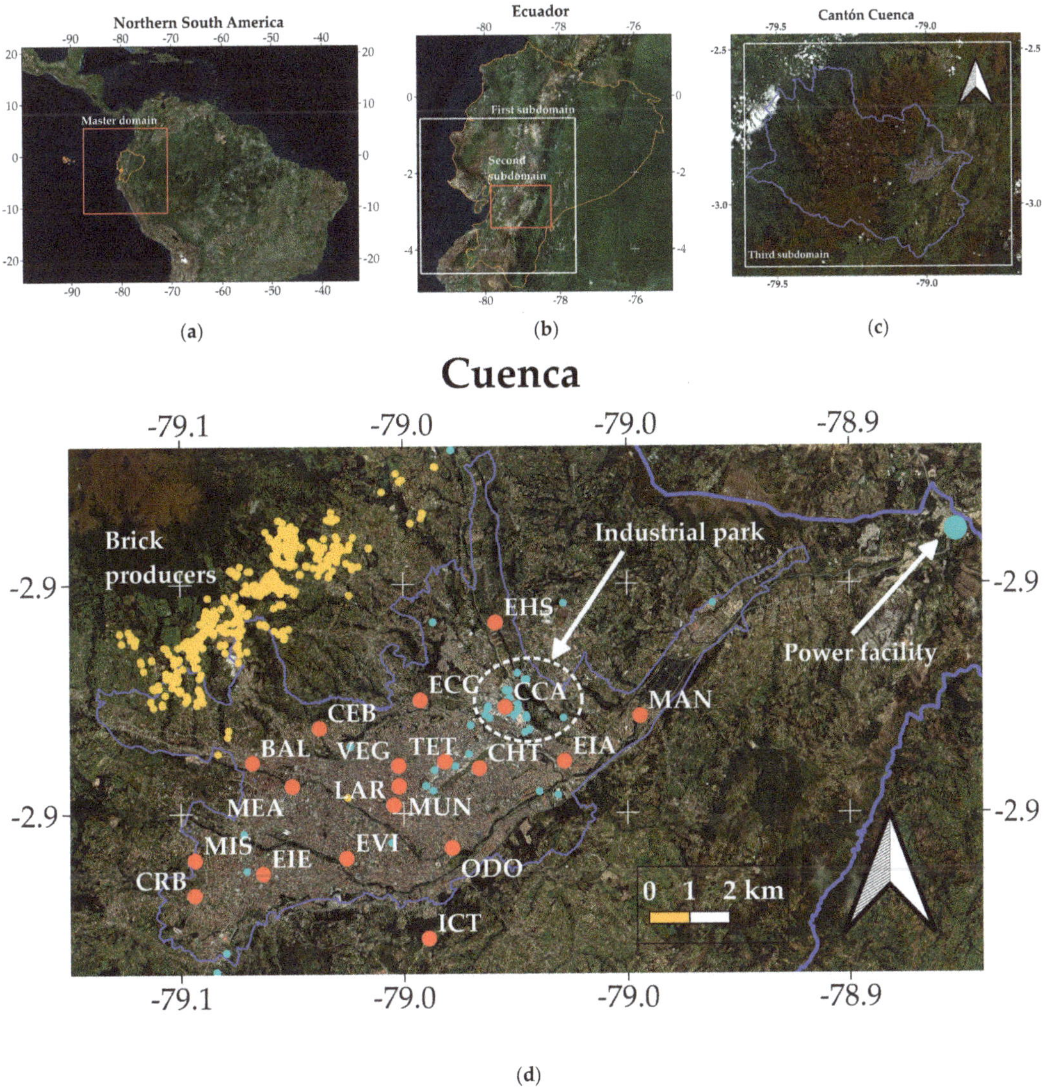

Figure 1. Location of: (**a**) Ecuador and (**b**,**c**) Cantón Cuenca. (**d**) The urban area of Cuenca (blue border) and the air quality network from Cuenca (red dots). Nomenclature of stations: MUN, Municipio; MAN, Machángara; CEB, Cebollar; BCB, and Bomberos; ICT, Ictocruz; CRB, and Colegio Rafael Borja; MIS, Misicata; EIE, Escuela Ignacio Escandón; EVI, and Escuela Velasco Ibarra; ODO, Facultad de Odontología; LAR, Calle Larga; BAL, Balzay; MEA, and Mercado El Arenal; CHT, Colegio Herlinda Toral; VEG, Vega Muñoz; TET, and Terminal Terrestre; EIA, Escuela Ignacio Andrade; ECC, Escuela Carlos Crespi; CCA, and Colegio Carlos Arízaga; and EHS, Escuela Héctor Sempértegui. Yellow dots indicate the artisanal brick producers. Cyan dots indicate industries.

The WRF-Chem V3.2 model was used in Cuenca, which is a region with a complex topography and land use configuration, where atmospheric modeling is challenging. Emissions from the inventory of 2014 were used to assess the influence of planetary boundary layer schemes [20], as well as global atmospheric datasets [21] for modeling meteorology and air quality. For these studies, the direct effects between aerosols and meteorology were activated. This was performed because this feedback improved the performance when modeling the air quality in Cuenca [20]. However, a preliminary modeling assessment with direct and indirect effects showed no benefits for Cuenca [22]. Moreover, the 2014 emission inventory and the WRF-Chem model were used to assess the impact on air quality due to the future replacement of diesel by electrical buses [23], and to estimate the benefit on $PM_{2.5}$ levels, due to the advance of combustion activities on 31 December [24]. Although the modeling performances for these studies were promising, they require improvements. For this purpose, other components, such as the land surface and urban canopy model, deserve a dedicated evaluation to improve the configuration of schemes and options for atmospheric modeling in this region.

At the time of writing this manuscript, the Municipality of Cuenca is updating the emission inventory, taking the year 2021 as its basis. One way to assess the quality of the new emission inventory will be using these results in WRF-Chem, for the purposes of modeling both meteorology and air quality during selected periods in 2021.

Therefore, the goals of this contribution are to address the following questions:

- What land surface scheme from WRF-Chem V3.2 provides the best performance for modeling the meteorological variables and air quality in Cuenca?
- Is there a benefit when considering the influence of the urban area through the urban canopy option?
- What are the recommendation for a land surface and urban canopy model to assess the quality of the in-progress emission inventory of Cuenca?

2. Methods

2.1. *The Air Quality Network*

To monitor meteorology and air quality, since 2012 the city has had an automatic station in the historic center (MUN station, Figure 1), at coordinates 2.89° S, 79.0° W, and 2500 masl. In addition, the air quality network has around twenty passive stations for monitoring the monthly mean concentrations of NO_2 and O_3 (Figure 1d). The municipality of Cuenca is the accredited entity to operate the air quality network; further, it is mandated to apply the required methods of the Ecuadorian regulations.

The study of records provided by the MUN station allowed the identification of the weekend effect (WE) in the historic center of Cuenca [23]. The WE indicates counterintuitive behavior regarding O_3, which increases during weekends, although the emissions of its precursors (NO_x and NMVOC) decrease.

2.2. *Emission Inventory of 2014*

Cuenca's last emission inventory was built for 2014 [25]. This inventory indicates that on-road traffic was the primary source of pollutants (94.9% of CO, 71.2% of NOx, 42.4% of $PM_{2.5}$, and 39.6% of NMVOC). Other NMVOC-relevant sources were found to be solvents (29.7%) and vegetation (19.5%). Industries, most of them located in the industrial park–which is in the northeast urban area (Figure 1d)–were the most significant source of SO_2 (60.1%). Approximately 600 artisan brick producers (northwest, out of the urban area) generated 38.5% of the $PM_{2.5}$ emissions. In addition, a thermoelectric facility (northeast, out of the urban area) generated 35.1% of SO_2 and 18.5% of NO_x emissions.

2.3. *Modeling Approach*

We used the WRF-ChemV3.2 for modeling the meteorological and air quality variables in Cuenca during September 2014. The WRF-Chem is a non-hydrostatic and Eulerian 3-D model that is used for atmospheric research and forecasting purposes [26]. It can optionally include the chemical transport of air pollutants in its modeling, thus, allowing for the use of an online approach for modeling interactions between air pollutants and meteorological variables.

Firstly, a master domain and two nested subdomains (Figure 1) were used to run the meteorological simulations. The inner subdomain is a grid of 100 × 82 cells, 1 km on each side, which covers the territory of Cuenca (Figure 1c). Table 2 details the main features of the domains.

Table 2. Configuration of the domains for modeling the meteorology and air quality in Cuenca.

Domains	Nomenclature	Size (Cells)	Spatial Resolution (km)
Master (Figure 1a)	D01	70 × 70	27
First subdomain (Figure 1b)	D02	52 × 52	9
Second subdomain (Figure 1b)	D03	61 × 42	3
Third (inner) subdomain (Figure 1c)	D04	100 × 82	1

The land use maps were defined based on the US Geological Survey (USGS) categories. Figure 2 depicts the land use map for the inner domain, which indicates that cropland grassland mosaic, dryland cropland, pasture, and shrubland were the most important in terms of coverage surrounding the urban and built-up land area. Each cell was assigned a unique category based on the most abundant land use range.

Based on the performance of the global meteorological datasets over this region, we used the FNL Global Operational Analysis (Final), produced by the National Centers for Environmental Prediction (NCEP) [27], for generating the initial and boundary conditions. The chemical transport of pollutants option was activated for the third subdomain for the purposes of reading the hourly emissions coming from the emission inventory of Cuenca in 2014. The simulation period covers 1 to 27 September 2014, and it was focused on accounting for the high O_3 levels that are typically measured during this month. The Carbon Bond Mechanism Z (CBMZ) [28] and the model for simulation aerosol chemistry and interaction (MOSAIC) were used to speciate and represent, respectively, the corresponding hourly emissions [29]. Table 3 summarizes the primary schemes and options selected for the modeling approach.

Table 3. Main schemes and options selected for modeling the meteorology and air quality in Cuenca (WRF-Chem V3.2).

Component and WRF-Chem Option Nomenclature	WRF-Chem Option	Model, Description, and References
Microphysics (mp_physics)	4	WRF Single–moment 5–class (Hong et al., 2004) [30]
Longwave Radiation (ra_lw_physics)	1	RRTM (Mlawer et al., 1997) [31]
Shortwave Radiation (ra_sw_physics)	2	Goddard (Chou and Suarez, 1999) [32]
Surface Layer (sf_clay_physics)	1	MM5 similarity (Paulson, 1970) [33]
Planetary Boundary Layer (bl_pbl_physics)	1	Yonsei University (Hong et al., 2006) [34]
Cumulus Parameterization (cu_physics)	5	Grell 3D Ensemble (Grell, 1993) [35]
Options of chemical mechanisms and aerosol modules (chem_opt)	7	CBMZ (Zaveri and Peters, 1999) and MOSAIC (4 sectional aerosol bins) (Zaveri et al., 2008) [28,29]
Land Surface (sf_surface_physics)	1	(Dudhia, 1996) [10]
	2	(Chen and Dudhia, 2001) [11]
	3	(Smirnova et al., 1997, 2000) [12,13]
	7	(Pleim and Xiu, 1995; Xiu and Pleim, 2001) [14,15]
Urban surface (sf_urban_physics)	0	No urban physics
	1	Single-layer UCM (Kusaka et al., 2001; Kusaka and Kimura, 2004; Chen et al., 2006) [36–38]. This option can be used with the Noah land surface scheme

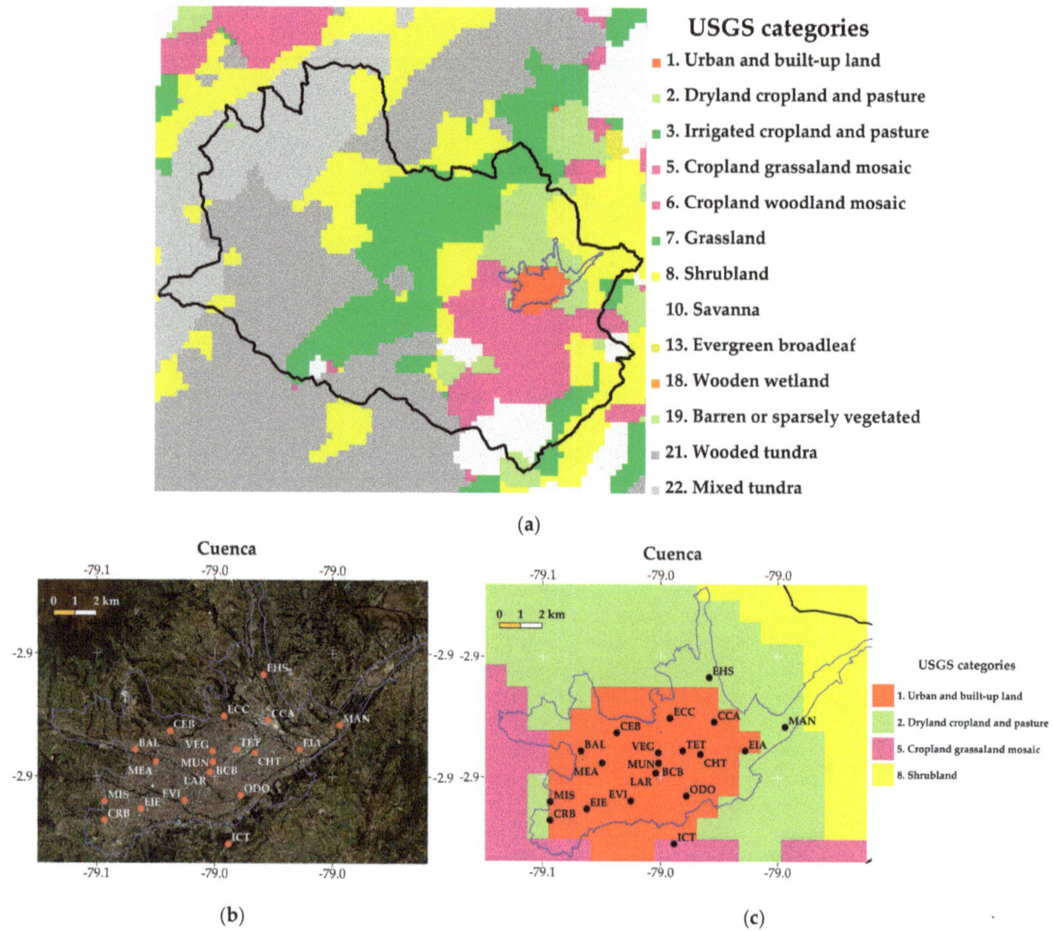

Figure 2. (**a**) Land use coverage (1 km of resolution) for the inner domain. Based on the US Geological Survey land use categories. (**b**) Zoom of the urban area. (**c**) Zoom of the urban area and the land use coverage for modeling.

2.4. Modeling Performance

The performance for modeling surface temperature was assessed by the gross error (GE), mean (MB), and index of agreement (IOA) (Table 4). The performance for modeling wind speed was assessed by the root mean square error (RMSA), MB, and IOA. For wind direction, GE and MB were used. The expressions of these metrics are described in the EEA (2011) [39] and in the study of Simon et al. (2012) [40].

For short-term air quality modeling performance, we evaluated the CO (maximum mean in 1 h and 8 h, per day), $PM_{2.5}$ (mean in 24 h), and O_3 (maximum mean in 8 h, per day) concentrations during periods consistent with the Ecuadorian air quality legislation and the WHO guidelines [17,41]. For these variables, we used MB, RMSE, fractional bias (FB), the mean normalized bias (MNB), and the correlation coefficient (r) [40].

In addition, we obtained the percentages of records, which were captured by modeling. It was considered that a record was adequately modeled if the maximum deviation between the observed and modeled value agreed with the accuracies of Table 4. For long-term air quality (NO_2 and O_3), the modeling performance was considered as the percentage of the passive stations with a maximum difference of 30%.

Table 4. Metrics for modeling meteorological and air quality variables [39,40].

Variable	Metric	Benchmark or Ideal Value	Accuracy
Hourly surface temperature	GE MB IOA	<2 °C <±0.5 °C ≥0.8	±2 °C
Hourly wind speed (10 m above the surface)	RMSE MB IOA	<2 m s^{-1} $<\pm0.5$ m s^{-1} ≥0.6	±1 m s^{-1}
Hourly wind direction (10 m above the surface)	GE MB	<30° <±10°	±30°
Short-term air quality (daily concentrations): Maximum 1 h CO mean, maximum 8 h CO mean, 24 h $PM_{2.5}$ mean, maximum 8 h O_3 mean	MB RMSE FB MNB r	0 0 0 0 1	±50%
Long-term air quality (monthly concentrations): NO_2 and O_3			±30%

3. Results

3.1. Meteorology

The 2 Noah land surface scheme was better when modeling surface temperature, followed by the 4 Pleim–Xiu scheme (Table 5). These options received the GE, MB, and IOA metrics into the corresponding value benchmark ranges. The 2 Noah option reproduced, in the best way, the mean daily profile (Figure 3a). On average, the 4 Pleim–Xiu option underestimated the surface temperature between 09:00 to 16:00 local time (LT) (MB = −0.5 °C). In addition, a strong relationship between measured and modeled temperatures (IOA = 0.9 and 1.0) was established for these two schemes. The 2 Noah scheme captured the highest percentage (78.0%) of the hourly temperature records (Table 6). The 1 5-Layer and 3 RUC options underestimated the temperature during the first hours of the day and then overestimated it during most of the daylight hours. The 5 Noah UCM option modeled the mean temperature profile well during the morning, as well as during the first hours of the afternoon (Figure 3a). However, this option underestimated the temperature during the hours without solar radiation.

Table 5. Metrics for the meteorological modeling. Bold numbers indicate values in the benchmark ranges.

Land Surface Scheme:	1 5-Layer	2 Noah	3 RUC	4 Pleim–Xiu	5 Noah UCM	Benchmark
			Hourly surface temperature:			
GE	2.7	**1.3**	3.6	**1.8**	2.4	**<2 °C**
MB	−0.9	**0.1**	1.0	**−0.5**	−1.9	**<±0.5 °C**
IOA	**0.9**	**0.9**	**0.8**	**0.8**	**0.8**	**≥0.8**
			Hourly wind speed:			
RMSE	**1.1**	**0.9**	**1.1**	**1.6**	**1.4**	**<2 m s^{-1}**
MB	**0.2**	**0.2**	**0.0**	**0.0**	**0.5**	**$<\pm0.5$ m s^{-1}**
IOA	**0.8**	**0.9**	**0.8**	**0.6**	**0.8**	**≥0.6**
			Hourly wind direction:			
GE	69.1	61.9	67.8	73.3	73.3	**<30°**
MB	12.0	−20.8	**6.1**	**−8.9**	22.3	**<±10°**

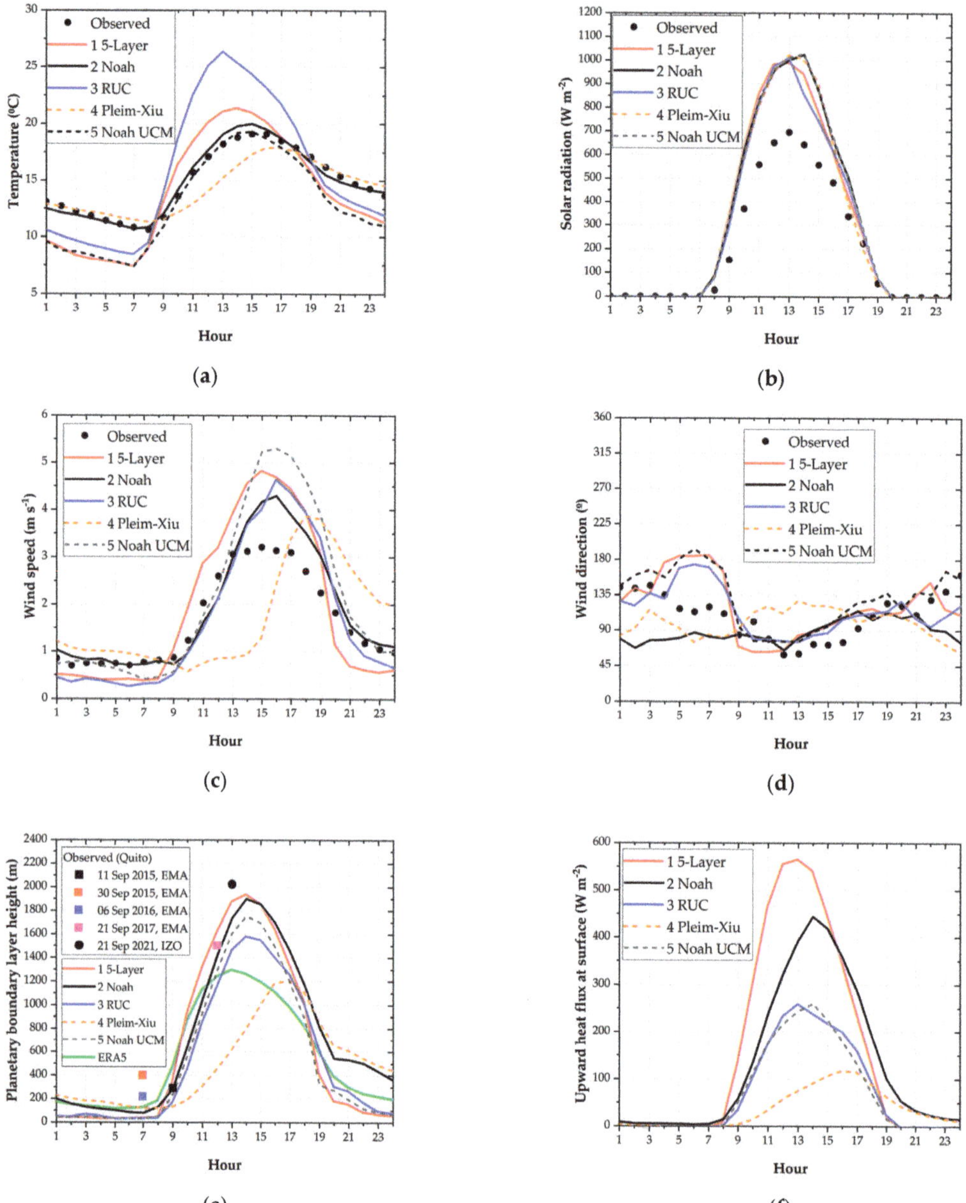

Figure 3. Daily mean profile during September 2014 at the location of the MUN station: (**a**) Temperature. (**b**) Global solar radiation. (**c**) Wind speed. (**d**) Wind direction. (**e**) Planetary boundary layer height (PBLH). Squares in (**e**) show PBLH observations at the EMA station [42]. The black dot indicates the observation at the IZO station [43]. (**f**) The modeled upward heat flux at the surface by the schemes.

Table 6. Percentages of the records that were captured by modeling. Meteorological variables. Bold numbers highlight the schemes with the best performances.

Land Surface Scheme:	1 5-Layer	2 Noah	3 RUC	4 Pleim–Xiu	5 Noah UCM	Number of Records
Hourly surface temperature	38.0	**78.0**	34.8	63.8	50.9	644
Hourly wind speed	66.9	**76.9**	71.0	49.5	62.4	644
Hourly wind direction	32.9	**37.1**	31.8	25.8	31.4	644

Figure 4 shows the spatial distribution of the modeled temperature on 22 September 2014, at 07:00 LT. Over the urban area, the 2 Noah option provided temperatures mostly between 11 and 13 °C. The 1 5-Layer option modeled temperatures mostly between 9 and 11 °C. The 3 RUC and 4 Pleim-Xiu schemes obtained temperatures ranging from 15–17 °C and 13–15 °C, respectively, in the complete urban area. The 5 Noah UCM option provided temperatures between 9 and 11 °C over the entire urban area. Figure 3 also shows the corresponding computed values of the planetary boundary layer heights. Over the urban area, the 2 Noah option calculated values between 70 and 300 m. The other options produced lower values, between 50 and 120 m for the 1 5-Layer and 4 Pleim–Xiu models, and between 50 and 100 m for 3 RUC and 5 Noah UCM options.

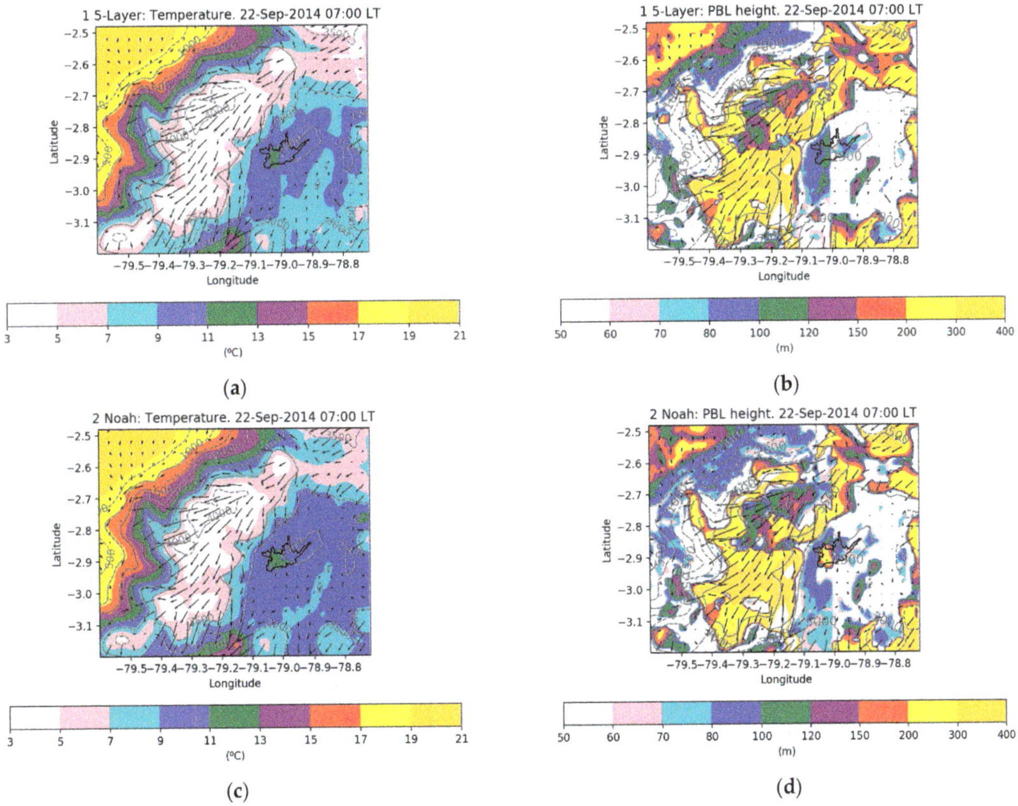

Figure 4. *Cont.*

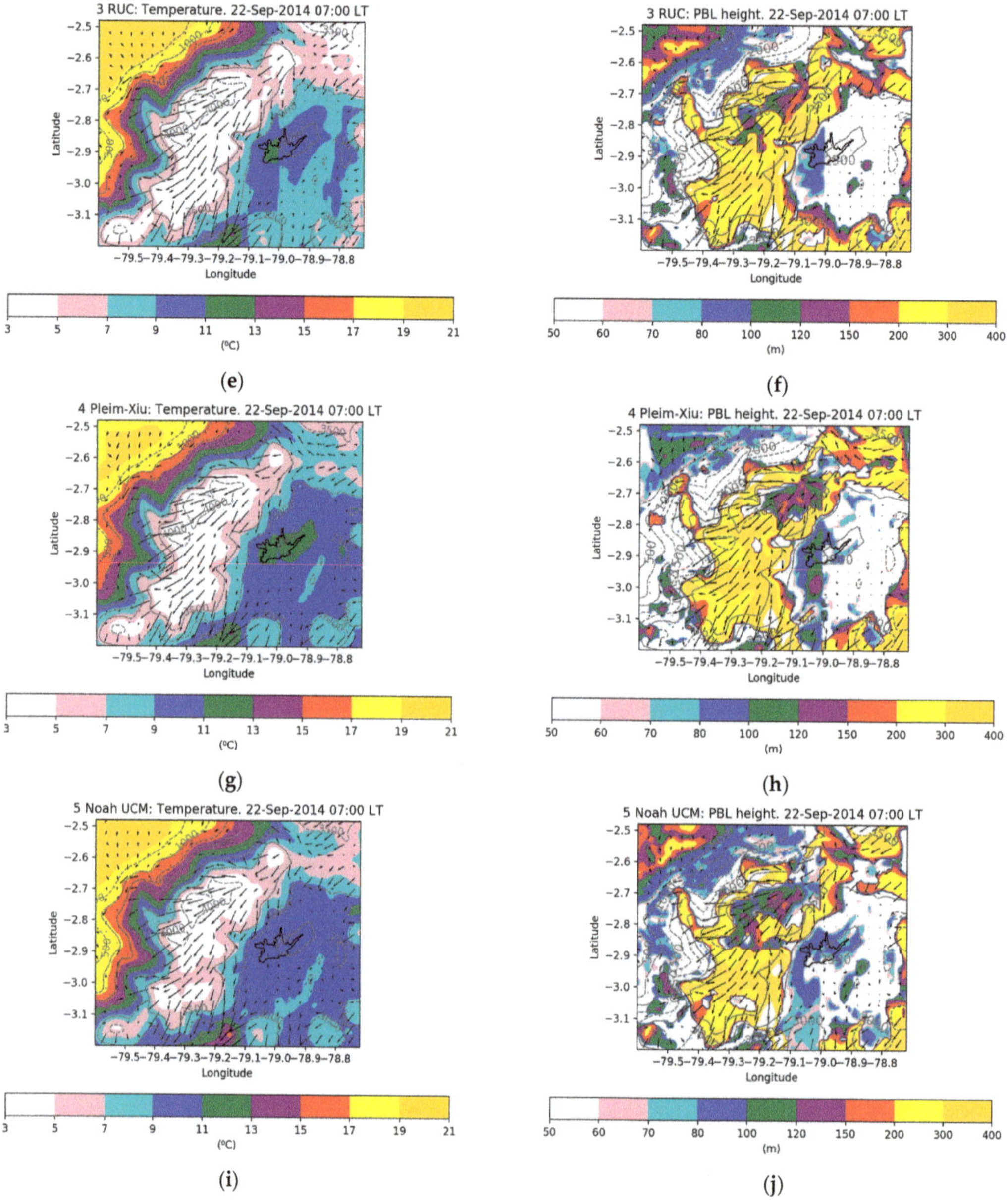

Figure 4. Modeled temperature and planetary boundary layer height on 22 September 2014 (07:00 LT): (**a**,**b**) 1 5-Layer; (**c**,**d**) 2 Noah; (**e**,**f**) 3 RUC; (**g**,**h**) 4 Pleim–Xiu; (**i**,**j**) 5 Noah UCM.

Figure 5 depicts the modeled temperature on 22 September 2014 at 13:00 LT. Over the urban area, the 2 Noah option provided temperatures mostly between 15 and 17 °C. The 1 5-Layer option modeled values mostly between 17 and 19 °C. In the complete urban area, the 3 RUC scheme obtained temperatures ranging from 15 and 17 °C. The 4 Pleim–Xiu option provided values between 13 and 15 °C for the entire urban area. The 5 Noah UCM option provided temperatures consistent with the 2 Noah scheme.

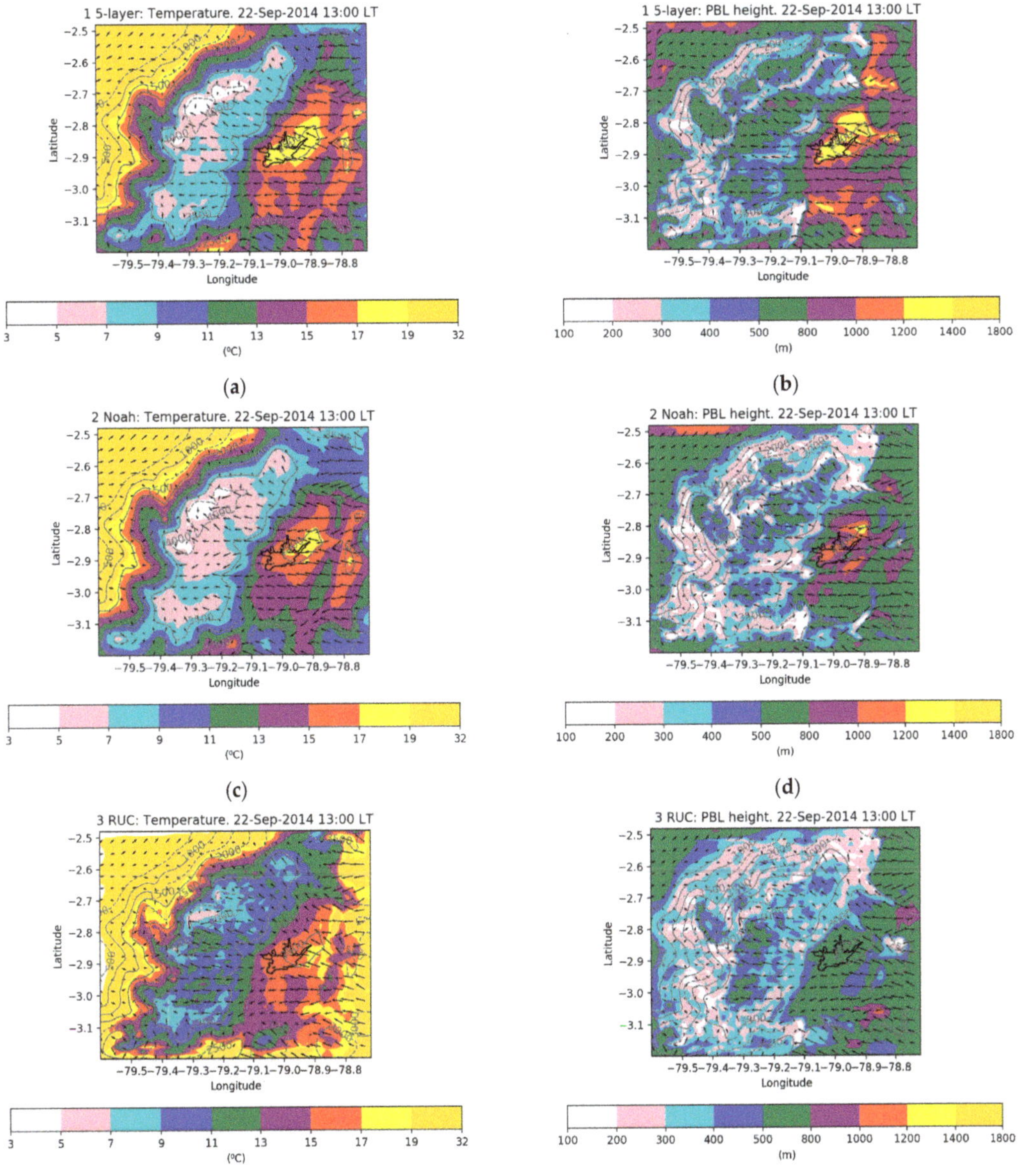

Figure 5. *Cont.*

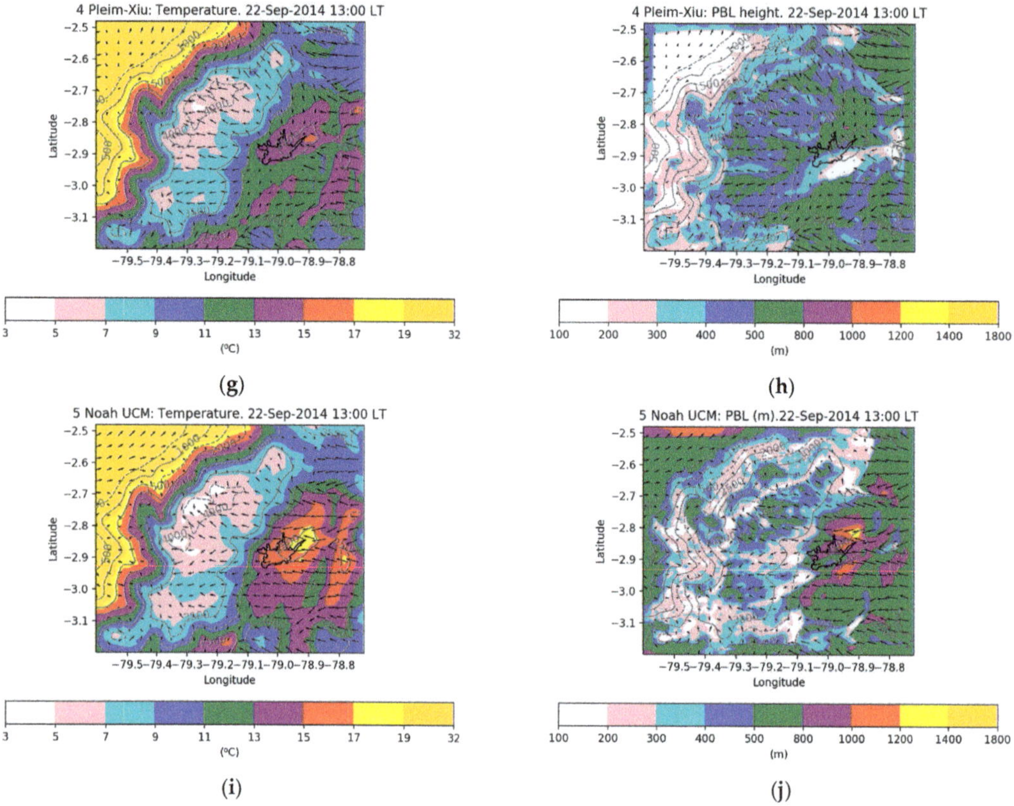

Figure 5. Modeled temperature and planetary boundary layer height on 22 September 2014 (13:00 LT): (**a**,**b**) 1 5-Layer; (**c**,**d**) 2 Noah; (**e**,**f**) 3 RUC; (**g**,**h**) 4 Pleim–Xiu; (**i**,**j**) 5 Noah UCM.

Although all the metrics, when using all the options, were in the corresponding benchmark ranges for modeling wind speed, the 2 Noah scheme achieved the best performance (Table 5). Although this scheme, on average, captured the values of the first hours of the day, it overestimated wind speed during the afternoon. This scheme captured the highest percentage (76.9%) of the hourly wind speed records (Table 6). The 4 Pleim–Xiu was the unique scheme that underestimated wind speed between 10:00 to 16:00 LT. The 5 Noah UCM was the option with the highest overestimation (on average, up to 2.2 m s^{-1} at 15:00 LT) during the afternoon (Figure 3c).

Although the 2 Noah scheme was better in terms of GE (61.9°), none of the options reached values for wind direction that was into the benchmark range (<30°). On average, the wind direction was correctly modeled between 09:00 to 21:00 LT for most of the land surface schemes (Figure 3d). Although found to be the best, the 2 Noah scheme captured only 37.1% of the hourly wind speed records (Table 6).

In all the land surface schemes, the levels of global solar radiation were overestimated (Figure 3b).

We did not find records of Cuenca's planetary boundary layer height (PBLH). However, for a preliminary comparison, Figure 3e shows five records during the Septembers' of different years measured in Quito, which is the capital of Ecuador and is an Andean city with similar features to Cuenca in terms of height and topography. Four of these records were taken at the Universidad San Francisco de Quito's Atmospheric Measurement Station (EMA) at coordinates 0.19° S, 78.4° W, and 2414 masl [42]. The other observation was taken in Izobamba (IZO), which is a station near Quito, at coordinates 0.36° S, 78.55°W, and 3048 masl, and is operated by the National Institute of Meteorology and Hydrology of Ecuador (INAMHI) [43]. In addition, we added the mean PBLH profile of September 2014, deduced from the hourly data of the fifth generation of the European Centre for Medium-Range Weather Forecasts (ECMWF) atmospheric reanalysis dataset (ERA5) [44], specifically for the location of Cuenca.

PBLH observations at 07:00 LT were 218 and 401 m. The closest mean modeled PBLH values were 123 and 79 m, as provided by the 4 Pleim–Xiu and 2 Noah schemes, respectively. At 07:00 LT, the ERA5 dataset indicated 128 m. At 09:00 LT, the observed PBLH (292 m) was near the modeled height (282 m), as per the 2 Noah option. At noon, the observed PBLH (1505 m), was consistent with the results of the 2 Noah (1425 m) and the 1 5-Layer (1622 m) schemes. The PBLH record at 13:00 LT (2025 m) was consistent with the modeled value of the 1 5-Layer (1877 m), and the 2 Noah schemes (1736 m). However, the ERA5 PBLH profile showed values at noon (1245 m) and 13:00 LT (1299 m) that were lower than the records.

3.2. Air Quality

The 2 Noah scheme was best fitted to model the maximum CO daily 1 h mean and the CO daily 8 h mean concentrations (Figure 6b). This scheme also correctly described the mean daily profile during peak CO concentrations (Figure 6f), whereas the other land surface options overestimated the maximum CO levels. On average, the 5 Noah UCM overestimated up to 3.6 mg m^{-3} of the CO level at 08:00 LT. The 2 Noah scheme received the best metrics for modeling CO concentrations (Table 7). However, this option received no solid linear relationship between the recorded and modeled values (r = 0.41 and r = 0.38, for the maximum CO daily 1 h mean and CO daily 8 h mean, respectively). The 2 Noah option presented the highest percentages of captured records (92.6% for the maximum CO daily 1 h mean, and 100.0% for the maximum CO daily 8 h mean (Table 8). The 5 Noah UCM option showed the lowest performance due to its overestimating the peak CO levels.

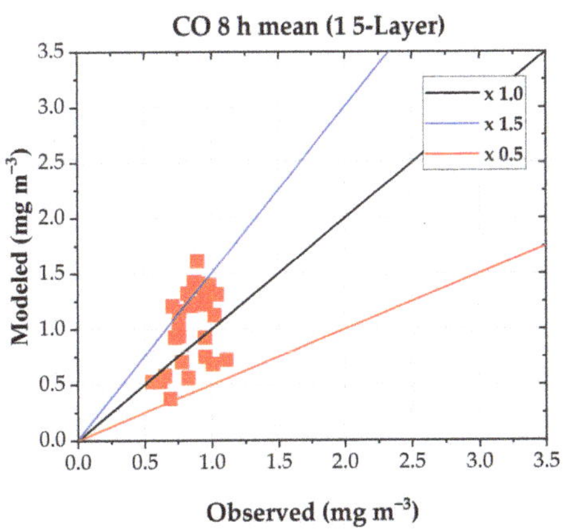

(a)

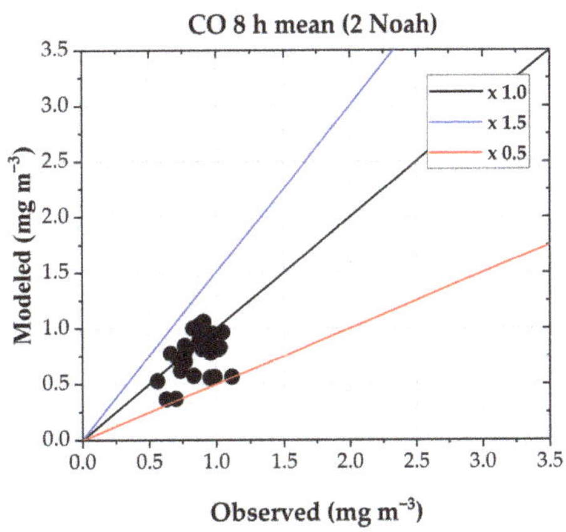

(b)

Figure 6. *Cont.*

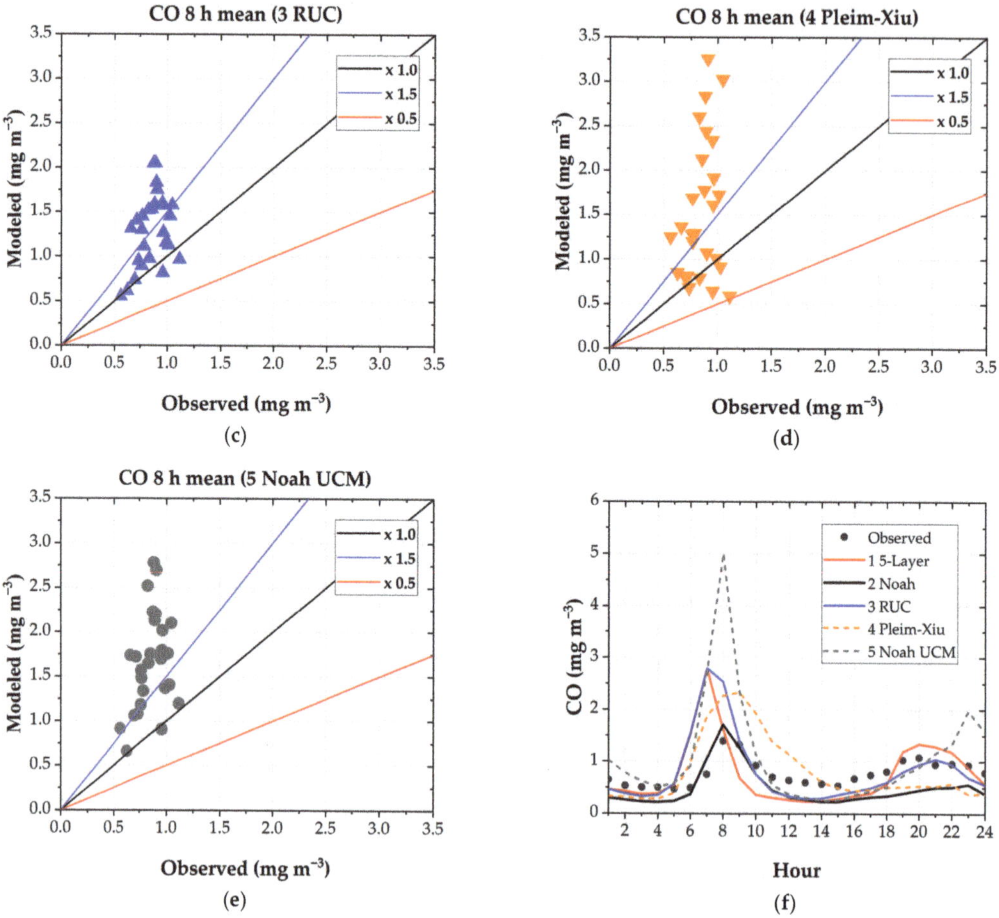

Figure 6. Observed versus modeled daily CO 8 h maximum mean at the MUN station: (**a**) 1 5-Layer; (**b**) 2 Noah; (**c**) 3 RUC; (**d**) Pleim–Xiu; (**e**) 5 Noah UCM. (**f**) The mean daily profile of hourly CO concentrations.

Although somewhat overestimated, the 2 Noah option showed the best performance for modeling the 24 h $PM_{2.5}$ mean concentrations (Figure 7), which were overestimated by the other options. All the land surface schemes show a weak linear correlation between the recorded and modeled values. The 2 Noah option presented the highest percentages of captured records (63.0%, Table 8). The 5 Noah UCM showed poor performance, whereby it overestimated, to a high degree, the $PM_{2.5}$ levels (Figure 6e,f) (MB = 26.7 $\mu g\ m^{-3}$).

For modeling the maximum 8 h O_3 daily mean, the 4 Pleim–Xiu option showed better performance (Table 7, Figure 8). All the schemes delivered no solid linear relationship between the recorded and modeled values (r between 0.09 and 0.25). The 4 Pleim–Xiu option, however, captured 96.3% of records, whereas the other options captured 85.2%.

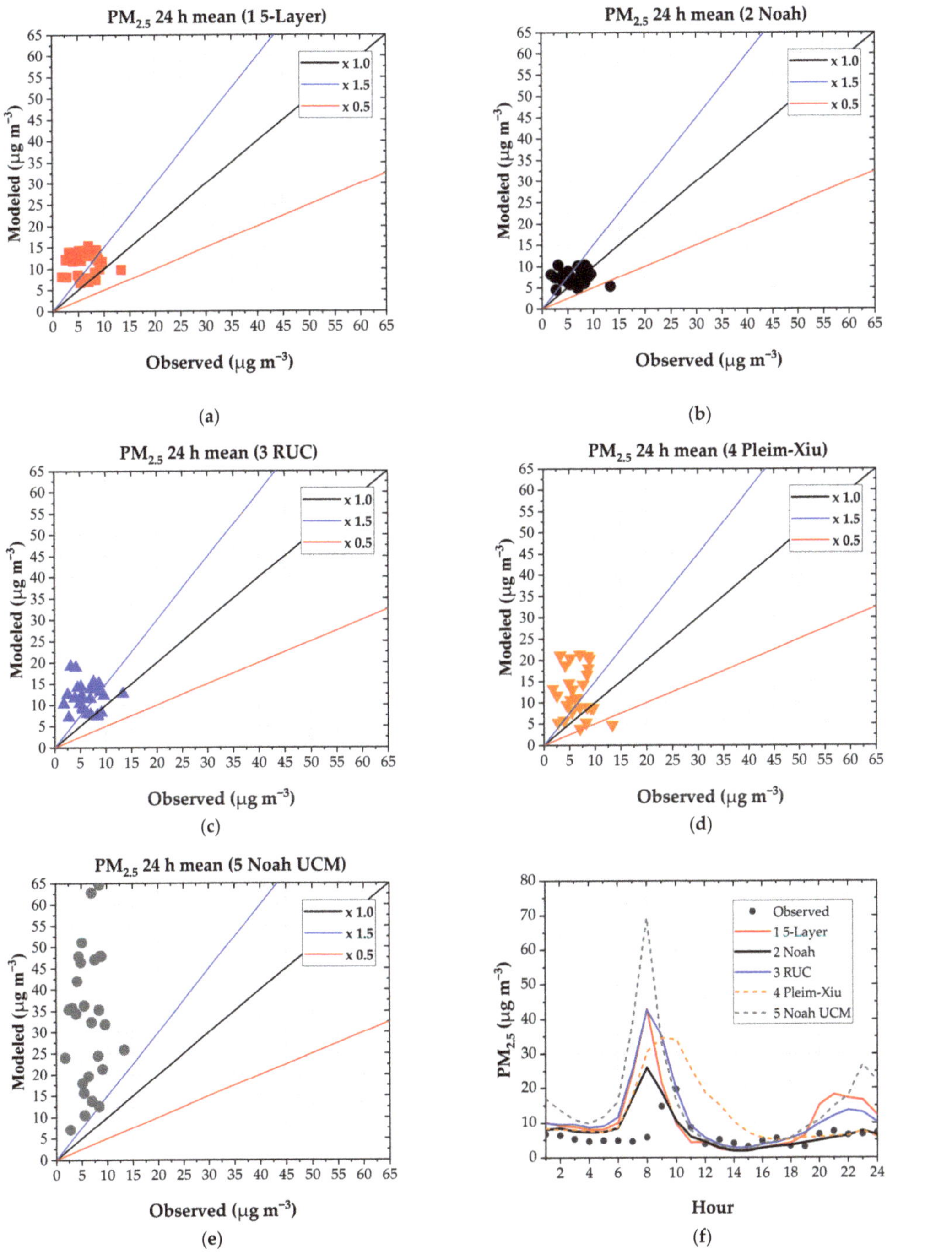

Figure 7. Observed versus modeled daily $PM_{2.5}$ 24 h mean at the MUN station: (**a**) 1 5-Layer; (**b**) 2 Noah; (**c**) 3 RUC; (**d**) Pleim–Xiu; (**e**) 5 Noah UCM. (**f**) The mean daily profile of hourly $PM_{2.5}$ concentrations.

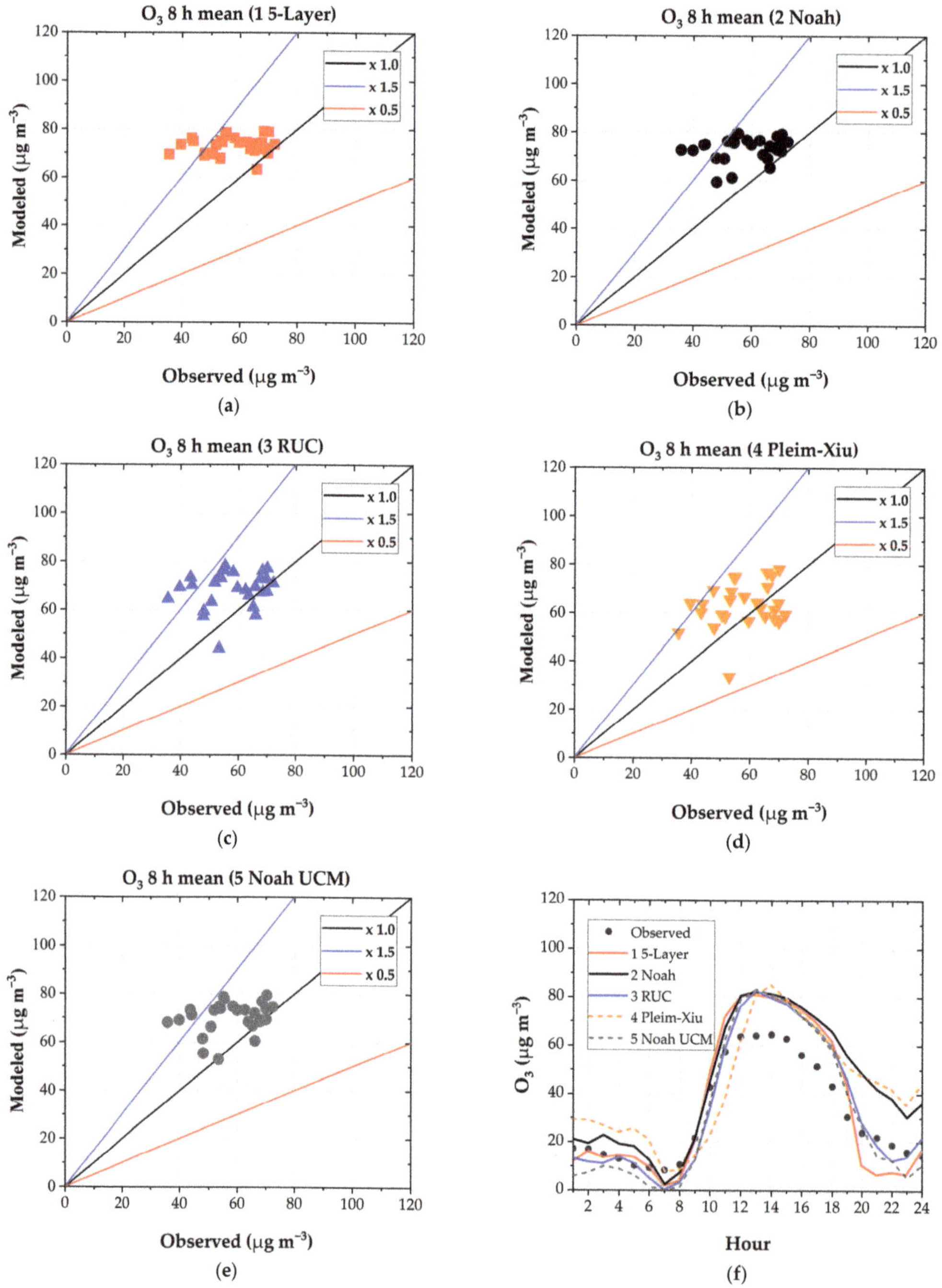

Figure 8. Observed versus modeled daily O_3 8 h maximum mean at the MUN station: (**a**) 1 5-Layer; (**b**) 2 Noah; (**c**) 3 RUC; (**d**) Pleim–Xiu; (**e**) 5 Noah UCM. (**f**) The mean daily profile of hourly O_3 concentrations.

Table 7. Metrics for short-term air quality modeling. Bold numbers highlight the schemes with the best performances.

Land Surface Scheme:	**1 5-Layer**	**2 Noah**	**3 RUC**	**4 Pleim–Xiu**	**5 Noah UCM**	**Ideal Value**
Maximum 1 h CO mean:						
MB	1.41	**0.10**	1.41	0.87	3.47	**0**
RMSE	1.74	**0.56**	1.73	1.60	3.91	**0**
FB	59.7	**6.1**	59.9	41.6	102.5	**0**
MNB	88.44	**8.42**	89.38	55.50	216.89	**0**
r	**0.41**	**0.41**	0.37	0.22	0.33	**1**
Maximum 8 h CO mean:						
MB	0.17	**−0.09**	0.45	0.70	0.81	**0**
RMSE	0.35	**0.21**	0.58	1.02	0.96	**0**
FB	18.1	**−10.7**	41.5	57.8	64.31	**0**
MNB	19.70	**−9.52**	52.52	81.73	95.43	**0**
r	**0.43**	0.38	0.40	0.25	0.34	**1**
24 h $PM_{2.5}$ mean:						
MB	4.52	**1.17**	5.50	5.92	26.65	**0**
RMSE	5.78	**3.25**	6.94	8.68	30.75	**0**
FB	52.4	**16.8**	60.2	63.4	135.28	**0**
MNB	107.59	**45.73**	130.72	139.34	512.89	**0**
r	0.06	0.01	−0.06	−0.09	0.09	**1**
Maximum 8 h O_3 mean:						
MB	15.31	15.29	10.54	**5.55**	12.35	**0**
RMSE	18.57	18.51	15.59	**13.11**	16.31	**0**
FB	23.4	23.3	16.7	**9.1**	19.28	**0**
MNB	30.84	30.61	21.99	**12.62**	25.16	**0**
r	0.09	0.20	0.18	**0.25**	0.24	**1**

Table 8. Percentages of the records that were captured by modeling. Air quality variables. Bold numbers highlight the schemes with the best performances.

Land Surface Scheme:	**1 5-Layer**	**2 Noah**	**3 RUC**	**4 Pleim–Xiu**	**5 Noah UCM**	**Number ofRecords**
Short-term air quality:						
Maximum 1 h CO mean	29.6	**92.6**	25.9	40.7	7.4	27
Maximum 8 h CO mean	74.1	**100.0**	48.1	37.0	22.2	27
24 h $PM_{2.5}$ mean	44.4	**63.0**	29.6	33.3	3.7	27
Maximum 8 h O_3 mean	85.2	85.2	85.2	**96.3**	85.2	27
Long-term air quality:						
NO_2, monthly mean	80.0	**93.3**	80.0	73.3	73.3	15
O_3, monthly mean	**93.8**	56.3	**93.8**	56.3	**93.8**	16

Regarding the NO_2 monthly mean concentrations, the 2 Noah scheme captured 93.3% of the records from all passive stations, followed by the 1 5-Layer and 3 RUC options, which captured 80.0% of the records (Table 8, Figure 9a). The 1 5-Layer, 3 RUC, and 5 Noah UCM options adequately modeled 93.8% of the O_3 monthly mean concentrations, whereas the 2 Noah and 4 Pleim–Xiu schemes captured 56.3% of the records from the passive stations.

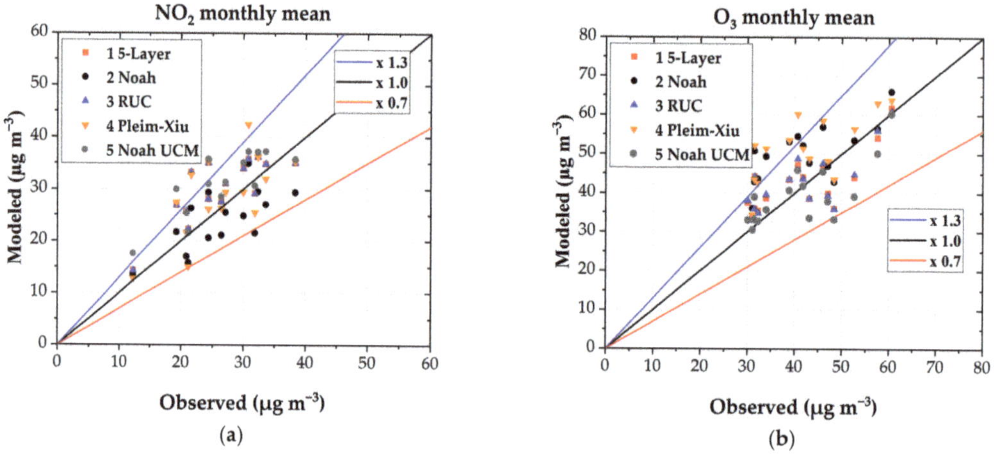

Figure 9. Passive stations: Observed versus modeled monthly means: (**a**) NO_2 and (**b**) O_3.

4. Discussion and Conclusions

We used a last-generation tool for atmospheric modeling in Cuenca, an Andean city in Southern Ecuador. We assessed the influence of the land surface options from the WRF-Chem V3.2, which is a 3D-Eulerian model used worldwide for the purposes of research and forecasting, to identify the scheme with the best performance for modeling meteorology and air quality. Results indicated that, on average, the 2 Noah scheme was the better option when compared to the 1 5-Layer, 3 RUC, and 4 Pleim–Xiu schemes (which were the other land surface schemes available from the WRF-Chem V3.2). The 2 Noah scheme was the option that best modeled the entire daily mean profile of the surface temperature (Figure 3a), which is a fundamental parameter directly related to the planetary boundary layer height and atmospheric stability. Therefore, this implied that the 2 Noah scheme, with respect to modeling the concentrations of primary and secondary pollutants, was the most successful. The 2 Noah scheme also adequately modeled the daily mean profile of the wind speed from the beginning of the daily cycle up to 13:00 LT (Figure 3c), which was found to be better than the other options. The 4 Pleim–Xiu scheme underestimated the surface temperature and wind speed from 10:00 to 17:00 LT (Figure 3a,c), thus resulting in computing low values regarding the planetary boundary layer height (Figure 3e). Therefore, this increased the computed CO levels, which were higher than the records (Figure 6d,f).

Atmospheric monitoring is scarce in Ecuador, especially for parameters such as the planetary boundary layer height [42,43,45]. We did not find PBLH records from Cuenca. However, five observations of this parameter were measured in Quito, which we preliminary used as a reference. Regarding this, the 2 Noah scheme also provided consistent modeled values of the planetary boundary layer height. In short, more surface and vertical observations will be required in the Andean region of Ecuador for complete assessments of numerical models.

Data from global datasets, such as the ERA5 reanalysis [44], a fifth-generation and one of the most advanced products, need to be assessed for their use in the tropical Andean region of South America [21]. In this sense, Dias-Júnior et al. [46] reported that the data from the ERA5 dataset underestimated the planetary boundary layer height between 15 and 30% when compared to daytime observations in a station downwind from the city of Manaus (Central Amazonia). This underestimation is consistent with the data depicted in Figure 3e, which suggested that the ERA5 data for the location of Cuenca underestimated the PBLH around midday, between 17.3 and 35.9%.

Modeled upward heat fluxes at the surface (Figure 3f) were consistent with the corresponding PBLH profiles. The peak fluxes around midday (between 444 to 566 W m^{-2})

were computed by the 1 5-Layer and 2 Noah, which were the schemes that calculated the highest PBLH values at 14:00 LT (1899–1938 m) (Figure 3e). The 3 RUC and 5 Noah UCM options computed the maximum fluxes of 260 W m^{-2}, with peak PBLH values of 1582 and 1753 m, respectively. The 4 Pleim–Xiu scheme estimated the lowest peak heat flux at the surface (118 W m^{-2}), but at 16:00 LT, rather than at 13:00 or 14:00 LT as the other four schemes had delivered. The Pleim–Xiu option estimated the maximum PBLH in 1209 m, at 17:00 LT. Further, it showed the peaks in temperature and wind speed as displaced to the right when compared to the other schemes. The modeled upward heat fluxes at the surface showed a clear relationship and consistency with the corresponding planetary boundary layer heights.

Our results suggested that the 2 Noah land surface scheme's better modeling performance would result from its better ability to model the upward heat fluxes at the surface. This preliminary explanation needs to be tested based on observations of heat fluxes and PBLH.

The highest heat fluxes from the 1 5-layer scheme would be related to its approach, which sets the soil moisture with the land use, and the season-dependent constant value without vegetation effects. The other schemes considered the presence of vegetation and delivered more sophisticated treatments for moisture. A detailed analysis of why each option reached its corresponding performance is out of the scope of this study, as this analysis would be challenging due to the complex interactions between the land surface scheme and the other modeling parameterizations.

Our finding is consistent with Rizaa et al. (2018) [4], who reported that the 2 Noah scheme was better than the 3 RUC when modeling (WRF-Chem V3.6.1) the daily mean PM_{10} levels in Southern Italy, which were produced by a severe dust episode. Our finding is also consistent with the conclusion of Constantinidou et al. (2020) [47], who reported that the 2 Noah scheme was better than the 3 RUC and other schemes for modeling temperature (WRF V3.8.1) in climate simulations in the Middle East-North Africa region. Similarly, Lee et al. [48], especially when modeling the meteorological parameters during a campaign monitoring in South Korea, reported that the 2 Noah scheme produced the best agreement between observations and simulations when compared to the 1 5-Layer, 3 RUC, and 4 Pleim–Xiu schemes.

Our results indicated that modeling meteorological and air quality parameters in Cuenca were sensitive to land surface schemes. Similar conclusions were reported in the literature. Misenis and Zhang (2010) [49] indicated that both meteorological and chemical predictions were sensitive to land surface schemes, when modeling (WRF-Chem V3.6.1) a 5-day summer episode in Houston, United States. Other studies also reported that surface temperature was sensitive to land surface schemes [5,50]. Table 9 compares the main results of this contribution with other assessments on the influence of land surface schemes. Most of the evaluations cover the modeling of meteorological variables.

Table 9. Comparison with other assessments on the influence of land surface schemes.

Region	Period	Model	Main Results	Reference
Andean region of Ecuador	September 2014	WRF-Chem V3.2	Noah provided better performances for modeling meteorological and air quality variables than the 5-Layer, RUC, and Pleim–Xiu schemes. The combination of the 2 Noah land surface scheme and the Urban Canopy Model option was not better than modeling with the 2 Noah scheme alone. Moreover, this combination produced the poorest results for CO and $PM_{2.5}$	This contribution
Oregon, United States	22 to 28 September 2014	WRF V3.7.1	The Pleim–Xiu scheme produced lower and more reliable sensible heat fluxes than Noah. However, Noah's latent heat fluxes improved compared to Pleim–Xiu, when North American Regional Reanalysis forcing data was used.	Sun et al. (2017) [2]
South of Chile	One year	WRF V4.0	The 5-Layer was better than an improved version of the Noah scheme	Somos and Manquehual (2020) [3]
Southern Italy	March 2016	WRF-Chem V3.6.1	The Noah and an improved Noah scheme were better, especially for modeling daily average PM_{10} concentrations, compared to RUC	Rizza et al. (2018) [4]

Table 9. *Cont.*

Region	Period	Model	Main Results	Reference
Tibetan Plateau	March 2017	WRF V3.7.1	The near-surface air temperature was sensitive to land surface schemes. The Community land surface model was better for modeling a snow event compared to Noah, and an improved Noah scheme	Liu et al. (2019) [5]
East China	23 July 2003	WRF V3.0	The 5-Layer and Noah schemes had approximately the same performance. RUC produced the maximum differences with records	Zeng et al. (2015) [8]
Middle East-North Africa	2000 to 2010	WRF V3.8.1	The Noah land surface scheme was better for modeling temperature and rainfall compared to an improved Noah scheme, RUC, and the Community land surface scheme	Constantinidou et al. (2020) [47]
Haean basin-South Korea	23 to 26 September 2010	WRF V3.3	The significant impact of the land surface scheme was shown in meteorological simulations. The best agreement between observation and simulation was obtained for Noah, compared to 5-Layer, RUC, and Pleim–Xiu	Lee et al. (2016) [48]
Houston, United States	Five-day summer episode during 2000	WRF-Chem V3.6.1	Both meteorological and chemical predictions at the surface and aloft show stronger sensitivity to the land surface compared to planetary boundary layers schemes	Misenis and Zhang (2010) [49]
Lake Tana Basin, Ethiopia	March to August 2015	WRF V3.8	Temperature and rainfall were sensitive to land surface schemes and land use data choice. RUC and updated USGS land use data were better for temperature. Noah and updated USGS land use were better for rainfall	Teklay et al. (2019) [50]
Northern Italy	2006 to 2015	WRF V3.8	An improved Noah version simulated a dry soil event close to records. There were no differences for rainfall modeling using Noah, improved Noah, and a third land surface scheme	Zhuo et al. (2019) [51]
Eastern Italian Alps	12 to 15 February 2006	WRF V3.8.1	Noah and an improved Noah scheme improved their modeling performance for near-surface temperature over snow-covered terrain after modification of the mean grid cell albedo	Tomasi et al. (2017) [52]
Central Asia	Summers from 2000 to 2018	WRF V4.0	Modeled variables (2 m temperature and rainfall) were sensitive to land surface schemes. The Community land surface scheme was better than Pleim–Xiu and an improved Noah scheme	Lu et al. (2021) [53]
Western United States	1 October 1995 to 30 September 1996	WRF V3.0	Land surface schemes strongly affect temperature simulations. The Community land surface model was better for modeling compared to the 5-Layer, Noah, and RUC schemes	Jin et al. (2010) [54]

Contrary to the finding of this study, other studies have reported that other land surface schemes were better than the 2 Noah scheme. Lu et al. (2021) [53] indicated the community land surface scheme was better than the 3 RUC scheme and an improved version of the 2 Noah scheme when modeling temperature and rainfall (WRF V4.0) during summer in Central Asia. The 1 5-Layer scheme, has been reported as the simplest land surface scheme in terms of structure and physics and was found to be better than an improved version of the 2 Noah scheme for modeling (WRF V4.0) meteorological variables over the south of Chile [3]. The 4 Pleim–Xiu was better than the 2 Noah scheme for modeling sensible heat fluxes (WRF V3.7.1) in Oregon, United States [2]. However, 2 Noah's improved version provided better results than the 4 Pleim–Xiu scheme, specifically when the North American Regional Reanalysis forcing dataset was used.

Combining the 2 Noah scheme and the urban canopy model option did not show benefits in Cuenca when compared to results with the Noah scheme alone. Moreover, this combination produced the poorest computed results due to overestimating CO and $PM_{2.5}$ levels. These results are partly explained by the underestimation of surface temperature at hours of peak emissions of the primary pollutants, which were accompanied by lower planetary boundary layer heights (Figure 3e) when compared to the 2 Noah option, thereby limiting the volume of the atmosphere with respect to dispersing primary pollutants. The results suggested that, although the urban canopy model can consider the urban effects at the subgrid-scale, this feature does not necessarily improve the modeling performance. The decrease in performance when modeling with the 2 Noah scheme and the urban canopy model is consistent with the conclusion by Liao et al. (2014) [55], who reported that when using the urban canopy model, higher PM_{10} concentrations were computed (WRF-Chem V3.2.1), specifically when modeling climate and air quality in the Yangtze River Delta, China. Although the urban canopy models describe the effects of urban areas at a subgrid-scale (less than 1 km for the inner domain of Cuenca), it seems that this option from WRF-Chem V3.2 is unsuitable or has properties that do not represent the urban area of Cuenca. Due to its nature, urban canopy models are expected to improve the modeling performance.

Interestingly all the options, including the 5 Noah UCM, showed good performances for modeling O_3 concentrations, capturing between 85.2 and 96.3% of the maximum O_3 8 h mean daily records (Table 8). Except for the 4 Pleim–Xiu scheme, the other options tend to overestimate O_3 levels (Figure 8). The presence of the WE in the historic center of Cuenca indicates that around the MUN station, the O_3 abundance is promoted by decreasing NO_x emissions and that an increase of NO_x will thus reduce the O_3 levels. On the other hand, an increase or decrease in NMVOC emissions will increase or decrease the production of O_3, respectively. This feature needs to be considered for the new emissions inventory for Cuenca, where the ratio between NMVOC to NO_x should be compatible with a regime for O_3 production that is limited by NMVOC emissions.

Our results and conclusions from other studies suggest that dedicated assessments should be conducted to identify an appropriate land surface option for each study region.

Our results suggest that for the purposes of modeling meteorology and air quality with the WRF-Chem V3.2 in the Andean region of Ecuador, the 2 Noah scheme would provide properly computed values of surface temperature, wind speed, and air quality variables. Therefore, we suggest that the 2 Noah land surface scheme be utilized for assessing the quality of Cuenca's new atmospheric emission inventory, which is currently in preparation.

Based on this and previous contributions, we preliminary recommend the following options for modeling with the WRF-Chem V3.2, both for modeling the meteorology and air quality in the Andean region of Ecuador:

- Land use categories: Based on the USGS dataset and categories (as per this contribution);
- Global atmospheric dataset: FNL [21];
- Land surface scheme: Noah (as per this contribution);
- Urban Canopy Model: None (as per this contribution);
- Planetary Boundary Layer: Yonsei University option [20];
- Chemical mechanisms and aerosol modules: CBMZ and MOSAIC with direct effects [20,23].

Our contribution provided insights for atmospheric studies in the Andean region of Ecuador, which is a complex area in which atmospheric modeling is challenging. This is due to the influence of the Andes chain, the intertropical converge zone, and the poor coverage of atmospheric monitoring, both at the surface and especially in terms of vertical sounding [21,44]. Our findings help to understand the influence of land surface schemes in modeling meteorology and air quality, which are valuable features because, in most of the studies, we found that only meteorological variables were assessed.

In the future, other options need to be explored, such as using an updated version of WRF-Chem, as well as determining the corresponding versions of the land surface and urban canopy model options. In addition, although the land uses a map of this assessment that acceptably described the current situation in Cuenca, it can be updated. For this purpose, it is necessary to update the data in the USGS or to assess the data coming from the Moderate Resolution Imaging Spectroradiometer (MODIS), or even from dedicated local studies in Cuenca to describe in the best way the land uses categories and their properties.

Although temperature and wind speed were adequately modeled in Cuenca, other meteorological variables, such as solar radiation and wind direction, require improved modeling performance. For solar radiation, dedicated studies for assessing the cumulus, land surface, and microphysics options can be evaluated. One limitation of modeling in the Andean region is the lack of dedicated schemes. We have used options developed and tested mainly in the Northern Hemisphere. Therefore, the development or updating of the parameterization options is strongly desirable, specifically by considering the features of the Andean region. Moreover, updated topography can improve the modeling performance for wind direction. Additionally, the potential benefits from data assimilation need to be evaluated.

Funding: This research was funded by the USFQ Poli-Grants 2022–2023 program.

Informed Consent Statement: Not applicable.

Data Availability Statement: Not applicable.

Acknowledgments: This research is part of the "Emisiones atmosféricas y Calidad del Aire en el Ecuador 2022–2023" project. Simulations were done at the High-Performance Computing system at the Universidad San Francisco de Quito. Publication of this article was funded by the Universidad San Francisco de Quito USFQ Research Publication Fund.

Conflicts of Interest: The author declares no conflict of interest.

References

1. Warner, T.T. *Numerical Weather and Climate Prediction*; Cambridge University Press: Cambridge, UK, 2011.
2. Sun, X.; Holmes, H.; Osibanjo, O.; Sun, Y.; Ivey, C. Evaluation of Surface Fluxes in the WRF Model: Case Study for Farmland in Rolling Terrain. *Atmosphere* **2017**, *8*, 197. [CrossRef]
3. Somos-Valenzuela, M.; Manquehual-Cheuque, F. Evaluating Multiple WRF Configurations and Forcing over the Northern Patagonian Icecap(NPI) and Baker River Basin. *Atmosphere* **2020**, *11*, 815. [CrossRef]
4. Rizza, U.; Miglietta, M.M.; Mangia, C.; Ielpo, P.; Morichetti, M.; Iachini, C.; Virgili, S.; Passerini, G. Sensitivity of WRF-Chem Model to Land Surface Schemes: Assessment in a Severe Dust Outbreak Episode in the Central Mediterranean(Apulia Region). *Atmos. Res.* **2018**, *201*, 168–180. [CrossRef]
5. Liu, L.; Ma, Y.; Menenti, M.; Zhang, X.; Ma, W. Evaluation of WRF Modeling in Relation to Different Land Surface Schemes and Initial and Boundary Conditions: A Snow Event Simulation Over the Tibetan Plateau. *J. Geophys. Res. Atmos.* **2019**, *124*, 209–226. [CrossRef]
6. Stensrud, D. *Parameterization Schemes: Keys to Understanding Numerical Weather Prediction Models*; Cambridge University Press: New York, NY, USA, 2009.
7. Fisher, R.A.; Koven, C.D. Perspectives on the Future of Land Surface Models and the Challenges of Representing Complex Terrestrial Systems. *J. Adv. Model. Earth Syst.* **2020**, *12*, e2018MS001453. [CrossRef]
8. Zeng, X.-M.; Wang, N.; Wang, Y.; Zheng, Y.; Zhou, Z.; Wang, G.; Chen, C.; Liu, H. WRF-Simulated Sensitivity to Land Surface Schemes in Short and Medium Ranges for a High-Temperature Event in East China: A Comparative Study: WRF Sensitivity to Land Surface Schemes. *J. Adv. Model. Earth Syst.* **2015**, *7*, 1305–1325. [CrossRef]
9. Skamarock, W.C.; Klemp, J.B.; Dudhia, J.; Gill, D.O.; Barker, D.M.; Duda, M.G.; Huang, X.-Y.; Wang, W.; Powers, J.G. A Description of the Advanced Research WRF Version 3. 2008. Available online: https://opensky.ucar.edu/islandora/object/technotes:500 (accessed on 12 February 2023).
10. Dudhia, J. A multi-layer soil temperature model for MM5. Workshop Boulder. In Proceedings of the Sixth Annual PSU/NCAR Mesoscale Model Users' Workshop, Bouilder, CO, USA, 22–24 July 1996; pp. 49–50.
11. Chen, F.; Dudhia, J. Coupling an Advanced Land Surface—Hydrology Model with the Penn State–NCAR MM5 Modeling System. Part II: Preliminary Model Validation. *Mon. Weather Rev.* **2001**, *129*, 587–604. [CrossRef]
12. Smirnova, T.G.; Brown, J.M.; Benjamin, S.G. Performance of Different Soil Model Configurations in Simulating Ground Surface Temperature and Surface Fluxes. *Mon. Weather Rev.* **1997**, *125*, 1870–1884. [CrossRef]
13. Smirnova, T.G.; Brown, J.M.; Benjamin, S.G.; Kim, D. Parameterization of Cold-Season Processes in the MAPS Land-Surface Scheme. *J. Geophys. Res. Atmos.* **2000**, *105*, 4077–4086. [CrossRef]
14. Pleim, J.E.; Xiu, A. Development and Testing of a Surface Flux and Planetary Boundary Layer Model for Application in Mesoscale Models. *J. Appl. Meteorol.* **1995**, *34*, 16–32. [CrossRef]
15. Xiu, A.; Pleim, J.E. Development of a Land Surface Model. Part I: Application in a Mesoscale Meteorological Model. *J. Appl. Meteorol.* **2001**, *40*, 192–209. [CrossRef]
16. Espinoza Claudia, C. *Informe de Calidad Del Aire Cuenca 2021*; EMOV EP: Cuenca, Ecuador, 2022; Available online: https://www.researchgate.net/publication/362194795_Informe_de_Calidad_del_Aire_Cuenca_2021 (accessed on 12 February 2023).
17. World Health Organization. *WHO Global Air Quality Guidelines: Particulate Matter(PM2.5 and PM10), Ozone, Nitrogen Dioxide, Sulfur Dioxide and Carbon Monoxide*; World Health Organization: Geneva, Switzerland, 2021.
18. World Health Organization (Ed.) *Air Quality Guidelines: Global Update 2005: Particulate Matter, Ozone, Nitrogen Dioxide, and Sulfur Dioxide*; World Health Organization: Copenhagen, Denmark, 2006.
19. Parra, R.; Cadena, E.; Flores, C. Maximum UV Index Records(2010–2014) in Quito(Ecuador) and Its Trend Inferred from Remote Sensing Data (1979–2018). *Atmosphere* **2019**, *10*, 787. [CrossRef]
20. Parra, R. Performance Studies of Planetary Boundary Layer Schemes in WRF-Chem for the Andean Region of Southern Ecuador. *Atmos. Pollut. Res.* **2018**, *9*, 411–428. [CrossRef]
21. Parra, R. Effect of Global Atmospheric Datasets in Modeling Meteorology and Air Quality in the Andean Region of Ecuador. *Aerosol Air Qual. Res.* **2022**, *22*, 210292. [CrossRef]
22. Parra, R. *Effects of Aerosols Feedbacks in Modeling Meteorology and Air Quality in the Andean Region of Southern Ecuador*; WIT Press: Santiago de Compostela, Spain, 2021; pp. 39–50. [CrossRef]

23. Parra, R.; Espinoza, C. Insights for Air Quality Management from Modeling and Record Studies in Cuenca, Ecuador. *Atmosphere* **2020**, *11*, 998. [CrossRef]
24. Parra, R.; Saud, C.; Espinoza, C. Simulating PM2.5 Concentrations during New Year in Cuenca, Ecuador: Effects of Advancing the Time of Burning Activities. *Toxics* **2022**, *10*, 264. [CrossRef]
25. EMOV-EP. Inventario de Emisiones Atmosféricas Del Cantón Cuenca 2014. 2016. [CrossRef]
26. WRF Model Users Site. Available online: https://www2.mmm.ucar.edu/wrf/users/ (accessed on 30 December 2022).
27. CISL RDA: NCEP FNL Operational Model Global Tropospheric Analyses, Continuing from July 1999. Available online: https://rda.ucar.edu/datasets/ds083.2/ (accessed on 30 December 2022).
28. Zaveri, R.A.; Peters, L.K. A New Lumped Structure Photochemical Mechanism for Large-Scale Applications. *J. Geophys. Res. Atmospheres* **1999**, *104*, 30387–30415. [CrossRef]
29. Zaveri, R.A.; Easter, R.C.; Fast, J.D.; Peters, L.K. Model for Simulating Aerosol Interactions and Chemistry(MOSAIC). *J. Geophys. Res.* **2008**, *113*, D13204. [CrossRef]
30. Hong, S.-Y.; Dudhia, J.; Chen, S.-H. A Revised Approach to Ice Microphysical Processes for the Bulk Parameterization of Clouds and Precipitation. *Mon. Weather Rev.* **2004**, *132*, 103–120. [CrossRef]
31. Mlawer, E.J.; Taubman, S.J.; Brown, P.D.; Iacono, M.J.; Clough, S.A. Radiative Transfer for Inhomogeneous Atmospheres: RRTM, a Validated Correlated-k Model for the Longwave. *J. Geophys. Res. Atmos.* **1997**, *102*, 16663–16682. [CrossRef]
32. Chou, M.-D.; Suarez, M.J. *A Solar Radiation Parameterization for Atmospheric Studies*; NASA/TM-1999-104606/VOL15; NASA: Washington, DC, USA, 1999. Available online: https://ntrs.nasa.gov/citations/19990060930 (accessed on 30 December 2022).
33. Paulson, C.A. The Mathematical Representation of Wind Speed and Temperature Profiles in the Unstable Atmospheric Surface Layer. *J. Appl. Meteorol.* **1970**, *9*, 857–861. [CrossRef]
34. Hong, S.-Y.; Noh, Y.; Dudhia, J. A New Vertical Diffusion Package with an Explicit Treatment of Entrainment Processes. *Mon. Weather Rev.* **2006**, *134*, 2318–2341. [CrossRef]
35. Grell, G.A. Prognostic Evaluation of Assumptions Used by Cumulus Parameterizations. *Mon. Weather Rev.* **1993**, *121*, 764–787. [CrossRef]
36. Kusaka, H.; Kondo, H.; Kikegawa, Y.; Kimura, F. A Simple Single-Layer Urban Canopy Model For Atmospheric Models: Comparison With Multi-Layer And Slab Models. *Bound. Layer Meteorol.* **2001**, *101*, 329–358. [CrossRef]
37. Kusaka, H.; Kimura, F. Coupling a Single-Layer Urban Canopy Model with a Simple Atmospheric Model: Impact on Urban Heat Island Simulation for an Idealized Case. *J. Meteorol. Soc. Jpn. Ser. II* **2004**, *82*, 67–80. [CrossRef]
38. Chen, F.; Tewari, M.; Kusaka, H.; Warner, T. Current Status of Urban Modeling in the Community Weather Research and Forecast (WRF) Model (2006—Annual2006_6urban). Available online: https://ams.confex.com/ams/Annual2006/techprogram/paper_98678.htm (accessed on 30 December 2022).
39. The Application of Models under the European Union's Air Quality Directive: A Technical Reference Guide—European Environment Agency. Available online: https://www.eea.europa.eu/publications/fairmode (accessed on 30 December 2022).
40. Simon, H.; Baker, K.R.; Phillips, S. Compilation and Interpretation of Photochemical Model Performance Statistics Published between 2006 and 2012. *Atmos. Environ.* **2012**, *61*, 124–139. [CrossRef]
41. World Health Organization, Regional Office for Europe. *Air Quality Guidelines for Europe*; World Health Organization, Regional Office for Europe: Geneva, Switzerland, 2000.
42. Cazorla, M.; Juncosa, J. Planetary Boundary Layer Evolution over an Equatorial Andean Valley: A Simplified Model Based on Balloon-Borne and Surface Measurements. *Atmos. Sci. Lett.* **2018**, *19*, e829. [CrossRef]
43. Muñoz, L.E.; Campozano, L.V.; Guevara, D.C.; Parra, R.; Tonato, D.; Suntaxi, A.; Maisincho, L.; Páez, C.; Villacís, M.; Córdova, J.; et al. Comparison of Radiosonde Measurements of Meteorological Variables with Drone, Satellite Products, and WRF Simulations in the Tropical Andes: The Case of Quito, Ecuador. *Atmosphere* **2023**, *14*, 264. [CrossRef]
44. ERA5 Hourly Data on Pressure Levels from 1959 to Present. Available online: https://cds.climate.copernicus.eu/cdsapp#!/dataset/reanalysis-era5-pressure-levels?tab=form (accessed on 12 February 2023).
45. Cazorla, M.; Parra, R.; Herrera, E.; da Silva, F.R. Characterizing Ozone throughout the Atmospheric Column over the Tropical Andes from in Situ and Remote Sensing Observations. *Elem. Sci. Anthr.* **2021**, *9*, 00019. [CrossRef]
46. Dias-Júnior, C.Q.; Carneiro, R.G.; Fisch, G.; D'Oliveira, F.A.F.; Sörgel, M.; Botía, S.; Machado, L.A.T.; Wolff, S.; dos Santos, R.M.N.; Pöhlker, C. Intercomparison of Planetary Boundary Layer Heights Using Remote Sensing Retrievals and ERA5 Reanalysis over Central Amazonia. *Remote Sens.* **2022**, *14*, 4561. [CrossRef]
47. Constantinidou, K.; Hadjinicolaou, P.; Zittis, G.; Lelieveld, J. Performance of Land Surface Schemes in the WRF Model for Climate Simulations over the MENA-CORDEX Domain. *Earth Syst. Environ.* **2020**, *4*, 647–665. [CrossRef]
48. Lee, C.B.; Kim, J.-C.; Belorid, M.; Zhao, P. Performance Evaluation of Four Different Land Surface Models in WRF. *Asian J. Atmos. Environ.* **2016**, *10*, 42–50. [CrossRef]
49. Misenis, C.; Zhang, Y. An Examination of Sensitivity of WRF/Chem Predictions to Physical Parameterizations, Horizontal Grid Spacing, and Nesting Options. *Atmos. Res.* **2010**, *97*, 315–334. [CrossRef]
50. Teklay, A.; Dile, Y.T.; Asfaw, D.H.; Bayabil, H.K.; Sisay, K. Impacts of Land Surface Model and Land Use Data on WRF Model Simulations of Rainfall and Temperature over Lake Tana Basin, Ethiopia. *Heliyon* **2019**, *5*, e02469. [CrossRef]
51. Zhuo, L.; Dai, Q.; Han, D.; Chen, N.; Zhao, B. Assessment of Simulated Soil Moisture from WRF Noah, Noah-MP, and CLM Land Surface Schemes for Landslide Hazard Application. *Hydrol. Earth Syst. Sci.* **2019**, *23*, 4199–4218. [CrossRef]

52. Tomasi, E.; Giovannini, L.; Zardi, D.; de Franceschi, M. Optimization of Noah and Noah_MP WRF Land Surface Schemes in Snow-Melting Conditions over Complex Terrain. *Mon. Weather Rev.* **2017**, *145*, 4727–4745. [CrossRef]
53. Lu, S.; Guo, W.; Xue, Y.; Huang, F.; Ge, J. Simulation of Summer Climate over Central Asia Shows High Sensitivity to Different Land Surface Schemes in WRF. *Clim. Dyn.* **2021**, *57*, 2249–2268. [CrossRef]
54. Jin, J.; Miller, N.L.; Schlegel, N. Sensitivity Study of Four Land Surface Schemes in the WRF Model. *Adv. Meteorol.* **2010**, *2010*, 1–11. [CrossRef]
55. Liao, J.; Wang, T.; Wang, X.; Xie, M.; Jiang, Z.; Huang, X.; Zhu, J. Impacts of Different Urban Canopy Schemes in WRF/Chem on Regional Climate and Air Quality in Yangtze River Delta, China. *Atmos. Res.* **2014**, *145–146*, 226–243. [CrossRef]

atmosphere

Article

A Coupled CH_4, CO and CO_2 Simulation for Improved Chemical Source Modeling

Beata Bukosa [1,*], Jenny A. Fisher [2], Nicholas M. Deutscher [2] and Dylan B. A. Jones [3]

1 National Institute of Water and Atmospheric Research, Wellington 6021, New Zealand
2 Centre for Atmospheric Chemistry, School of Earth, Atmospheric and Life Sciences, Faculty of Science, Medicine and Health, University of Wollongong, Wollongong, NSW 2422, Australia
3 Department of Physics, University of Toronto, Toronto, ON M5S 1A7, Canada
* Correspondence: beata.bukosa@niwa.co.nz

Abstract: Understanding greenhouse gas–climate processes and feedbacks is a fundamental step in understanding climate variability and its links to greenhouse gas fluxes. Chemical transport models are the primary tool for linking greenhouse gas fluxes to their atmospheric abundances. Hence, accurate simulations of greenhouse gases are essential. Here, we present a new simulation in the GEOS-Chem chemical transport model that couples the two main greenhouse gases—carbon dioxide (CO_2) and methane (CH_4)—along with the indirect greenhouse gas carbon monoxide (CO) based on their chemistry. Our updates include the online calculation of the chemical production of CO from CH_4 and the online production of CO_2 from CO, both of which were handled offline in the previous versions of these simulations. In the newly developed coupled (online) simulation, we used consistent hydroxyl radical (OH) fields for all aspects of the simulation, resolving biases introduced by inconsistent OH fields in the currently available uncoupled (offline) CH_4, CO and CO_2 simulations. We compare our coupled simulation with the existing v12.1.1 GEOS-Chem uncoupled simulations run the way they are currently being used by the community. We discuss differences between the uncoupled and coupled calculation of the chemical terms and compare our results with surface measurements from the NOAA Global Greenhouse Gas Reference Network (NOAA GGGRN), total column measurements from the Total Carbon Column Observing Network (TCCON) and aircraft measurements from the Atmospheric Tomography Mission (ATom). Relative to the standard uncoupled simulations, our coupled results suggest a stronger CO chemical production from CH_4, weaker production of CO_2 from CO and biases in the OH fields. However, we found a significantly stronger chemical production of CO_2 in tropical land regions, especially in the Amazon. The model–measurement differences point to underestimated biomass burning emissions and secondary production for CO. The new self-consistent coupled simulation opens new possibilities when identifying biases in CH_4, CO and CO_2 source and sink fields, as well as a better understanding of their interannual variability and co-variation.

Keywords: carbon dioxide; carbon monoxide; chemical production; modeling; GEOS-Chem; carbon cycle

check for updates

Citation: Bukosa, B.; Fisher, J.A.; Deutscher, N.M.; Jones, D.B.A. A Coupled CH_4, CO and CO_2 Simulation for Improved Chemical Source Modeling. *Atmosphere* **2023**, *14*, 764. http://doi.org/10.3390/atmos14050764

Academic Editor: Daniel Viúdez-Moreiras

Received: 11 March 2023
Revised: 20 April 2023
Accepted: 20 April 2023
Published: 22 April 2023

1. Introduction

Accurate simulations of greenhouse gases are vital for climate predictions. Carbon dioxide (CO_2) and methane (CH_4) are the two main anthropogenic greenhouse gases and have a significant impact on our climate. Due to human activities, the atmospheric amounts of CO_2 and CH_4 have increased globally by 40% and 150%, respectively, since the industrial revolution [1]. Carbon monoxide (CO) is less abundant than CO_2 and CH_4; however, through its indirect effects on CH_4, ozone and CO_2, it can also have a climate impact [2]. Changes in the atmospheric amounts of these gases, driven by changes in their sources and sinks, largely control our future climate, but uncertainties about these processes and their budgets still remain [3–5]. All three carbon greenhouse gases are chemically dependent, and a change in one can affect the others.

In the GEOS-Chem model, each of these gases have their own stand-alone simulation, decoupled from one another. All three simulations are widely used for carbon gas flux inversion and source attribution [6–12]. Previous studies have emphasized the importance of the inclusion of the 3-D chemical production of CO_2 from the collective oxidation of CO, CH_4 and non-methane volatile organic compounds (NMVOCs) [13,14], but this chemical production, together with the secondary production of CO from CH_4, is handled offline in the stand-alone carbon gas simulations of the GEOS-Chem model [15–17]. Moreover, the chemical production and hydroxyl radical (OH) fields used by each of these individual simulations are calculated from different model versions (based on the model version in use when the stand-alone simulation was developed), introducing inconsistencies between the simulations. Here, we present a new simulation in GEOS-Chem that couples CH_4, CO and CO_2 with an online calculation of their chemical production using consistent and updated OH fields for a more accurate simulation of these gases.

The dominant loss process of CH_4, the second most important anthropogenic greenhouse gas, is through a reaction with OH:

$$CH_4 + OH \rightarrow CH_3 + H_2O \quad (1)$$

that eventually leads to the formation of CO after a series of intermediate steps [18]:

$$CHO + O_2 \rightarrow CO + HO_2 \quad (2)$$

Both CH_4 and CO have a common sink in the atmosphere through the reaction with OH. The role of CO in determining tropospheric OH indirectly affects the atmospheric burden of CH_4 [19]. Along with CH_4, it is one of the principal sinks of OH. Through the reaction with OH, CO can also lead to the chemical formation of CO_2 [13,14]:

$$CO + OH \rightarrow CO_2 + H \quad (3)$$

The oxidation of both primary CO, from direct anthropogenic and biomass burning emissions, and secondary CO, as an intermediate in the oxidation of CH_4 and NMVOCs, leads to the formation of CO_2. CO_2 can also be produced from the oxidation of carboxy–peroxy radical (RCO_3) and alkenoid ozonolysis (reaction of ethene with ozone; C_2H_4+O_3) [20], but this is thought to only be a minor contributor.

In regions that are not dominated by strong anthropogenic point emissions or biomass burning emissions, the major source of CO is CH_4 oxidation by OH through Reactions (1) and (2) and the intermediate reactions. Studies have found the yield of CO from CH_4 oxidation to range from 0.70–1 [4,21–26]. The CO chemical production from CH_4 is estimated to be 760–1086 Tg CO yr^{-1}, with CO also chemically produced from NMVOCs, with estimates of 320–820 Tg CO yr^{-1} [4,17,25,27–30]. The combined CO chemical production represents more than half of the total CO source.

The reaction of CO with OH radicals represents its largest sink, removing 2325–2630 Tg CO yr^{-1} [15,17,25,28,29,31]. The total chemical CO_2 source is estimated to be around 1.04–1.1 Pg C yr^{-1} [14,15], which is approximately 12% of the annual anthropogenic CO_2 source (9.4 Pg C yr^{-1}, averaged for 2008–2017) [32]. Around 90–94% of the CO_2 chemical production is from CO oxidation [20,33]. In contrast to the majority of the CO_2 sources that are emitted at the surface, CO_2 from the oxidation of CO is produced throughout the atmosphere. Although significant efforts have been made to constrain the total budgets of CO_2, CH_4 and CO, discrepancies in the chemical terms between studies suggest that these terms are still subject to uncertainties that can impact our understanding of the total budgets [4,14,15,17,30].

In this study, we introduced a new simulation in the GEOS-Chem model that couples the chemistry of CH_4, CO and CO_2. The coupling of the carbon greenhouse gases represents an important modeling improvement and capability when studying these gases [34,35]. With the new coupled GEOS-Chem carbon simulation, we eliminated the previous offline handling of the chemical production terms and OH inconsistency between the three species [15–17], enabling us to have (i) better estimates of the chemical terms, (ii) consistent OH fields between species, and (iii) simultaneous and consistent simulations of CH_4, CO and CO_2 that can help when constraining their fluxes based on their covariation [10,36,37]. Moreover, the coupled simulation removes the need to run the individual simulations separately if interested in all three gases, and it requires fewer computational resources than running three independent simulations.

We first describe the method for the online calculation of the chemical production and the difference between the existing (uncoupled) and new (coupled) versions of these simulations (Section 2). We then compare the stand-alone simulations of all three gases, each run the way they are currently being used by the community, with the coupled simulation. For both versions, we analyze their annual budgets and the contribution of chemical production to the total amount of each gas (Sections 3.1 and 3.2) as well as their global spatial and vertical distribution (Sections 3.3 and 3.4). We validate the new coupled simulations against global surface flask measurements at sites that are part of the NOAA Global Greenhouse Gas Reference Network (NOAA GGGRN), column measurements from the Total Carbon Column Observing Network (TCCON) and aircraft in situ measurements from the Atmospheric Tomography Mission (ATom) (Section 3.5). Finally, we discuss a sensitivity simulation designed to test the impact of using inconsistent OH fields between the three uncoupled simulations, an issue that impacts the existing uncoupled simulations but is resolved in our new coupled simulation (Section 3.6).

2. Methods

2.1. Uncoupled Geos-Chem Carbon Gas Simulations

The uncoupled CH_4, CO and CO_2 simulations used here are based on version 12.1.1 of the GEOS-Chem 3-D global chemical transport model. The existing uncoupled simulations are described in Nassar et al. [15] and Nassar et al. [38] for CO_2, Wecht et al. [16] and Maasakkers et al. [39] for CH_4 and Fisher et al. [17] for CO. Each of these simulations are used routinely and independently for evaluating new emission inventories, estimating and resolving emissions, analyzing spatial and temporal changes of CH_4, CO and CO_2, source/sink attribution and inversion studies [6–12].

These simulations are decoupled from other gases and from one another; hence, they require input fields, including chemical production rates and OH losses. GEOS-Chem can also perform a full chemistry simulation, known as coupled aerosol–oxidant chemistry in the troposphere and stratosphere simulation. Of the three species, only CO is simulated online in the full chemistry simulation. CO_2 and CH_4 are not modeled as active species in the full chemistry, and their response to sources and sinks can only be modeled via the currently uncoupled simulations. The full chemistry simulation is required for the functionality of some of the stand-alone simulations because it provides input fields for those simulations. Various versions of the full chemistry simulation were run previously by the developers of each uncoupled simulation to archive the production rates and oxidant fields that are currently used in the carbon gas simulations (Table 1). Both the production and oxidant fields are computed using 3-D archives of monthly average values. All three carbon gas simulations are linear, and each includes a suite of tracers tagged by source type and/or region.

The equations below describe the changes in the emission, deposition, production and loss terms that occur within each grid box for the stand-alone CH_4, CO and CO_2 simulations. Note that advective transport fluxes between grid boxes (including between the troposphere and the stratosphere) are in addition to the terms described in each equation. The GEOS-Chem model dynamically calculates the tropopause height at every timestep and uses this information to assign each grid box to either the troposphere or the stratosphere. The simulated CH_4 in the troposphere is based on Equation (4):

$$\frac{d[CH_{4Trop}]}{dt} = E_{CH_4} - S_{CH_4} - k_{CH_4,OH}[OH][CH_4] - k_{CH_4,Cl}[Cl][CH_4] \quad (4)$$

where E_{CH_4} represents the surface emissions (gas, oil, coal, livestock, landfills, wastewater, biofuel, rice, biomass burning, wetlands, seeps, termites and other anthropogenic emissions; see Table S1), S_{CH_4} is the sink from soil absorption, [OH], [Cl] and [CH_4] are the atmospheric concentrations of OH, Cl and CH_4, and $k_{CH_4,OH}$ and $k_{CH_4,Cl}$ are the pressure- and temperature-dependent rate constants for oxidation of CH_4 by OH and Cl, respectively. While [CH_4] is calculated at each model timestep, in the uncoupled simulation, [OH] and [Cl] are provided as monthly mean values archived from a prior full chemistry simulation.

In the stratosphere, Equation (4) becomes:

$$\frac{d[CH_{4Strat}]}{dt} = E_{CH_4} - L(CH_4) \tag{5}$$

where $L(CH_4)$ represents the stratospheric CH_4 sink based on stratospheric CH_4 loss frequencies archived from the NASA Global Modeling Initiative model [40,41] as described by Murray et al. [42].

Simulated CO in the troposphere is based on Equation (6):

$$\frac{d[CO_{Trop}]}{dt} = E_{CO} + P(CO) - k_{CO}[OH][CO] \tag{6}$$

where E_{CO} represents the surface emissions (fossil fuel, biofuel and biomass burning), P(CO) accounts for the chemical production of CO from CH_4 and NMVOC oxidation and k_{CO} is the pressure- and temperature-dependent rate constant for oxidation of CO by OH from the Jet Propulsion Laboratory (JPL) data evaluation [43]. As in the CH_4 simulation, [OH] is provided as a monthly mean value archived from a prior full chemistry simulation.

The chemical production of CO (P(CO)) can be further separated into the production from CH_4 ($P(CO)_{CH_4}$) and the production from NMVOCs ($P(CO)_{NMVOC}$):

$$P(CO) = P(CO)_{CH_4} + P(CO)_{NMVOC} \tag{7}$$

The $P(CO)_{CH_4}$ and $P(CO)_{NMVOC}$ terms are monthly averaged archived fields that were obtained with the v9-01-03 GEOS-Chem 2° × 2.5° full chemistry simulation from the simulated monthly CO chemical production rates (P(CO)) as described by Fisher et al. [17]. In brief, the simulated P(CO) is split offline to the $P(CO)_{CH_4}$ and $P(CO)_{NMVOC}$ terms based on the CH_4 loss rates (L(CH_4)) that are also simulated and saved from a full chemistry simulation. A 100% CO yield from CH_4 is assumed; hence, the production of CO from CH_4 is equal to the CH_4 loss:

$$P(CO)_{CH_4} = L(CH_4) \tag{8}$$

The remaining $P(CO)_{NMVOC}$ contribution is then calculated as the difference between the total CO production and the production of CO from CH_4:

$$P(CO)_{NMVOC} = P(CO) - P(CO)_{CH_4} \tag{9}$$

Since the 100% yield may overestimate the production of CO from the oxidation of CH_4, the simulation caps the $P(CO)_{CH_4}$ to the total P(CO), where it is greater than P(CO) [17]. Hence, this assumption will retain consistency in the P(CO) terms between the full chemistry and uncoupled simulations.

In the v9-01-03 full chemistry simulation used to calculate the L(CH_4) and $P(CO)_{CH_4}$ fields used by the uncoupled CO simulation [17], CH_4 mixing ratios were prescribed as fixed values. One annual value for each of four latitude bands (30–90° S, 0–30° S, 0–30° N, 30–90° N) was applied throughout the troposphere, defined as averages of surface observations from NOAA carbon cycle surface flasks.

Table 1. GEOS-Chem production (*P*) and loss (*L*) terms used for the uncoupled and coupled carbon gas simulations, along with the full chemistry model versions used to create the chemical terms and hydroxyl radical (*OH*) fields. The chemical terms are for the troposphere unless otherwise specified. Other source and sink fields are shown in Table S1.

	CH_4	CO	CO_2
Fields used by both uncoupled and coupled simulations			
Stratospheric L(CH_4)	Archived fields [1]	-	-
Stratospheric L(CO)	-	GMI [2]	-
Stratospheric P(CO)	-	GMI [2]	-
$P(CO)_{NMVOC}$	-	$P(CO)_{NMVOC}$ = P(CO) − $P(CO)_{CH_4}$	-
	-	archived, full chemistry v9-01-03 [3]	-
Fields used by uncoupled simulations only			
L(CH_4) [4]	online [5]	archived, full chemistry v9-01-03 [3]	-
Time resolution	Every model timestep, 20 min	Monthly mean, 2009–2011 average	-
$P(CO)_{CH_4}$	-	archived, $P(CO)_{CH_4}$ = Trop. L(CH_4)	-
Time resolution	-	Monthly mean, 2009–2011 average	-
L(CO) [4,6]	-	online, v9-01-03 [OH] [3]	archived, full chemistry v8-02-01 [7]
Time resolution	-	Every model timestep, 20 min	Monthly mean, 2004–2010
P(CO_2) [6]	-	-	archived, P(CO_2) = L(CO)
Time resolution	-	-	Monthly mean, 2004–2010
Fields used by coupled simulation only			
L(CH_4)	online, v9-01-03 [OH] [3,8]	-	-
Time resolution	Every model timestep, 20 min	-	-
$P(CO)_{CH_4}$	-	online, $P(CO)_{CH_4}$ = L(CH_4)	-
Time resolution	-	Every model timestep, 20 min	-
L(CO)	-	online, v9-01-03 [OH] [3]	-
Time resolution	-	Every model timestep, 20 min	-
P(CO_2) [9]	-	-	online, P(CO_2) = L(CO)
Time resolution	-	-	Every model timestep, 20 min

[1] Murray et al. [42], [2] NASA Global Modeling Initiative model, [3] Fisher et al. [17], [4] Note that, in the uncoupled simulations, there are two entities for L(CH_4) and L(CO) because there is a different treatment based on whether they are being used to calculate the concentration of the species itself or as a proxy for the production of another species, [5] Note that the public uncoupled CH_4 simulation uses v5-07-08 OH fields [44]; however, the uncoupled CH_4 simulation is not used in our analysis, [6] Troposphere and stratosphere, [7] Nassar et al. [15], [8] Note that the sensitivity simulation described in Section 3.6 instead uses v5-07-08 OH fields as described in the text, [9] The stratospheric P(CO_2) is calculated online (every model timestep, 20 min) from stratospheric L(CO) (P(CO_2) = L(CO)).

In the stratosphere, Equation (6) becomes:

$$\frac{d[CO_{Strat}]}{dt} = P(CO) - L(CO) \tag{10}$$

where P(CO) represents the stratospheric production rates of CO and L(CO) represents the stratospheric CO sink based on stratospheric CO loss frequencies. Both quantities are from the NASA Global Modeling Initiative model.

The simulated CO_2 throughout the atmosphere is based on Equation (11):

$$\frac{d[CO_2]}{dt} = E_{CO_2} + P(CO_2) + D_{CO_2} \tag{11}$$

where E_{CO_2} represents the surface (fossil fuel, biomass burning, biofuel, shipping) and 3-D (aviation) emissions, P(CO_2) accounts for the 3-D chemical production from the oxidation of CO, and D_{CO_2} represents the net contribution from ocean exchange, balanced and net annual terrestrial exchange. Note that D_{CO_2} can be positive or negative since these processes have negative values in regions where they act as a net sink and positive values where they act as a net source. Analogous to the CO simulation, the P(CO_2) term is a monthly averaged archived field that was obtained with the v8-02-01 GEOS-Chem 4° × 5° full chemistry simulation from the simulated monthly CO loss rates (L(CO)) as described by Nassar et al. [15], assuming that the CO_2 production is equal to the CO loss:

$$P(CO_2) = L(CO) \tag{12}$$

Some of the emission inventories used in the CO_2 simulation already include CO_2 from CO oxidation (effectively assuming a prompt oxidation of precursors at the point of emission), but these amounts are only in the form of surface emissions rather than distributed throughout the atmosphere, leading to a bias in the model [14]. With the inclusion of a 3-D chemical source in the CO_2 simulation, this bias needs to be corrected by subtracting the CO_2 chemical production "emitted" at the surface (in the emission inventories) from the total CO_2. Nassar et al. [15] quantified a 0.825 Pg C yr^{-1} global annual value for this surface correction based on emissions of all reactants that undergo oxidation to CO_2 and are included in emission inventories. This includes emissions from fossil fuel, biospheric CH_4 (wetlands, ruminants, rice, termites, landfill) and biospheric NMVOC emissions (isoprene and monoterpene). The emission inventories used for biofuel and biomass burning explicitly account for CO_2, CO, CH_4 and NMVOC separately; hence, no surface correction is applied.

2.2. *Coupled Geos-Chem Simulation*

Our updates couple CH_4, CO and CO_2 based on the chemical loss and production reactions between these species, providing a single, self-consistent simulation. A schematic diagram of the coupling is shown in Figure 1.

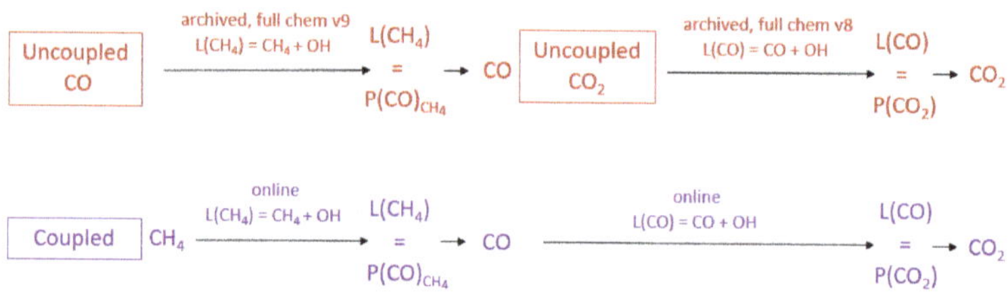

Figure 1. Schematic diagram of the tropospheric uncoupled carbon monoxide (CO) and carbon dioxide (CO_2 simulations (top) versus the coupled CH_4–CO–CO_2 simulation (CH_4 methane, bottom). The diagram also shows the model version (i.e., full chemistry simulation) used for the creation of the loss and production fields in the uncoupled simulation. Colors correspond to simulations shown in subsequent sections (see text for details). Note that both simulations use the same CO production from non-methane volatile organic compounds (P(CO)$_{NMVOC}$) field described in Sections 2.1 and 2.2 (not shown on diagram).

The starting point of the coupled simulation is the calculation of CH_4 based on Equations (4) and (5). The tropospheric CH_4 loss rates are calculated from the oxidation of tropospheric CH_4 by OH at every time step. As before, a 100% yield of CO from CH_4 oxidation is assumed [4], and the tropospheric CH_4 loss is passed to the CO part of the simulation at every timestep as the chemical production of CO from CH_4 (P(CO)$_{CH_4}$) in the troposphere. The calculation of the CO production in the stratosphere and from NMVOCs uses the same method as in the uncoupled CO-only simulation. In the troposphere, the total chemical production of CO (P(CO)) is equal to the sum of the archived P(CO)$_{NMVOC}$ field and the now-online calculated P(CO)$_{CH_4}$. The global tropospheric P(CO)$_{NMVOC}$ term is equal to 480 Tg CO yr^{-1} in both the coupled and uncoupled CO simulations for each simulation year.

The chemical production of CO_2 (P(CO_2)) is then calculated from the simulated CO loss from the oxidation of CO by OH in the troposphere and from the archived CO loss in the stratosphere. As in the uncoupled version, a 100% yield of CO_2 from CO is assumed [15]. For the chemical surface correction, due to the inclusion of the chemically produced CO_2 in other emission inventories, we retained the same correction method and values as in the uncoupled simulation [15].

We used consistent archived OH fields (from v9-01-03) for all aspects of the simulation. We also note that our coupled simulation does not require running additional full chemistry simulations to create the chemical production fields, and therefore allows users to easily update the OH fields used for the oxidation of the three species (and online calculation of the chemical production terms) as new model versions become available.

The new coupling now allows for time-specific changes in and tracking of the chemical production terms. This is an improvement of the uncoupled simulations, where the prescribed fields were based on simulations of specific prior years and therefore could not capture the year-specific variations and dependencies between these gases. In the uncoupled simulations, all of the prescribed chemical production and loss fields are monthly mean values, whereas, with the coupled simulation, these fields are calculated online at every timestep (i.e., 20 min), allowing us to track the day-to-day and diurnal variability of the simulated chemical production terms.

2.3. Experimental Design

Our aim in this work was to compare the newly coupled simulation to the default (public) v12.1.1 uncoupled simulations currently used by the GEOS-Chem community. All aspects of the coupled and uncoupled simulations not associated with the chemical coupling were kept as consistent as possible with the public versions of the uncoupled simulations. The only two exceptions were: (i) the inclusion of a diurnal scaling to the OH field used for CH_4 oxidation in the troposphere and (ii) the use of the Quick Fire Emissions Dataset (QFEDv2, Darmenov and da Silva [45]) for CO_2 emissions (further details in Section S1). The former provided consistency between carbon gases in terms of the treatment of diel OH variability (which was already included for CO but not for CH_4), and the latter ensured that we could use consistent biomass burning emissions for all three species.

We ran both the uncoupled and coupled simulations from January 2005 through December 2017. The meteorological inputs for GEOS-Chem come from the Modern-Era Retrospective analysis for Research and Applications, Version 2 (MERRA2) reanalysis developed by the NASA Global Modelling and Assimilation Office (GMAO). The native horizontal resolution of MERRA2 is $0.5° \times 0.625°$. We ran the simulations at $2° \times 2.5°$ horizontal resolution with 47 vertical levels. We used 10 min as the transport and convection timestep and 20 min for the chemistry and emissions timestep. The production and loss terms used by each simulation are shown in Table 1, with additional common emission fields (i.e., source and sink processes) in Table S1. For simulation periods that are outside of the specified inventory time range, the model re-used the data from the closest year.

Based on the recommendation from the GEOS-Chem carbon cycle working group, both the uncoupled and coupled simulations were initialized with a 10-year spinup for CO_2 and CH_4 using 2005 as a base spinup year, whereas, for CO, the model was spun up for 6 months in 2005. The spinup was carried out with the uncoupled v11-01 simulations described in Bukosa et al. [10]. The initial fields prior to the spinup were based on year 2005 for CO_2 and 2010 for CH_4. Due to the increasing trend of CO_2 and CH_4 in the atmosphere, each spinup year (repeating year 2005) adds the yearly growth rate of 2005 to the modeled CO_2 and CH_4 values, leading to globally higher simulated values relative to measurements. The global modeled growth of CO_2 and CH_4 in 2005 at the surface is 1.41 ppm yr^{-1} and 0.96 ppb yr^{-1}, respectively. We quantified the overall offset by calculating the difference between the modeled CO_2 and CH_4 values at the end of the 10-year spin-up (calculated for 1 January 2005) and measurements at baseline NOAA GGGRN sites (Barrow, Mauna Loa, American Samoa (Tutuila) and South Pole, average value for January 2005). The resulting offset was 14 ppm for CO_2 and 45.8 ppb for CH_4, and we subtracted this offset from the

CO_2 and CH_4 initial fields prior to the simulations. Due to differences between emission inventories used in Bukosa et al. [10] and those used here, we used the first simulation year (2005) as an additional spinup year for all three gases.

3. Results and Discussions

3.1. Chemical Production Budgets

The main terms impacted by the coupling of CH_4, CO and CO_2 are the production of CO from CH_4 ($P(CO)_{CH_4}$) in the troposphere and the production of CO_2 from CO ($P(CO_2)$). Furthermore, the changes in these terms also impact the total source budgets for CO and CO_2 and the sink term for CO (loss of CO by OH (L(CO))). The global and hemispheric budgets for the chemical production for both coupled and uncoupled versions of the model along with known literature values are shown in Table 2. The annual global budgets of the chemical production terms $P(CO)_{CH_4}$ and $P(CO_2)$ from the uncoupled and coupled simulations are shown in Figure 2, with their regional distributions in Figures S5 and S6.

Table 2. Global and hemispheric budgets (SH—Southern Hemisphere, NH—Northern Hemisphere) for CO production from CH_4 ($P(CO)_{CH_4}$ in Tg CO yr^{-1}) and CO_2 production from CO ($P(CO_2)$ in Pg C yr^{-1}) from the uncoupled (U) and coupled (C) simulations, as well as literature values for the global budgets. The budgets from the simulations are shown as a multi-year mean based on years 2006–2017. The range of values for individual years is shown in the parentheses.

		Global		NH		SH	
Chemical Terms	**Prior Work**	U	C	U	C	U	C
$P(CO)_{CH_4}$	760–1086 [1,2,3,4,5,6]	902 [7] (901–905) [7]	937 (913–960)	521 [7] (520–522) [7]	536 (522–549)	381 [7] (380–382) [7]	401 (390–411)
$P(CO_2)$	1.04–1.1 [8,9]	1.1 [10] (1.08–1.11) [10]	1.03 (1.01–1.05)	0.67 [10] (0.63–0.68) [10]	0.62 (0.62–0.63)	0.43 [10] (0.43–0.46) [10]	0.40 (0.39–0.42)

[1] Holloway et al. [27], base year: not defined; [2] Bergamaschi et al. [25], base year: 1993–1995; [3] Duncan et al. [4], base year: 1988–1997; [4] Arellano Jr. and Hess [28], base year: 2000–2001; [5] Stein et al. [29], base year: 2008; [6] Zeng et al. [30], base year: 2004; range based on different model simulations. [7] Fisher et al. [17], base year: 2009–2011 average; [8] Nassar et al. [15], base year: 2000–2009; [9] Suntharalingam et al. [14], base year: 1988–1997; [10] Nassar et al. [15], base year: 2006–2010.

The results from our coupled simulation remain consistent with the range of values found in prior work. The coupled simulation shows stronger $P(CO)_{CH_4}$ than the fields used by the uncoupled simulation. The stronger chemical production is mainly driven by different CH_4 levels between the coupled simulation and the full chemistry simulation (used as the input in the uncoupled CO simulation). The $P(CO)_{CH_4}$ calculation in both simulations is based on the same OH version (v9-01-03), although minor differences exist due to the temporal resolution of the OH fields (monthly OH fields with diurnal scaling for coupled, hourly for uncoupled). The coupled $P(CO)_{CH_4}$ values are stronger than the uncoupled values for all years and for both hemispheres (11–55 Tg CO yr^{-1} difference). This difference represents 0.5–2.3% of the total CO source in the coupled simulation. The CO_2 chemical source shows weaker values in the coupled simulation relative to the uncoupled one (0.04–0.09 Pg C yr^{-1} difference). This difference represents 0.3–0.7% of the total CO_2 source in the coupled simulation. The stronger uncoupled $P(CO_2)$ values are a result of different CO amounts used for the CO loss calculation between the coupled and full chemistry (used as the input for the uncoupled) simulations, as well as more abundant OH used to calculate L(CO) for the uncoupled simulation (v8-02-01, Figure S2) relative to the OH field used in the coupled simulation (v9-01-03).

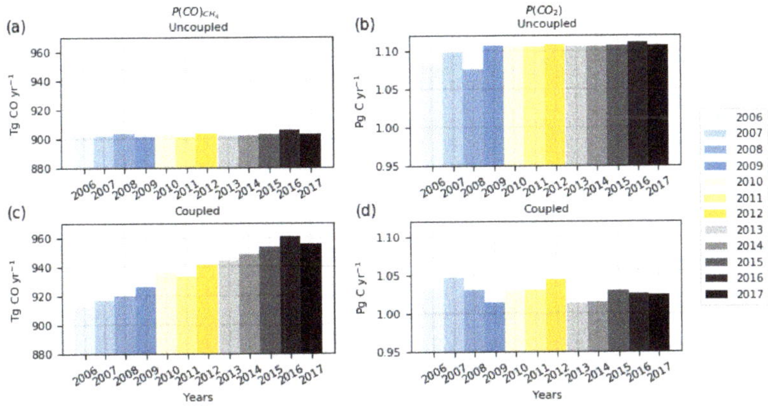

Figure 2. Annual values of the global chemical production term budgets for CO production from CH_4 (**a**,**c**) and CO_2 production from CO (**b**,**d**) from the uncoupled (**a**,**b**) and coupled (**c**,**d**) simulations. Regional distributions are shown in Figures S5 and S6.

Figure 2a,b show that there is some very minor inter-annual variability in the chemical production fields simulated by the uncoupled simulation despite the fact that the input chemical production fields used in this simulation do not vary inter-annually for CO (based on 2009–2011 average values) and do not vary inter-annually after year 2010 for CO_2. This small variability in the uncoupled simulation is exclusively driven by: (1) leap years in 2008, 2012 and 2016 that lead to a larger total annual production and (2) interannual variability in the meteorological fields (e.g., pressure levels, tropopause height) affecting the calculation of the total tropospheric budget.

The inter-annual variability of the chemical fields is one of the key benefits of the coupled simulation. We found a consistent increasing trend in $P(CO)_{CH_4}$ over the 2006–2017 period in the coupled simulation (Figure 2c) due to increasing atmospheric CH_4 concentrations leading to an increased CH_4 loss and associated CO production. The $P(CO)_{CH_4}$ increase is the most pronounced in tropical regions (Figure S5). For P(CO_2), we do not observe a trend in the coupled results. The 2006–2010 P(CO_2) results do not entirely match the inter-annual variability shown in the uncoupled fields. The year-to-year change of the chemical fields in the coupled simulation is driven by the inter-annual variability of the emission fields used to simulate CH_4 and CO (Tables 1 and S1). Some of the emission-driven variability may potentially be linked to El Niño Southern Oscillation (ENSO). For example, we observe the strongest growth in $P(CO)_{CH_4}$ during 2009/2010 and 2015/2016, which coincide with moderate and strong El Niño years, while we find no growth during 2010/2011, a strong La Niña year, highlighting the potential impact of climate anomalies on the chemical terms. However, we note that 2016 is also a leap year, which will also impact the production increase. Different ENSO-triggered CH_4 processes lead to opposite changes in CH_4: during El Niño events, wetland emissions are reduced, whereas biomass burning emissions are enhanced [46–49]. Our coupled simulation shows that these changes can have an imprint on the chemical production of CO that is not captured in the uncoupled simulation. The availability of OH via CO also impacts the CH_4 interannual variability; however, we were unable to quantify the OH-driven changes here as none of our simulations included OH inter-annual variability or OH-feedbacks. We recommend that future updates to the coupled simulation prioritize the inclusion of a CO–OH–CH_4 feedback in the calculation [34,50].

Figure 3 shows the budgets throughout the year for each chemical term in different latitudinal bands. Figure 3a shows that both the uncoupled (red) and coupled (indigo) $P(CO)_{CH_4}$ have a similar annual cycle, with an overall stronger production in the coupled simulation. The coupled simulation also shows more variability due to the year-specific

CH_4 loss, with the most variability in tropical regions. Although covering a shorter time period with year-specific fields, from 2006 to 2010, the uncoupled $P(CO_2)$ values show more variability than in the coupled version (Figure 3b). The uncoupled simulation also shows stronger $P(CO_2)$ values in all latitudinal bands. The largest difference between simulations is during December–March, mostly in Northern Hemisphere (NH) tropical and Southern Hemisphere (SH) mid-latitude regions.

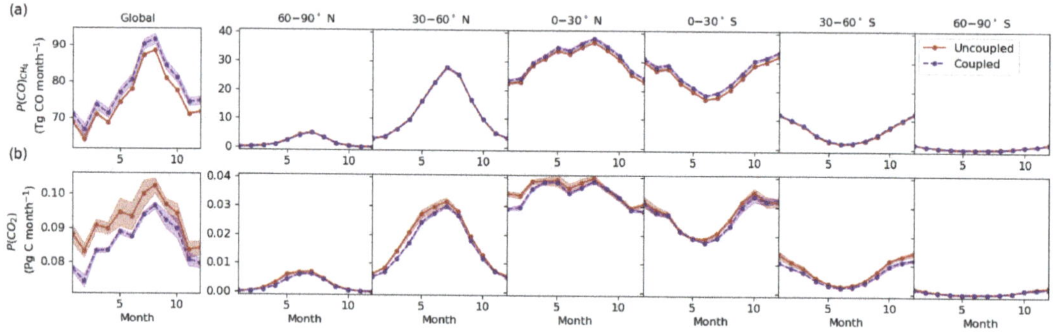

Figure 3. Monthly total atmospheric column CO production from CH_4 (**a**) and CO_2 production from CO (**b**), with 1 standard deviation (shaded region), from the uncoupled (red) and coupled (indigo) simulation averaged for 2006–2017 and summed globally (left) and over different regions (right).

3.2. Chemical Source Contributions

Due to the linearity of the GEOS-Chem carbon greenhouse gas simulations, in addition to simulating the total amount of each gas, we can also quantify the mole fractions of individual processes (referred to as tracers). These include the CO_2 mole fraction from CO_2 chemical production (CO_{2CO}) and the CO mole fraction from CO production from CH_4 (CO_{CH_4}). Figure 4 shows these chemical production tracers (Figure 4a,b), as well as the total CO and CO_2 mole fractions (Figure 4c,d) at the surface for different latitudinal bands. Note that, in contrast to the CO source tracers, where the atmospheric sink terms (e.g., OH) are applied to each tracer, for CO_2, there is no sink applied to the different source tracers. This leads to a trend in CO_{2CO} and its accumulation in the atmosphere. To highlight differences in the seasonal cycle, we detrended the CO_{2CO} data shown in Figure 4b,d and added the mean 2006–2017 yearly growth rates.

Implementing the online calculation of the chemical terms results in higher CO_{CH_4} values in the coupled simulation relative to the uncoupled simulation, along with a stronger variability (Figure 4a), similar to the production rates (Table 2, Figure 3a). An average 0.8 ± 0.5 ppb difference is present across the NH between the coupled and uncoupled results, whereas, in the SH, we find a larger difference of 1.4 ± 0.5 ppb. Both the coupled and uncoupled simulations show similar seasonal cycles. The difference in CO_{CH_4} is also reflected in the total amounts of CO (Figure 4c), leading to slightly higher global surface CO values in the coupled simulation. $P(CO)_{CH_4}$ and CO_{CH_4} show a seasonal cycle, with the maximum production during NH summer and minimum during winter, which is the opposite of the seasonal cycle of the total CO mole fractions.

The CO_{2CO} mole fractions also show a similar seasonal cycle between the two simulations in both hemispheres (Figure 4b). The uncoupled simulation shows a stronger yearly global surface growth rate of 0.52 ppm yr^{-1} due to the stronger chemical production, whereas the coupled simulation shows a weaker growth rate of 0.48 ppm yr^{-1} due to weaker production. Overall, the coupling does not significantly impact the resulting CO_2 mole fractions between simulations. The differences between the coupled and uncoupled simulations are too small to be reflected in the total CO_2 surface values (Figure 4d). As already highlighted, $P(CO_2)$ is a 3-D source; hence, the signal of this source in the surface mole fractions is small relative to the other more dominant CO_2 surface fluxes.

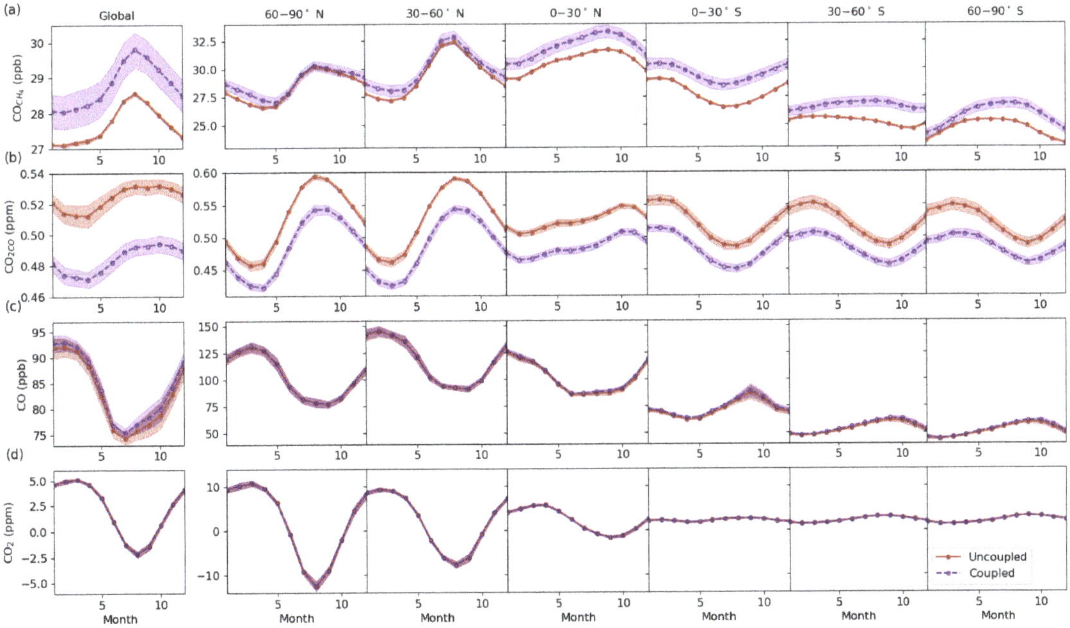

Figure 4. Surface mole fractions from chemical production of CO_{CH_4} (**a**) and CO_{2CO} (**b**) and total CO (**c**) and CO_2 (**d**) mole fractions from the uncoupled (red) and coupled (indigo) simulations with 1 standard deviation (based on 2006–2017 average values). Note that the CO_2 values are detrended and added to the mean 2006–2017 yearly growth rates.

3.3. Global Distribution

Figure 5 shows the total column chemical production of CO from CH_4 with corresponding CO_{CH_4} mole fractions at the surface and a 500 hPa altitude from the coupled simulation, as well as the difference between the coupled and uncoupled simulations. Figure 6 is the same as Figure 5 but for the chemical production of CO_2 and the CO_{2CO} mole fractions. For CO_2, we additionally removed the long-term trend from the CO_{2CO} mole fractions. The seasonal changes of both the production terms and mole fractions are shown in Figures S7–S10 for CO and Figures S11–S14 for CO_2.

The online calculation of $P(CO)_{CH_4}$ has a small impact on its global spatial distribution; both the coupled and uncoupled simulations show similar distributions (Figure 5a,d). The simulations used the same OH fields; hence, the differences in $P(CO)_{CH_4}$ are driven by the different handling of the CH_4 values before the OH loss is applied. The main difference between the two simulations is the stronger $P(CO)_{CH_4}$ over tropical ocean regions and weaker $P(CO)_{CH_4}$ over NH land regions in the coupled version. On a yearly scale, the surface CO_{CH_4} mole fractions from the coupled simulation show higher values above both ocean and land regions (Figure 5e), as a result of the stronger $P(CO)_{CH_4}$ over tropical ocean regions. A similar behavior is observed at 500 hPa; however, the differences are smaller and more diffuse. We find the same differences throughout the seasons (Figures S7–S10). We further discuss the simulated mole fractions and the impact of the coupling on total *CO* in Section 3.5.

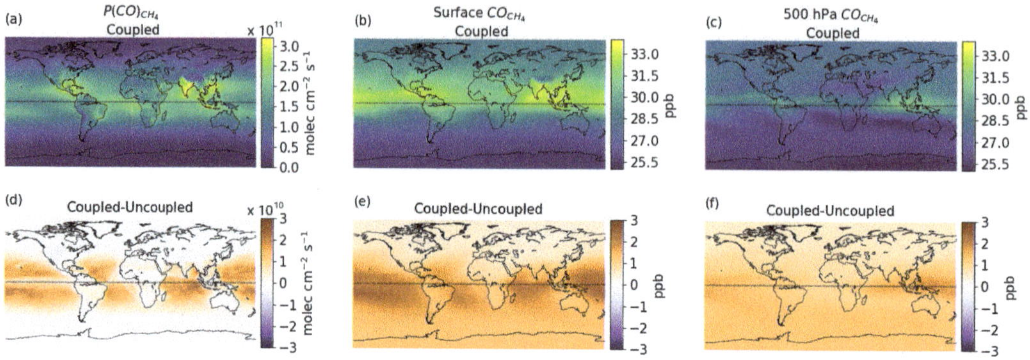

Figure 5. Average 2006–2017 total column CO chemical production from CH_4 (**a**) and corresponding mole fractions (i.e., CO_{CH_4}) at the surface (**b**) and at 500 hPa (**c**) from the coupled simulation, along with the difference in each field between the coupled and uncoupled simulation (**d**–**f**).

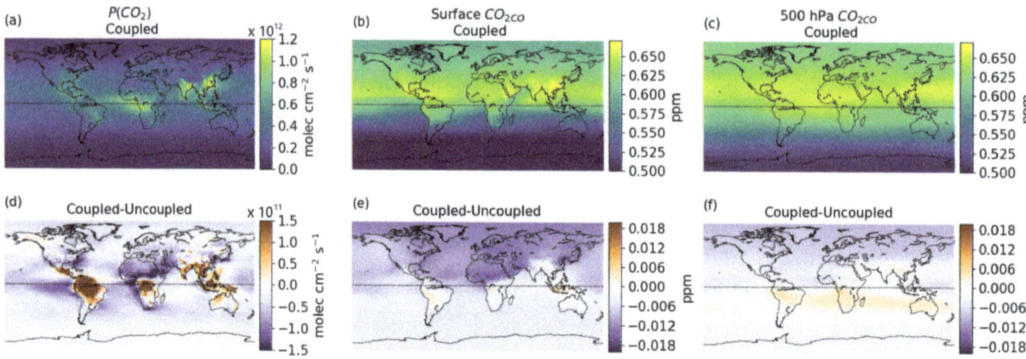

Figure 6. Average 2006–2017 total column CO_2 chemical production from CO (**a**) and corresponding mole fractions (i.e., CO_{2CO}) at the surface (**b**) and at 500 hPa (**c**) from the coupled simulation, along with the difference in each field between the coupled and uncoupled simulation (**d**–**f**).

The coupled simulation shows stronger P(CO_2) (Figure 6d) in certain land regions (South America, Central Africa, Indonesia, parts of East Asia and Australia) despite the annual global chemical source being weaker than in the uncoupled simulation. Moreover, in the uncoupled simulation, there is almost no P(CO_2) observed above the Amazon [15]; however, our results suggest substantial P(CO_2) in this region. The difference patterns appear to be mostly independent of season (Figures S11–S14). The chemical production is overall weaker above the ocean in the coupled simulation for all seasons; however, the coupled simulation does show stronger P(CO_2) during certain periods in tropical and NH mid-latitude regions. The stronger P(CO_2) in the coupled simulation above South America, Central Africa, Indonesia, parts of East Asia and Australia is present in all seasons, but with the strongest contribution during September–November. South America, Central Africa and northern Australia are characterized by strong biomass burning, especially during the SH dry season, when frequent fires are observed (September–November), emitting large amounts of CO into the atmosphere [51]. Our coupled model simulates the P(CO_2) in these regions during the fire season to be stronger than the previous fields used in the uncoupled simulation. The stronger P(CO_2) from the coupled simulation in other regions such as East Asia and North America points to enhanced anthropogenic CO emissions that lead to a stronger chemical production of CO_2. In addition to the primary CO emissions, the secondary production of CO from NMVOC could also have a significant impact on the P(CO_2) in regions where we observe differences. Different model versions were used to save the P(CO)$_{NMVOC}$ and P(CO_2) in the uncoupled simulations, as discussed in Section 2. The updated chemistry between model versions would additionally impact the P(CO_2) through the CO production from NMVOC in regions where we expect a significant contribution from this production term (e.g., the Amazon). The spatial distribution of the surface mole fractions is similar between simulations, with lower values overall in the coupled simulation, especially in the NH, due to the globally stronger P(CO_2). However, the coupled simulation does show more abundant CO_{2CO} over tropical land regions, with stronger P(CO_2), and over the SH during June-November (Figures S13 and S14). The higher coupled simulation CO_{2CO} values over tropical land regions are more pronounced at 500 hPa.

3.4. Vertical Latitudinal Distribution

Figure 7 shows the vertical latitudinal distribution of $P(CO)_{CH_4}$ and P(CO_2) for different months, averaged for 2006–2017, as well as the difference between the uncoupled and coupled simulations. The strongest $P(CO)_{CH_4}$ in the coupled simulation occurs between the surface and 3 km altitude. For most months, this chemical production is stronger in the NH than in the SH; however, around November we observe stronger production in the SH, potentially due to biomass burning and wetland activity, that leads to enhanced CH_4 levels and its subsequent loss driving CO production. Although the strongest production occurs between 50° S–50° N, we also observe production in Antarctic regions in December–January and in Arctic regions in May–July, corresponding to their summer periods. Production is stronger in the Arctic than in the Antarctic due to higher CH_4 levels and the stronger loss in the NH. As for the global spatial distribution results, the coupling has a small impact on the vertical distribution (Figure 7c, stronger production in the coupled than in the uncoupled).

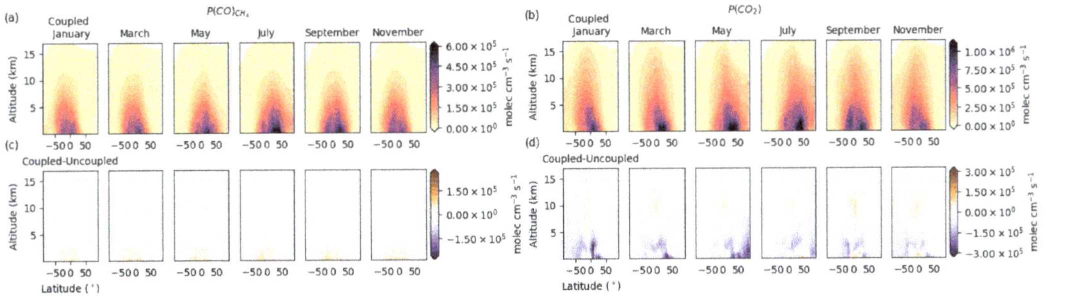

Figure 7. Vertical latitudinal distribution of the coupled CO chemical production from CH_4 (**a**) and CO_2 chemical production from CO (**b**) and the coupled–uncoupled differences (**c**,**d**) over different months, averaged over 2006–2017.

The strongest CO_2 chemical production in the coupled simulation occurs between the surface and 4 km altitude, and CO_2 is produced chemically up to 15 km (Figure 7b). From January–July, we observe stronger production in the NH, with the strongest production in tropical regions at the beginning of the year, moving toward higher latitudes by July. Based on the distribution of this source in the NH (Figure 6), strong production occurs over China and India from anthropogenic CO, with mixed biomass burning influence from other regions. For the remaining months, both hemispheres show strong P(CO_2), with the SH showing stronger production in September, presumably due to additional biomass burning in the tropics (e.g., Indonesia, Australia, Africa, S America). Using an uncoupled version of the CO_2 simulation, Nassar et al. [15] did not find a biomass burning contribution over the Amazon; however, our coupled simulation, as already discussed, suggests a significant contribution from this region. Relative to the uncoupled simulation, the coupled simulation shows weaker production in mid-latitude and polar regions, with a stronger contribution in the tropics at surface levels and above 5 km. The Arctic and Antarctic regions show weaker production in the coupled simulation.

3.5. Model Evaluation with Column, Surface and Aircraft Measurements

We validated the new coupled simulation against global column retrievals and calibrated surface flask and aircraft in situ measurements (Figure 8, Table A1 in Appendix A). Long-term time series of column-averaged dry-air mole fractions of CO and CO_2 were measured by TCCON [52]. In addition, long-term time series of surface mole fractions exist at different sites across the globe as part of NOAA GGGRN (Dlugokencky et al. [53,54]). For a vertical profile comparison, we used aircraft measurements from the ATom campaigns [55]; for both CO_2 and CO, we used the merged ATom data product collected from the NOAA-Picarro and Harvard Quantum Cascade Laser System instrument.

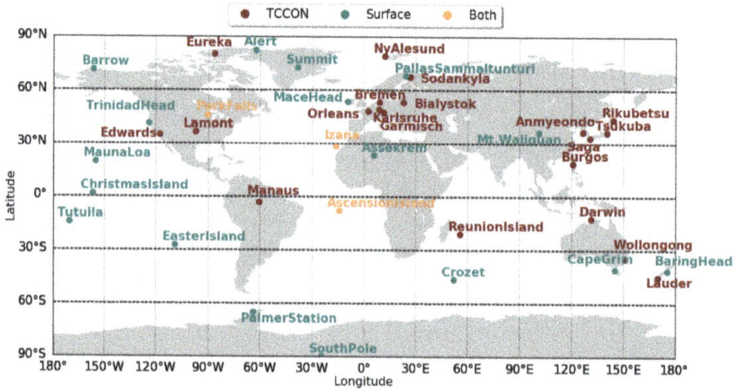

Figure 8. Locations of the flask surface sites from the NOAA Global Greenhouse Gas Reference Network (NOAA GGGRN, turquoise, Dlugokencky et al. [53,54]) along with sites that measure column-averaged dry-air mole fractions as part of the Total Carbon Column Observing Network: (TCCON, red, https://tccondata.org/, accessed on 10 march 2023) and sites that are both part of TCCON and NOAA GGGRN (orange). For site details, see Table A1 in Appendix A.

We used column measurements from TCCON as the main data product to highlight the differences between the coupled and uncoupled simulations. Both the CO and CO_2 chemical sources are produced throughout the column; hence, relative to surface measurements, these measurements are more representative of the impact of chemical production on the total amounts of the gases. In order to compare the total CO and CO_2 model output with the column-averaged measurements, we converted the modeled mole fractions to column-averaged dry-air mole fractions (X_{gas}) by dividing the vertical column of the gas of interest (Ω_{gas}) with the total dry-air column (Ω_{O_2}) based on the method described by Wunch et al. [56]:

$$X_{gas} = 0.2095\frac{\Omega_{gas}}{\Omega_{O_2}} \quad (13)$$

and smoothed according to Equation (14) [57]:

$$c_s = c_a + h^T a^T (x_m - x_a) \quad (14)$$

where c_s represents the smoothed column model dry-air mole fraction, c_a is the TCCON a priori column dry-air mole fraction, h^T represents the vertical column summation, a^T is the TCCON averaging kernel and x_m and x_a are the model and a priori dry-air mole fraction profiles.

The modeled vertical profiles were saved at a daily temporal resolution and extracted for the closest grid box to each TCCON station. For the comparison with surface measurements, we extracted the grid points at the lowest level in the model. For the comparison with aircraft measurements, model outputs were saved for grid boxes corresponding to the measured time, latitude, longitude and level along the plane flight track. Both the aircraft measurements and modeled output were averaged to the model temporal (20 min) and spatial (2° × 2.5°) resolution to calculate one average value for each unique grid-box–time-step combination.

Both the column and surface measurements are impacted by data gaps. In order to minimize the impact of the non-continuous measurements and inconsistent measurement time periods on the analysis, we used a consistent time period (2010–2017) when analyzing the measurement–model differences. We found the fewest data gaps during this time period; however, a few sites are still subject to missing measurements (column: Ny Alesund, Rikubetsu, Edwards, Anmyeondo, Saga, Ascension Island, Reunion; surface: Trinidad, Easter Island, Christmas Island). Due to short timeseries at the Manaus and Burgos TCCON sites, we excluded them from the plots representing the measurement–model differences in the next section (Section 3.5.1); however, the full timeseries at all sites can be found in Figures S15–S18.

3.5.1. Comparison with Column Measurements

Figure 9 shows the differences between the modeled values (uncoupled and coupled) and measurements at different TCCON sites for CO (Figure 9a–e) and CO_2 (Figure 9f–j) plotted against the latitude of each site. We also show the normalized mean bias between the modeled and measured values on each plot. Mid-latitude European sites (Białystok, Bremen, Karlsruhe, Orléans and Garmisch, grouped into Other EU sites) show similar results; hence, we only present their mean value. The timeseries comparison of the total CO, CO_{CH_4}, CO_2 and CO_{2CO} mole fractions for each site can be found in Figures S15 and S16.

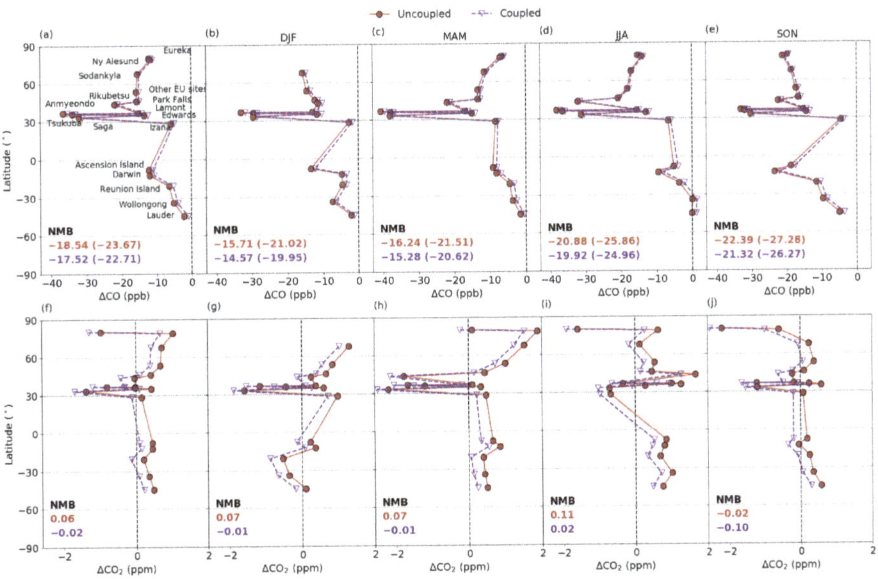

Figure 9. Column-averaged mole fraction model–measurement differences (uncoupled (red) and coupled (indigo)) for CO (**a**–**e**) and CO_2 (**f**–**j**) as a function of latitude, averaged for 2010–2017, with annual values (**a**,**f**) and for different seasons: December–January–February (DJF,: (**b**,**g**)) , March–April–May (MAM, (**c**,**h**)), June–July–August (JJA, (**d**,**i**)), September–October–November (SON, (**e**,**j**)). The numbers inset represent the normalized mean bias (NMB). For CO, we also show the NMB based on the unscaled CO values (shown in the parentheses).

The coupled CO results lead to a smaller bias between the modeled and measured values in both hemispheres and all seasons (except at SH mid-latitude sites, Lauder and Wollongong in June–July–August). Differences in the CO values are driven by differences in the CH_4 loss calculation, and we find that the stronger CH_4 loss in the coupled simulation leads to a smaller model–measurement bias, suggesting that this term was potentially underestimated in the uncoupled simulation. The distribution of the model–measurement differences between sites is consistent between the coupled and uncoupled simulation but with larger differences for sites in the SH. Previous studies showed that CO values in the SH are dominated by CH_4 and NMVOC oxidation [17,30,58]; hence, the larger SH offset between the coupled and uncoupled simulation is driven by the dominance of chemical production relative to other CO sources.

On a yearly scale and for all seasons, the smallest CO model–measurement bias is present at sites closest to the South Pole, with an increasing negative bias (i.e., underestimation of the modeled values compared to measurements) toward the NH. In the NH, the biases show a smaller latitudinal dependence than in the SH, presumably due to the larger differences in the CO sources between regions/sites. Differences in the modeled–measured values throughout the seasons highlight potential contributors to the observed biases. In the SH, we find a stronger negative bias during austral spring (September–October–November), while the seasonal dependence in the NH is more variable. The larger SH bias during austral spring suggests an underestimated biomass burning source, since this period aligns with the burning season in the SH. Due to uncertainty in the TCCON bias-correction to in situ scales for CO, we also compared our modeled CO with the unscaled TCCON CO values that are higher by approximately 7%. For all sites, we obtained the unscaled values by multiplying the column CO by 1.0672. The potential TCCON bias is apparent in comparison to MOPITT (Measurements of Pollution in the Troposphere) [59] and to NDACC (Network for Detection of Atmospheric Composition Change) [60]. The resultant Normalized Mean Bias (NMB) relative to the unscaled values is also shown in Figure 9a–e. A comparison with the unscaled values further increases the negative model–measurement bias.

Relative to CO, where we observe a consistent negative bias, the CO_2 biases are more variable between both sites and seasons. On a yearly scale and for most sites, the coupled CO_2 results show a smaller model–measurement bias than the uncoupled results. The main difference between the uncoupled and coupled CO_2 values is the weaker CO_2 chemical production in the coupled simulation, suggesting that this source term might have been overestimated in the uncoupled simulation (although there could be other compensating biases). For most sites and seasons, the simulated CO_2 values are higher than the measurements (Figure 9f–j). An exception in the NH is Eureka, where there is a consistent negative bias, and NH mid-latitude sites that show either a negative and positive bias depending on the season. We find the largest model–measurement bias for the NH mid-latitude sites (30–45° N). This bias is potentially driven by a combination of biases in the terrestrial and anthropogenic emissions that dominate CO_2 variability in these regions. The sites in the 30–45° N band also show the largest biases in the CO comparisons; hence, common CO and CO_2 anthropogenic emissions might be the dominant driver of this bias. Whether or not this bias extends to NH tropical regions cannot be determined due to a lack of TCCON sites in tropical regions. Relative to the NH, the SH biases are smaller and less variable, presumably due to less variable CO_2 sources/sinks.

Overall, our coupled simulation led to a smaller model–measurement bias than the original uncoupled simulations. However, we note that the reduced bias could potentially be compensating for biases in other emissions fields and that some of the differences in the biases are small when compared to other uncertainties in the system. Further bias reductions would come from reducing uncertainties in other fluxes and transport. The inclusion of an OH feedback between species would additionally impact the model–measurement bias, especially during enhanced localized emission events (i.e., fires). As an example, strong CO emissions would lead to depleted OH values, resulting in a weaker oxidation of CH_4 and the production of $P(CO)_{CH_4}$. This feedback is not captured in either of the simulations since the OH fields are fixed.

3.5.2. Comparison with Surface Measurements

We complemented the column measurements with surface measurements. Figure 10 shows the differences between the uncoupled and coupled simulations versus the measurements at surface sites for CO (Figure 10a–e) and CO_2 (Figure 10f–j) plotted against the latitude of each site. The timeseries comparison for each site can be found in Figures S17 and S18. Note that, relative to the column results, the surface comparison is more strongly impacted by the coarse model resolution (2° × 2.5°). The measured and modeled column values are more representative of regional and larger-scale processes so the impact of the model resolution is weaker.

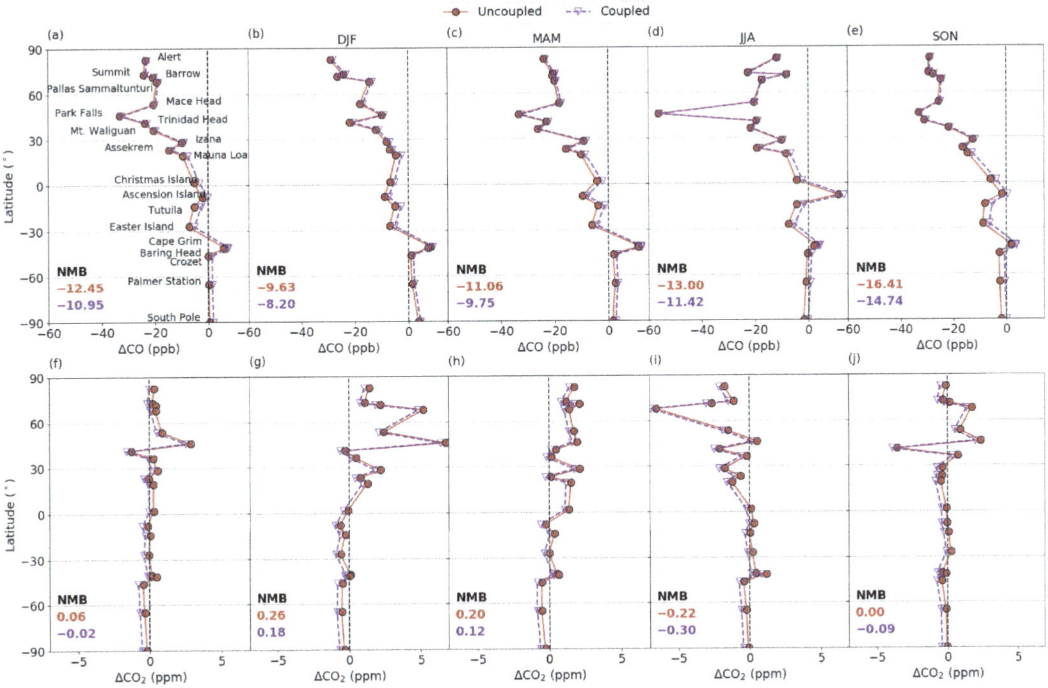

Figure 10. Surface mole fraction model–measurement differences (uncoupled (red) and coupled (indigo)) CO (**a–e**) and CO_2 (**f–j**) as a function of latitude, averaged for 2010–2017, with annual values (**a**,**f**) and for different seasons: December–January–Febuary (DJF, (**b**,**g**)) , March–April–May (MAM, (**c**,**h**)), June–July–August (JJA, (**d**,**i**)), September–October–November (SON, (**e**,**j**)).

For CO, the model–measurement biases in the surface data are similar to the column results. However, in the column data, we lacked measurements in NH tropical regions and 45–90° S, limiting us from identifying the biases. With available measurements in each latitudinal band, the surface comparison further amplifies the latitudinal dependence of the bias, increasing from the SH polar regions toward the North Pole. Relative to the column comparison, at the surface, the large bias at NH mid-latitude sites is less pronounced. The column CO values were consistently lower than the measurements; however, at the surface, we see an overestimation of the CO values for some SH sites (although a number of these sites are in regions where we lack column measurements). The overestimated values in the SH might be partially due to transport errors [61], such as the weaker vertical mixing [62,63] in the model leading to the buildup of CO in the planetary boundary layer.

Having column measurements in these regions would be beneficial when identifying the processes responsible for the observed model–measurement biases.

Differences between the surface and column comparison are more pronounced in the CO_2 data. We still find, on average, a smaller model–measurement bias in the coupled results; however, the latitudinal distribution of the biases is different relative to the column results. In the SH, we observe a consistent negative bias in the coupled results, whereas the column data pointed to a positive bias (i.e., overestimated modeled values compared to measurements). The negative bias tends to be larger for sites between 45–90° S, a region where we lack column measurements. The differences in the polar regions are potentially impacted by additional CO_2 exchange from air–sea ice interaction [64], a process that is not included in the simulation, and are still subject to large uncertainties [65]. The surface measurements in the NH tropics (where column measurements are lacking) show a positive bias during boreal winter/spring and negative bias during boreal summer/autumn.

3.5.3. Comparison with Aircraft Measurements

We further compared the simulations with aircraft measurements collected as part of ATom (campaign 1: July–August 2016, 2: January–February 2017, 3: September–October 2017 and 4: April–May 2018). Figure 11 shows the differences between the modeled and measured CO and CO_2 values during the four campaigns as a function of latitude and pressure. The spatial distributions of the differences between the modeled and measured values and CO_{CH_4} and CO_{2CO} cross-sections (Section S3) are shown in Figures S19 and S20.

The latitudinal change of the aircraft model–measurement differences for CO follows the pattern seen in both the column and surface data (a smaller model–measurement bias in the coupled simulation and, on average, underestimated modeled values compared to measurements). The negative CO bias is present during all seasons and latitudinal bands except during ATom 2 (austral summer) and ATom 4 (austral fall) south of 50–60° S. This negative bias is also present in the surface data, whereas, in the column data, we do not have sites south of 45° S. The model also underestimated the CO values compared to measurements at all vertical levels during all four campaigns/seasons (Figure 11c,g,k,o). Differences between the two simulations decrease with increasing altitudes in the model. Differences between the two simulations also decrease at higher latitudes during all ATom campaigns except ATom 2, where the difference between simulations is lowest in the SH mid-latitudes.

For CO_2, we find a consistent negative bias in all latitudinal bands except 65–70° N during ATom 2 (boreal winter). This is different from both the surface and column comparison, where we had a mixture of both negative and positive biases. Different biases in the surface, column and aircraft comparisons suggest that potential biases in vertical transport should also be explored. The coupled simulation shows a smaller model–measurement bias for both the surface and column comparison; however, for the aircraft comparison, we find a smaller bias in the uncoupled results. The model underestimated the CO_2 values compared to measurements in the uncoupled simulation, and the weaker chemical production in the coupled simulation further increased this bias. For CO_2 (in contrast to CO), the offset between the coupled and uncoupled simulations is consistent across all latitudinal bands and vertical levels.

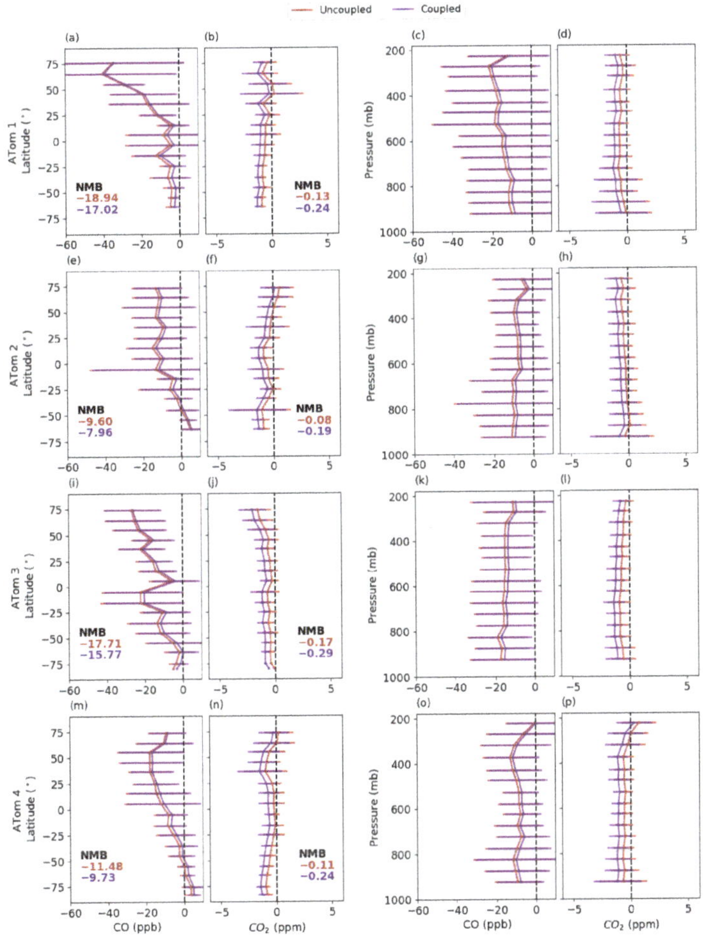

Figure 11. Aircraft model–measurement (uncoupled (red) and coupled (indigo)) CO and CO_2 differences, shown as their latitudinal (CO: (**a**,**e**,**i**,**m**) CO_2: (**b**,**f**,**j**,**n**)) and altitudinal distribution (CO: (**c**,**g**,**k**,**o**), CO_2: (**d**,**h**,**l**,**p**)) during the four Atmospheric Tomography Mission (ATom) campaigns in June–July 2016 (**a**–**d**), December–January 2017 (**e**–**h**), August–September 2017 (**i**–**l**) and March–April 2018 (**m**–**p**). Horizontal lines show standard deviation within each bin. The data are averaged into 10° latitudinal and 50 mb pressure bins.

3.6. The Importance of Consistent OH Fields

In this section, we explore the impact of inconsistent OH fields on the chemical production terms. In our coupled simulation, we used consistent OH fields (from v9-01-03) for all aspects of the simulation. However, as discussed in Section 2.1, the uncoupled simulations also rely on production and loss fields derived from the full chemistry simulation that was available at the time each capability was developed or updated, leading to differences in the resulting chemical fields. For the uncoupled simulations in the default v12.1.1 GEOS-Chem model, these fields were derived from GEOS-Chem v5-07-08 for the CH_4 simulation [44], v9-01-03 for the CO simulation [17] and v8-02-01 for the CO_2 simulation [15]. The global annual mean OH is largest in the v8-02-01 full chemistry simulation (11.8×10^5 molecules cm^{-3}) followed by v9-01-03 (11.4×10^5 molecules cm^{-3}) and v5-07-08 (10.8×10^5 molecules cm^{-3})

(http://wiki.seas.harvard.edu/geos-chem/index.php/Mean_OH_concentration, accessed on 10 March 2023). The yearly change and annual/seasonal global spatial patterns of the OH fields are shown in Figures S2–S4.

The OH disconnect inherent in the existing uncoupled simulations can introduce biases and inconsistencies in the simulated CH_4, CO and CO_2. Of most importance are inconsistencies in the archived OH fields used to calculate chemical loss rates for CH_4 and CO. Note that the uncoupled CO_2 simulation does not directly use OH (i.e., there is no CO_2 OH sink), so, for the CO_2 simulation, there is only a small and indirect influence from OH through OH-driven CO loss in the full chemistry simulation used to calculate the CO_2 chemical production fields. Here, we used the coupled model to perform an additional sensitivity simulation, retaining the default version of the OH used in the v12.1.1 uncoupled CH_4 simulation (i.e., v5-07-08 OH for the calculation of L(CH_4)) to highlight the impact of inconsistent OH fields currently in use in GEOS-Chem. We will refer to this sensitivity simulation as the *coupled-origOH* simulation.

We performed a 1 year simulation (2006) to analyze the impact of the OH disconnect. Both the coupled and *coupled-origOH* simulations were initialized with the same CH_4, CO and CO_2 initial conditions and both simulations used the same OH fields (v9-01-03) to calculate L(CO) and P(CO_2). The differences in the modeled values between the two simulations are exclusively driven by differences between the v5-07-08 and v9-01-03 OH fields used to calculate L(CH_4) and, by extension, $P(CO)_{CH_4}$. Differences between the OH fields are shown in Figures S2–S4. Briefly, the v9-01-03 OH shows higher values, with a peak during boreal summer (July); however, at the surface, the v5-07-08 OH has an earlier peak in June and also shows a second peak in October, when the v9-01-03 OH shows a decline. The seasonal cycles at higher altitudes are more consistent between the two OH versions. On both annual and seasonal scales, the v5-07-08 fields show lower surface OH above most land regions and NH ocean regions. A similar pattern is observed at higher altitudes (500 hPa), but with smaller and more diffuse differences.

Figure 12a shows the monthly global total column $P(CO)_{CH_4}$ from the coupled and *coupled-origOH* simulation. Using the default v5-07-08 OH fields for the $L(CH_{4Trop})$ and $P(CO)_{CH_4}$ calculations results in a 43 Tg CO yr^{-1} global decrease (≈5% change) relative to the coupled simulation (coupled 913 Tg CO yr^{-1}, *coupled-origOH* 870 Tg CO yr^{-1}), with weaker $P(CO)_{CH_4}$ in the *coupled-origOH* simulation due to lower OH values (Figure S2). The *coupled-origOH* $P(CO)_{CH_4}$ shows weaker production globally between May–September. The weaker $P(CO)_{CH_4}$ is present over land, whereas stronger production is observed over tropical and SH ocean regions (Figure 12b), following the differences in the spatial distribution of the OH fields (Figures S3 and S4).

The same OH field (v9-01-03) was used to calculate L(CO) and P(CO_2) in the coupled and *coupled-origOH* simulation; hence, the P(CO_2) is only impacted by differences in the component of the total CO loss that comes from CO_{CH_4}, which itself is only affected by $P(CO)_{CH_4}$ through L(CH_4). As a result, using the v5-07-08 OH fields for the $L(CH_{4Trop})$ calculation has a smaller impact on P(CO_2). Both the *coupled-origOH* and coupled simulations show similar P(CO_2) budgets but with stronger production in the coupled simulation between June–October (Figure 12e, 1.03 Pg C yr^{-1} coupled and 1.02 Pg C yr^{-1} in the *coupled-origOH*, ≈1% change). Similar to $P(CO)_{CH_4}$, the *coupled-origOH* shows weaker P(CO_2) everywhere except for tropical and SH ocean regions; however, the differences are more diffuse due to the indirect impact of the OH differences on P(CO_2).

Figure 12c shows the surface mole fractions of the chemical production of CO from CH_4. Using inconsistent OH fields between simulations leads to significant differences in the CO_{CH_4} seasonal cycle. In the *coupled-origOH* simulation, the mole fractions have an inverted seasonal cycle relative to the coupled results, showing a maximum in boreal winter and minimum in boreal summer. The inverted and incorrect seasonal cycle in the *coupled-origOH* simulation is driven by biases in the v5-07-08 OH fields. A detailed discussion of the seasonal cycle difference can be found in Section S2. For CO_{2CO} (Figure 12g), both simulations show a similar seasonal cycle, with slightly higher values in the coupled simulation due to stronger production. Similar to the total column production, the CO_{CH_4} mole fractions at the surface are highest in the tropical and SH ocean regions in the *coupled-origOH* simulation due to the stronger chemical production. The CO_{2CO} differences are more hemispheric, with lower mole fractions in the NH and higher mole fractions in the SH in the *coupled-origOH* simulation.

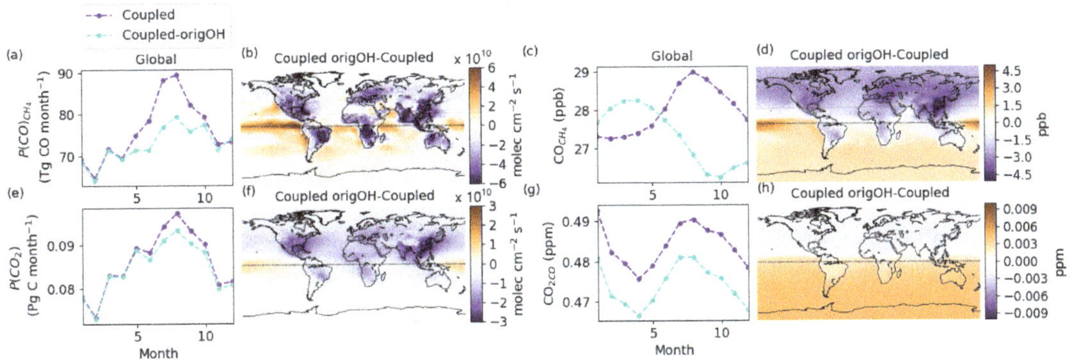

Figure 12. Monthly total atmospheric column CO production from CH_4 (**a**) and CO_2 production from CO (**e**), as well as surface mole fractions of CO_{CH_4} (**c**) and CO_{2CO} (**g**) from the coupled (indigo) and *coupled-origOH* (turquoise) simulation. Subplots (**b**,**f**) show the annual total column production difference between the *coupled-origOH* and coupled simulation, and (**d**,**h**) show the same but for the surface mole fractions of the chemical productions. All of the data are based on year 2006. Note that the CO_2 values are detrended and added to the 2006 growth rate.

We find that inconsistencies in the OH fields in the individual uncoupled simulations can have a significant impact on the production and loss terms, as well as the resulting mole fractions. Using the v5-07-08 OH fields that are currently the default in the v12.1.1 uncoupled CH_4 simulation showed an incorrect seasonal cycle of the CO_{CH_4} mole fractions, and, without the new coupling capability, this bias would not have been identified. By coupling CH_4, CO and CO_2, we therefore increase the consistency between GEOS-Chem simulations of these gases, not only by coupling their chemical production and loss terms but also by removing "hidden" inconsistencies between the individual simulations that arise through the use of different default OH fields.

4. Conclusions

We developed a coupled carbon greenhouse gas simulation in the GEOS-Chem chemical transport model that combines CH_4, CO and CO_2 through their chemical interdependence. The coupling between the three gases comes from the chemical production of CO from CH_4 loss ($P(CO)_{CH_4}$) and the chemical production of CO_2 from the oxidation of CO (P(CO_2)). In the uncoupled versions of these simulations that are currently widely used by the GEOS-Chem community, the chemical production calculations were handled offline based on monthly archived fields for specific years from older model versions. Moreover, the uncoupled simulations used inconsistent OH fields between the three gases. The new coupled simulation uses updated and consistent OH fields for all aspects of the simulation. We calculated $P(CO)_{CH_4}$ and P(CO_2) at every model timestep, enabling us to simulate the inter-annual variability of the chemical production fields and their follow-on effects.

Our budget estimates from the coupled simulation agree with known literature values. For the 2006–2017 time period, our coupled results show an increase in $P(CO)_{CH_4}$ with time and a dependence on climate anomalies (such as El Niño Southern Oscillation). We found differences between the coupled and uncoupled simulations ranging from 11–55 Tg CO yr^{-1}, with stronger production in the coupled simulation. Our P(CO_2) from the coupled simulation is weaker than in the uncoupled simulation, with a difference of 0.04–0.09 Pg C yr^{-1}.

Comparing the modeled values with three measurement products (TCCON total column measurements, NOAA GGGRN surface measurements and ATom aircraft data) led to (on average) a smaller model–measurement bias using the coupled simulation than using the original uncoupled simulations (except for CO_2 aircraft data); however, the reduced bias from the coupling could potentially be compensating for biases in other emissions fields. For CO, the remaining model–measurement biases in the SH can partially be explained by underestimated biomass burning emissions [10,58,66], especially during the dry season, and underestimated secondary CO production (CH_4 and NMVOC oxidation) [30]. Our coupled simulation suggests stronger P(CO_2) above tropical land regions than simulated previously and that the chemical production of CO_2 in the Amazon was significantly underestimated in previous P(CO_2) studies [15]. South America, Central Africa and northern Australia are characterized by strong biomass burning [67], and our coupled model simulates the P(CO_2) in these regions during the fire season to be stronger than in previous model versions, while the stronger P(CO_2) in regions such as East Asia [68] and North America points to enhanced anthropogenic CO emissions. For CO_2, the inclusion of the missing exchange from air–sea ice interaction could potentially contribute to better modeled values in the polar regions [64]. Our coupled model still excludes the OH feedback [34,50], which may be responsible for persistent biases in the modeled values, especially in regions where chemical production/loss is enhanced.

Our coupled simulation includes two major improvements relative to the default individual carbon gas simulations currently in use by the community: (i) the chemical coupling between species described above and (ii) consistent OH fields used for calculating CH_4 and CO loss. Using a sensitivity simulation, where we use the coupled simulation but retain the original (inconsistent) OH fields, we highlight the importance of using consistent and updated OH fields. We show that the default v5-07-08 OH fields currently used in the uncoupled v12.1.1 CH_4 simulation result in incorrect L(CH_4) and, by extension, $P(CO)_{CH_4}$ values, with an inverted seasonal cycle. In the coupled model, this has flow-on effects for CO and, to a lesser extent, CO_2.

The newly developed coupled simulation enables future investigations of the co-variations of CH_4, CO and CO_2, as well as their interannual variability, that will provide a better understanding of their interactions. We have shown that coupling the three gases improves model consistency, along with our ability to identify source and sink fields that are over- or underestimated in the model. The model–measurement differences are heavily influenced by the existing uncertainties in a variety of carbon gas sources and sinks [10,46,69]. The new coupled simulation paves the way for future improvements, including the inclusion of a CH_4–OH–CO feedback, additional source/sink fields, improvements to the CO yield estimates from CH_4 oxidation and implementation into the GEOS-Chem Adjoint used for inverse modeling, that will further improve our ability to constrain the fluxes of the carbon gases. With updates such as this simulation, we will be able to better highlight and identify the origin of the model–measurement differences and constrain the sources, sinks and budgets of CO_2, CH_4 and CO, crucial for future climate projections and mitigation policies.

Supplementary Materials: The following supporting information can be downloaded at: https://www.mdpi.com/article/10.3390/atmos14050764/s1, Figure S1: Global production of CO from CH_4 (a, d), its loss via OH (b, e) and their difference (c, f) in the coupled (indigo) and *coupled-origOH* (turquoise) simulations at the surface (a–c) and 500 hPa altitude (d–f) for year 2006.; Table S1: GEOS-Chem emission inventories used for both the uncoupled and coupled carbon gas simulations.; Figure S2: Globally averaged OH fields at the surface (a) and at 500 hPa (b) from the v9-01-03 (indigo, used by uncoupled CO and coupled simulation), v8-02-01 (red, uncoupled CO_2) and v5-07-08 (turquoise, uncoupled CH_4 and *coupled-origOH*) full chemistry simulations.; Figure S3: Surface (a–c) and 500 hPa (d–f) yearly averaged global spatial distribution of the OH fields based on the v9-01-03 (a, d) full chemistry simulation and the difference between v5-07-08–v9-01-03 (b, e) and v8-02-01–v9-01-03 (c, f); Figure S4: Surface yearly averaged global spatial distribution of the OH fields based on the v9-01-03 (a, d, g, j) full chemistry simulation and the difference between v5-07-08–v9-01-03 (b, e, h, k) and v8-02-01–v9-01-03 (c, f, i, l) for each season: December–January–February

(DJF, a–c), March–April–May (MAM, d–f), June–July–August (JJA, g–i), September–October–November (SON, j–l).; Figure S5: Annual budgets of the global and regional CO production from CH_4 from the uncoupled (a) and coupled (b) simulation.; Figure S6: Annual budgets of the global and regional CO_2 production from the uncoupled (a) and coupled (b) simulations.; Figure S7: Average December–January–February (DJF) 2006–2017 total column CO chemical production from CH_4 (a), corresponding mole fractions (i.e., CO_{CH_4}) at the surface (b) and at 500 hPa (c) from the coupled simulation along with the difference in each field between the coupled and uncoupled simulation (d–f).; Figure S8: Average March–April–May (MAM) 2006–2017 total column CO chemical production from CH_4 (a), corresponding mole fractions (i.e., CO_{CH_4}) at the surface (b) and at 500 hPa (c) from the coupled simulation along with the difference in each field between the coupled and uncoupled simulation (d–f).; Figure S9: Average June–July–August (JJA) 2006–2017 total column CO chemical production from CH_4 (a), corresponding mole fractions (i.e., CO_{CH_4}) at the surface (b) and at 500 hPa (c) from the coupled simulation along with the difference in each field between the coupled and uncoupled simulation (d–f).; Figure S10: Average September–October–November (SON) 2006–2017 total column CO chemical production from CH_4 (a), corresponding mole fractions (i.e., CO_{CH_4}) at the surface (b) and at 500 hPa (c) from the coupled simulation along with the difference in each field between the coupled and uncoupled simulation (d–f).; Figure S11: Average December–January–February (DJF) 2006–2017 total column CO_2 chemical production from CO (a) and corresponding mole fractions (i.e., CO_{2CO}) at the surface (b) and at 500 hPa (c) from the coupled simulation along with the difference in each field between the coupled and uncoupled simulation (d–f).; Figure S12: Average March–April–May (MAM) 2006–2017 total column CO_2 chemical production from CO (a) and corresponding mole fractions (i.e., CO_{2CO}) at the surface (b) and at 500 hPa (c) from the coupled simulation along with the difference in each field between the coupled and uncoupled simulation (d–f).; Figure S13: Average June–July–August (JJA) 2006–2017 total column CO_2 chemical production from CO (a) and corresponding mole fractions (i.e., CO_{2CO}) at the surface (b) and at 500 hPa (c) from the coupled simulation along with the difference in each field between the coupled and uncoupled simulation (d–f).; Figure S14: Average September–October–November (SON) 2006–2017 total column CO_2 chemical production from CO (a) and corresponding mole fractions (i.e., CO_{2CO}) at the surface (b) and at 500 hPa (c) from the coupled simulation along with the difference in each field between the coupled and uncoupled simulation (d–f).; Figure S15: Modelled CO (red-uncoupled, indigo-coupled) comparison with column measurements (black) at different TCCON sites (top plots), based on monthly average values. Note, the plots show the scaled CO TCCON values (see main text for details). The bottom plots represent the mixing ratios of the CO production from CH_4 from the different simulations. Note, Equation (14) (main text) cannot be directly used when calculating the column-averaged dry-air mole fractions of the chemical terms (i.e., CO_{2CO} and CO_{CH_4}) since the a priori (x_a) represents the profile of the total amount of each gas and has no information about the individual source contributions. The contribution of the a priori profiles is excluded for the calculation of CO_{CH_4}, which is converted to column-averaged dry-air mole fractions according to $c_s = h^T a^T x_m$.; Figure S16: Modelled CO_2 (red-uncoupled, indigo-coupled) comparison with column measurements (black) at different TCCON sites (top plots), based on monthly average values. The detrended values are shown in the middle plots. The bottom plots represent the detrended mixing ratios of the CO_2 production from CO from the different simulations. Note, Equation (14) (main text) cannot be directly used when calculating the column-averaged dry-air mole fractions of the chemical terms (i.e., CO_{2CO} and CO_{CH_4}) since the a priori (x_a) represents the profile of the total amount of each gas and has no information about the individual source contributions. The contribution of the a priori profiles is excluded for the calculation of CO_{2CO}, which is converted to column-averaged dry-air mole fractions according to $c_s = h^T a^T x_m$.; Figure S17: Modelled CO (red-uncoupled, indigo-coupled) comparison with surface measurements (black, top plots), based on monthly average values. The bottom plots represent the mixing ratios of the CO production from CH_4 from the different simulations.; Figure S18: Modelled CO_2 (red-uncoupled, indigo-coupled) comparison with surface measurements (black, top plots), based on monthly average values. The detrended values are shown in the middle plots. The bottom plots represent the detrended mixing ratios of the CO_2 production from CO from the different simulations.; Figure S19: Model-measurement differences for CO (a, b) and CO_2 (c, d) based on simulated values from the uncoupled (a,c) and coupled (b,d) simulations during the four ATom campaigns.; Figure S20: Altitude versus latitude cross-sections of chemically produced CO from CH_4 (CO_{CH_4}, a, b, e, f, i, j, m, n) and CO_2 (CO_{2CO}, c, d, g, h, k, l, o, p) mole fractions from the coupled simulation along with the uncoupled differences relative to the coupled simulation during the four ATom campaigns. References [70–82] are cited in the supplementary materials.

Author Contributions: Conceptualization, B.B., J.A.F., N.M.D. and D.B.A.J.; methodology, B.B., J.A.F., N.M.D. and D.B.A.J.; software, B.B.; validation, B.B.; formal analysis, B.B.; investigation, B.B.; data curation, B.B.; writing—original draft preparation, B.B., J.A.F., N.M.D. and D.B.A.J.; visualization, B.B.; supervision, N.M.D. and J.A.F.; funding acquisition, N.M.D. All authors have read and agreed to the published version of the manuscript.

Funding: This research was funded by the Australian Research Council grant number DE140100178, DP160101598 and FT180100327, Discovery Early Career Researcher (DECRA) University Postgraduate Award from the University of Wollongong, assistance of resources provided at the NCI National Facility systems at the Australian National University through the National Computational Merit Allocation Scheme supported by the Australian government grant number m19 and funding from New Zealand's Ministry of Business, Innovation and Employment through contract number C01X1817.

Institutional Review Board Statement: Not applicable.

Informed Consent Statement: Not applicable.

Data Availability Statement: GEOS-Chem is an open-source model and the coupled simulation is being implemented in an upcoming newer version of GEOS-Chem. The original v12.1.1 model is publicly available at https://doi.org/10.5281/zenodo.2249246 (accessed on 10 March 2023) and the implementation of the coupled simulation and results are available from the authors upon request. TCCON data are publicly available at https://tccondata.org/ (accessed on 10 March 2023). NOAA GGGRN surface data are publicly available at https://www.esrl.noaa.gov/gmd/dv/data/ (accessed on 10 March 2023). ATom data are publicly available at https://doi.org/10.3334/ORNLDAAC/1581. (accessed on 10 March 2023).

Acknowledgments: We thank Clare Paton-Walsh and Sara E. Mikaloff-Fletcher for valuable discussions about this paper. We thank the TCCON PIs and funding agencies for the TCCON measurements. We thank the NOAA ESRL Global Monitoring Laboratory, Boulder, Colorado, USA, for providing the different surface site data. We acknowledge the ATom Team for providing the aircraft data from all four campaigns.

Conflicts of Interest: The authors declare no conflict of interest.

Appendix A

Table A1. Column and surface stations used for the coupled simulation validation. Sites are ordered based on latitude, from highest to lowest.

Station	Latitude	Longitude	Elevation (m)
TCCON sites			
Eureka [1]	80.05° N	86.42° W	610
Ny Alesund [2]	78.90° N	11.89° E	20
Sodankyla [3]	67.37° N	26.63° E	188
Białystok [4]	53.23° N	23.02° E	180
Bremen [5]	53.10° N	8.85° E	27
Karlsruhe [6]	49.10° N	8.43° E	116
Orléans [7]	47.97° N	2.11° E	130
Garmisch [8]	47.48° N	11.06° E	740
Rikubetsu [9]	43.46° N	143.77° E	380
Lamont [10]	36.60° N	97.49° W	320
Anmyeondo [11]	36.54° N	126.33° E	30
Tsukuba [12]	36.05° N	140.12° E	30
Edwards [13]	34.96° N	117.88° W	699
Saga [14]	33.24° N	130.29° E	7
Burgos [15]	18.53° N	120.62° E	35

Table A1. *Cont.*

Station	Latitude	Longitude	Elevation (m)
Manaus [16]	3.21° S	60.60° W	50
Darwin [17]	12.43° S	130.89° E	30
Reunion Island [18]	20.90° S	55.48° E	87
Wollongong [19]	34.41° S	150.88° E	30
Lauder [20]	45.04° S	169.68° E	370
Both TCCON sites and surface [24,25,26]			
Park Falls [21]	45.94° N	90.27° W	440
Izana [22]	28.30° N	16.50° W	2370
Ascension Island [23]	7.91° S	14.33° W	10
Surface sites [24,25,26]			
Alert	82.45° N	62.51° W	185
Summit	72.50° N	38.42° W	3209
Barrow	71.32° N	156.61° W	11
Pallas Sammaltunturi	67.97° N	24.12° E	565
Mace Head	53.33° N	9.89° W	5
Trinidad Head	41.06° N	124.15° W	107
Mt. Waliguan	36.29° N	100.89° E	3810
Assekrem	23.26° N	5.63° E	2710
Mauna Loa	19.53° N	155.58° W	3397
Christmas Island	1.70° N	157.15° W	0
Tutuila	14.25° S	170.56° W	42
Easter Island	27.16° S	109.43° W	47
Cape Grim	40.67° S	144.69° E	94
Baring Head	41.41° S	174.87° E	85
Crozet	46.43° S	51.84° E	197
Palmer Station	64.77° S	64.05° W	10
South Pole	89.98° S	24.80° W	2810

[1] Stronget al. [83] [2] Notholt et al. [84] [3] Kivi et al. [85] [4] Deutscher et al. [86] [5] Notholt et al. [87] [6] Hase et al. [88] [7] Warneke et al. [89] [8] Sussmann and Rettinger [90] [9] Morino et al. [91] [10] Wennberg et al. [92] [11] Goo et al. [93] [12] Morino et al. [94] [13] Iraci et al. [95] [14] Kawakami et al. [96] [15] Morino et al. [97] [16] Dubey et al. [98] [17] Griffith et al. [99] [18] De Mazière et al. [100] [19] Griffith et al. [101] [20] Sherlock et al. [102] [21] Wennberg et al. [103] [22] Blumenstock et al. [104] [23] Feistet al. [105] [24] CO_2: Dlugokencky et al. [53] [25] CH_4: Dlugokencky et al. [53] [26] CO: Petron et al. [54].

References

1. Stocker, T.; Qin, D.; Plattner, G.K.; Tignor, M.; Allen, S.; Boschung, J.; Nauels, A.; Xia, Y.; Bex, V.; Midgley, P. *IPCC, 2013: Climate Change 2013: The Physical Science Basis. Contribution of Working Group I to the Fifth Assessment Report of the Intergovernmental Panel on Climate Change*; Cambridge University Press: Cambridge, UK, 2014.
2. Shindell, D.T.; Faluvegi, G.; Bell, N.; Schmidt, G.A. An emissions-based view of climate forcing by methane and tropospheric ozone. *Geophys. Res. Lett.* **2005**, *32*. [CrossRef]
3. Bousquet, P.; Ciais, P.; Miller, J.; Dlugokencky, E.; Hauglustaine, D.; Prigent, C.; Van der Werf, G.; Peylin, P.; Brunke, E.G.; Carouge, R.; et al. Contribution of anthropogenic and natural sources to atmospheric methane variability. *Nature* **2006**, *443*, 439–443. [CrossRef]
4. Duncan, B.N.; Logan, J.A.; Bey, I.; Megretskaia, I.A.; Yantosca, R.M.; Novelli, P.C.; Jones, N.B.; Rinsland, C.P. Global budget of CO, 1988–1997: Source estimates and validation with a global model. *J. Geophys. Res. Atmos.* **2007**, *112*. [CrossRef]
5. Liu, J.; Bowman, K.W.; Schimel, D.S.; Parazoo, N.C.; Jiang, Z.; Lee, M.; Bloom, A.A.; Wunch, D.; Frankenberg, C.; Sun, Y.; et al. Contrasting carbon cycle responses of the tropical continents to the 2015–2016 El Niño. *Science* **2017**, *358*, eaam5690. [CrossRef]
6. Bloom, A.A.; Bowman, K.W.; Lee, M.; Turner, A.J.; Schroeder, R.; Worden, J.R.; Weidner, R.; McDonald, K.C.; Jacob, D.J. A global wetland methane emissions and uncertainty dataset for atmospheric chemical transport models (WetCHARTs version 1.0). *Geosci. Model Dev.* **2017**, *10*, 2141–2156. [CrossRef]
7. Turner, A.J.; Jacob, D.J.; Wecht, K.J.; Maasakkers, J.D.; Lundgren, E.; Andrews, A.E.; Biraud, S.C.; Boesch, H.; Bowman, K.W.; Deutscher, N.M.; et al. Estimating global and North American methane emissions with high spatial resolution using GOSAT satellite data. *Atmos. Chem. Phys.* **2015**, *15*, 7049–7069. [CrossRef]

8. Wang, J.S.; Logan, J.A.; McElroy, M.B.; Duncan, B.N.; Megretskaia, I.A.; Yantosca, R.M. A 3-D model analysis of the slowdown and interannual variability in the methane growth rate from 1988 to 1997. *Glob. Biogeochem. Cycles* **2004**, *18*. [CrossRef]
9. Messerschmidt, J.; Parazoo, N.; Wunch, D.; Deutscher, N.M.; Roehl, C.; Warneke, T.; Wennberg, P.O. Evaluation of seasonal atmosphere–biosphere exchange estimations with TCCON measurements. *Atmos. Chem. Phys.* **2013**, *13*, 5103–5115. [CrossRef]
10. Bukosa, B.; Deutscher, N.M.; Fisher, J.A.; Kubistin, D.; Paton-Walsh, C.; Griffith, D.W.T. Simultaneous shipborne measurements of CO_2, CH_4 and CO and their application to improving greenhouse-gas flux estimates in Australia. *Atmos. Chem. Phys.* **2019**, *19*, 7055–7072. [CrossRef]
11. Kopacz, M.; Jacob, D.J.; Fisher, J.A.; Logan, J.A.; Zhang, L.; Megretskaia, I.A.; Yantosca, R.M.; Singh, K.; Henze, D.K.; Burrows, J.P.; et al. Global estimates of CO sources with high resolution by adjoint inversion of multiple satellite datasets (MOPITT, AIRS, SCIAMACHY, TES). *Atmos. Chem. Phys.* **2010**, *10*, 855–876. [CrossRef]
12. Palmer, P.I.; Jacob, D.J.; Jones, D.B.A.; Heald, C.L.; Yantosca, R.M.; Logan, J.A.; Sachse, G.W.; Streets, D.G. Inverting for emissions of carbon monoxide from Asia using aircraft observations over the western Pacific. *J. Geophys. Res. Atmos.* **2003**, *108*, 8828. [CrossRef]
13. Enting, I.G.; Mansbridge, J.V. Latitudinal distribution of sources and sinks of CO_2: Results of an inversion study. *Tellus B* **1991**, *43*, 156–170. [CrossRef]
14. Suntharalingam, P.; Randerson, J.T.; Krakauer, N.; Logan, J.A.; Jacob, D.J. Influence of reduced carbon emissions and oxidation on the distribution of atmospheric CO_2: Implications for inversion analyses. *Glob. Biogeochem. Cycles* **2005**, *19*, GB4003. [CrossRef]
15. Nassar, R.; Jones, D.B.A.; Suntharalingam, P.; Chen, J.M.; Andres, R.J.; Wecht, K.J.; Yantosca, R.M.; Kulawik, S.S.; Bowman, K.W.; Worden, J.R.; et al. Modeling global atmospheric CO_2 with improved emission inventories and CO_2 production from the oxidation of other carbon species. *Geosci. Model Dev.* **2010**, *3*, 689–716. [CrossRef]
16. Wecht, K.J.; Jacob, D.J.; Frankenberg, C.; Jiang, Z.; Blake, D.R. Mapping of North American methane emissions with high spatial resolution by inversion of SCIAMACHY satellite data. *J. Geophys. Res. Atmos.* **2014**, *119*, 7741–7756. [CrossRef]
17. Fisher, J.A.; Murray, L.T.; Jones, D.B.A.; Deutscher, N.M. Improved method for linear carbon monoxide simulation and source attribution in atmospheric chemistry models illustrated using GEOS-Chem v9. *Geosci. Model Dev. Discuss.* **2017**, *2017*, 4129–4144. [CrossRef]
18. Jacob, D.J. *Introduction to Atmospheric Chemistry*; Princeton University Press: Princeton, NJ, USA, 1999.
19. Isaksen, I.S.; Hov, Ø. Calculation of trends in the tropospheric concentration of O_3, OH, CO, CH_4 and NO_x. *Tellus B* **1987**, *39B*, 271–285. [CrossRef]
20. Folberth, G.; Hauglustaine, D.A.; Ciais, P.; Lathière, J. On the role of atmospheric chemistry in the global CO_2 budget. *Geophys. Res. Lett.* **2005**, *32*, L08801. [CrossRef]
21. Logan, J.A.; Prather, M.J.; Wofsy, S.C.; McElroy, M.B. Tropospheric chemistry: A global perspective. *J. Geophys. Res. Ocean.* **1981**, *86*, 7210–7254. [CrossRef]
22. Tie, X.; Jim Kao, C.Y.; Mroz, E. Net yield of OH, CO, and O_3 from the oxidation of atmospheric methane. *Atmos. Environ. Part A Gen. Top.* **1992**, *26*, 125–136. [CrossRef]
23. Manning, M.R.; Brenninkmeijer, C.A.M.; Allan, W. Atmospheric carbon monoxide budget of the southern hemisphere: Implications of 13C/12C measurements. *J. Geophys. Res. Atmos.* **1997**, *102*, 10673–10682. [CrossRef]
24. Novelli, P.C.; Lang, P.M.; Masarie, K.A.; Hurst, D.F.; Myers, R.; Elkins, J.W. Molecular hydrogen in the troposphere: Global distribution and budget. *J. Geophys. Res. Atmos.* **1999**, *104*, 30427–30444. [CrossRef]
25. Bergamaschi, P.; Hein, R.; Brenninkmeijer, C.A.M.; Crutzen, P.J. Inverse modeling of the global CO cycle: 2. Inversion of 13C/12C and 18O/16O isotope ratios. *J. Geophys. Res. Atmos.* **2000**, *105*, 1929–1945. [CrossRef]
26. Franco, B.; Blumenstock, T.; Cho, C.; Clarisse, L.; Clerbaux, C.; Coheur, P.F.; De Mazière, M.; De Smedt, I.; Dorn, H.P.; Emmerichs, T.; et al. Ubiquitous atmospheric production of organic acids mediated by cloud droplets. *Nature* **2021**, *593*, 233–237. [CrossRef] [PubMed]
27. Holloway, T.; Levy II, H.; Kasibhatla, P. Global distribution of carbon monoxide. *J. Geophys. Res. Atmos.* **2000**, *105*, 12123–12147. [CrossRef]
28. Arellano, A.F., Jr.; Hess, P.G. Sensitivity of top-down estimates of CO sources to GCTM transport. *Geophys. Res. Lett.* **2006**, *33*, L21807. [CrossRef]
29. Stein, O.; Schultz, M.G.; Bouarar, I.; Clark, H.; Huijnen, V.; Gaudel, A.; George, M.; Clerbaux, C. On the wintertime low bias of Northern Hemisphere carbon monoxide found in global model simulations. *Atmos. Chem. Phys.* **2014**, *14*, 9295–9316. [CrossRef]
30. Zeng, G.; Williams, J.E.; Fisher, J.A.; Emmons, L.K.; Jones, N.B.; Morgenstern, O.; Robinson, J.; Smale, D.; Paton-Walsh, C.; Griffith, D.W.T. Multi-model simulation of CO and HCHO in the Southern Hemisphere: Comparison with observations and impact of biogenic emissions. *Atmos. Chem. Phys.* **2015**, *15*, 7217–7245. [CrossRef]
31. Pétron, G.; Granier, C.; Khattatov, B.; Yudin, V.; Lamarque, J.F.; Emmons, L.; Gille, J.; Edwards, D.P. Monthly CO surface sources inventory based on the 2000–2001 MOPITT satellite data. *Geophys. Res. Lett.* **2004**, *31*, L21107. [CrossRef]
32. Le Quéré, C.; Andrew, R.M.; Friedlingstein, P.; Sitch, S.; Hauck, J.; Pongratz, J.; Pickers, P.A.; Korsbakken, J.I.; Peters, G.P.; Canadell, J.G.; et al. Global Carbon Budget 2018. *Earth Syst. Sci. Data* **2018**, *10*, 2141–2194. [CrossRef]
33. Ciais, P.; Borges, A.V.; Abril, G.; Meybeck, M.; Folberth, G.; Hauglustaine, D.; Janssens, I.A. The impact of lateral carbon fluxes on the European carbon balance. *Biogeosciences* **2008**, *5*, 1259–1271. [CrossRef]

34. Elshorbany, Y.F.; Duncan, B.N.; Strode, S.A.; Wang, J.S.; Kouatchou, J. The description and validation of the computationally Efficient CH_4–CO–OH (ECCOHv1.01) chemistry module for 3-D model applications. *Geosci. Model Dev.* **2016**, *9*, 799–822. [CrossRef]
35. Pison, I.; Bousquet, P.; Chevallier, F.; Szopa, S.; Hauglustaine, D. Multi-species inversion of CH_4, CO and H_2 emissions from surface measurements. *Atmos. Chem. Phys.* **2009**, *9*, 5281–5297. [CrossRef]
36. Wang, H.; Jacob, D.J.; Kopacz, M.; Jones, D.B.A.; Suntharalingam, P.; Fisher, J.A.; Nassar, R.; Pawson, S.; Nielsen, J.E. Error correlation between CO_2 and CO as constraint for CO_2 flux inversions using satellite data. *Atmos. Chem. Phys.* **2009**, *9*, 7313–7323. [CrossRef]
37. Pandey, S.; Houweling, S.; Krol, M.; Aben, I.; Röckmann, T. On the use of satellite-derived $CH_4 : CO_2$ columns in a joint inversion of CH_4 and CO_2 fluxes. *Atmos. Chem. Phys.* **2015**, *15*, 8615–8629. [CrossRef]
38. Nassar, R.; Napier-Linton, L.; Gurney, K.R.; Andres, R.J.; Oda, T.; Vogel, F.R.; Deng, F. Improving the temporal and spatial distribution of CO_2 emissions from global fossil fuel emission data sets. *J. Geophys. Res. Atmos.* **2013**, *118*, 917–933. [CrossRef]
39. Maasakkers, J.D.; Jacob, D.J.; Sulprizio, M.P.; Scarpelli, T.R.; Nesser, H.; Sheng, J.X.; Zhang, Y.; Hersher, M.; Bloom, A.A.; Bowman, K.W.; et al. Global distribution of methane emissions, emission trends, and OH concentrations and trends inferred from an inversion of GOSAT satellite data for 2010–2015. *Atmos. Chem. Phys.* **2019**, *19*, 7859–7881. [CrossRef]
40. Considine, D.B.; Logan, J.A.; Olsen, M.A. Evaluation of near-tropopause ozone distributions in the Global Modeling Initiative combined stratosphere/troposphere model with ozonesonde data. *Atmos. Chem. Phys.* **2008**, *8*, 2365–2385. [CrossRef]
41. Allen, D.; Pickering, K.; Duncan, B.; Damon, M. Impact of lightning NO emissions on North American photochemistry as determined using the Global Modeling Initiative (GMI) model. *J. Geophys. Res. Atmos.* **2010**, *115*, D22301. [CrossRef]
42. Murray, L.T.; Jacob, D.J.; Logan, J.A.; Hudman, R.C.; Koshak, W.J. Optimized regional and interannual variability of lightning in a global chemical transport model constrained by LIS/OTD satellite data. *J. Geophys. Res. Atmos.* **2012**, *117*, D20307. [CrossRef]
43. Burkholder, J.B.; Sander, S.P.; Abbatt, J.; Barker, J.R.; Huie, R.E.; Kolb, C.E.; Kurylo, M.J.; Orkin, V.L.; Wilmouth, D.M.; Wine, P.H. *Chemical Kinetics and Photochemical Data for Use in Atmospheric Studies, Evaluation No. 18*; JPL Publication 15-10; Jet Propulsion Laboratory: Pasadena, TX, USA, 2015.
44. Park, R.J.; Jacob, D.J.; Field, B.D.; Yantosca, R.M.; Chin, M. Natural and transboundary pollution influences on sulfate-nitrate-ammonium aerosols in the United States: Implications for policy. *J. Geophys. Res. Atmos.* **2004**, *109*, D15204. [CrossRef]
45. Darmenov, A.; da Silva, A. The quick fire emissions dataset (QFED)–documentation of versions 2.1, 2.2 and 2.4. *NASA Technical Report Series on Global Modeling and Data Assimilation*; NASA TM-2013-104606; NASA: Washington, DC, USA, 2015; Volume 32, p. 183.
46. Dlugokencky, E.J.; Nisbet, E.G.; Fisher, R.; Lowry, D. Global atmospheric methane: Budget, changes and dangers. *Philos. Trans. R. Soc. A Math. Phys. Eng. Sci.* **2011**, *369*, 2058–2072. [CrossRef] [PubMed]
47. Hodson, E.L.; Poulter, B.; Zimmermann, N.E.; Prigent, C.; Kaplan, J.O. The El Niño–Southern Oscillation and wetland methane interannual variability. *Geophys. Res. Lett.* **2011**, *38*, L08810. [CrossRef]
48. Schaefer, H.; Smale, D.; Nichol, S.E.; Bromley, T.M.; Brailsford, G.W.; Martin, R.J.; Moss, R.; Englund Michel, S.; White, J.W.C. Limited impact of El Niño–Southern Oscillation on variability and growth rate of atmospheric methane. *Biogeosciences* **2018**, *15*, 6371–6386. [CrossRef]
49. Rowlinson, M.J.; Rap, A.; Arnold, S.R.; Pope, R.J.; Chipperfield, M.P.; McNorton, J.; Forster, P.; Gordon, H.; Pringle, K.J.; Feng, W.; et al. Impact of El Niño–Southern Oscillation on the interannual variability of methane and tropospheric ozone. *Atmos. Chem. Phys.* **2019**, *19*, 8669–8686. [CrossRef]
50. Holmes, C.D. Methane Feedback on Atmospheric Chemistry: Methods, Models, and Mechanisms. *J. Adv. Model. Earth Syst.* **2018**, *10*, 1087–1099. [CrossRef]
51. Edwards, D.P.; Emmons, L.K.; Gille, J.C.; Chu, A.; Attié, J.L.; Giglio, L.; Wood, S.W.; Haywood, J.; Deeter, M.N.; Massie, S.T.; et al. Satellite-observed pollution from Southern Hemisphere biomass burning. *J. Geophys. Res. Atmos.* **2006**, *111*, D14312. [CrossRef]
52. Wunch, D.; Toon, G.C.; Blavier, J.F.L.; Washenfelder, R.A.; Notholt, J.; Connor, B.J.; Griffith, D.W.T.; Sherlock, V.; Wennberg, P.O. The Total Carbon Column Observing Network. *Philos. Trans. R. Soc. A Math. Phys. Eng. Sci.* **2011**, *369*, 2087–2112. [CrossRef]
53. Dlugokencky, E.J.; Mund, J.W.; Crotwell, A.M.; Crotwell, M.J.; Thoning, K.W. Atmospheric Carbon Dioxide Dry Air Mole Fractions from the NOAA GML Carbon Cycle Cooperative Global Air Sampling Network, 1968–2019. Version: 2020-07. *arXiv* **2020**, arXiv:2012.00839. https://doi.org/10.15138/wkgj-f215.
54. Petron, G.; Crotwell, A.M.; Crotwell, M.J.; Dlugokencky, E.J.; Madronich, M.; Moglia, E.; Neff, D.; Wolter, S.; Mund, J. Atmospheric Carbon Monoxide Dry Air Mole Fractions from the NOAA GML Carbon Cycle Cooperative Global Air Sampling Network, 1988–2020. Version: 2020-08. 2020. Available online: https://gml.noaa.gov/ccgg/arc/?id=132 (accessed on 10 March 2023).
55. Wofsy, S.; Afshar, S.; Allen, H.; Apel, E.; Asher, E.; Barletta, B.; Bent, J.; Bian, H.; Biggs, B.; Blake, D.; et al. ATom: Merged Atmospheric Chemistry, Trace Gases, and Aerosols. *ORNL DAAC* **2018**. [CrossRef]
56. Wunch, D.; Toon, G.C.; Wennberg, P.O.; Wofsy, S.C.; Stephens, B.B.; Fischer, M.L.; Uchino, O.; Abshire, J.B.; Bernath, P.; Biraud, S.C.; et al. Calibration of the Total Carbon Column Observing Network using aircraft profile data. *Atmos. Meas. Tech.* **2010**, *3*, 1351–1362. [CrossRef]
57. Deutscher, N.M.; Sherlock, V.; Mikaloff Fletcher, S.E.; Griffith, D.W.T.; Notholt, J.; Macatangay, R.; Connor, B.J.; Robinson, J.; Shiona, H.; Velazco, V.A.; et al. Drivers of column-average CO_2 variability at Southern Hemispheric Total Carbon Column Observing Network sites. *Atmos. Chem. Phys.* **2014**, *14*, 9883–9901. [CrossRef]

58. Té, Y.; Jeseck, P.; Franco, B.; Mahieu, E.; Jones, N.; Paton-Walsh, C.; Griffith, D.W.T.; Buchholz, R.R.; Hadji-Lazaro, J.; Hurtmans, D.; et al. Seasonal variability of surface and column carbon monoxide over the megacity Paris, high-altitude Jungfraujoch and Southern Hemispheric Wollongong stations. *Atmos. Chem. Phys.* **2016**, *16*, 10911–10925. [CrossRef]
59. Hedelius, J.K.; He, T.L.; Jones, D.B.A.; Baier, B.C.; Buchholz, R.R.; De Mazière, M.; Deutscher, N.M.; Dubey, M.K.; Feist, D.G.; Griffith, D.W.T.; et al. Evaluation of MOPITT Version 7 joint TIR–NIR X_{CO} retrievals with TCCON. *Atmos. Meas. Tech.* **2019**, *12*, 5547–5572. [CrossRef]
60. Zhou, M.; Langerock, B.; Vigouroux, C.; Sha, M.K.; Hermans, C.; Metzger, J.M.; Chen, H.; Ramonet, M.; Kivi, R.; Heikkinen, P.; et al. TCCON and NDACC X_CO measurements: Difference, discussion and application. *Atmos. Meas. Tech.* **2019**, *12*, 5979–5995. [CrossRef]
61. Schuh, A.E.; Jacobson, A.R.; Basu, S.; Weir, B.; Baker, D.; Bowman, K.; Chevallier, F.; Crowell, S.; Davis, K.J.; Deng, F.; et al. Quantifying the impact of atmospheric transport uncertainty on CO2 surface flux estimates. *Glob. Biogeochem. Cycles* **2019**, *33*, 484–500. [CrossRef] [PubMed]
62. Schuh, A.E.; Byrne, B.; Jacobson, A.R.; Crowell, S.M.; Deng, F.; Baker, D.F.; Johnson, M.S.; Philip, S.; Weir, B. On the role of atmospheric model transport uncertainty in estimating the Chinese land carbon sink. *Nature* **2022**, *603*, E13–E14. [CrossRef]
63. Stanevich, I.; Jones, D.B.A.; Strong, K.; Parker, R.J.; Boesch, H.; Wunch, D.; Notholt, J.; Petri, C.; Warneke, T.; Sussmann, R.; et al. Characterizing model errors in chemical transport modeling of methane: Impact of model resolution in versions v9-02 of GEOS-Chem and v35j of its adjoint model. *Geosci. Model Dev.* **2020**, *13*, 3839–3862. [CrossRef]
64. Graham, K.A.; Holmes, C.D.; Friedrich, G.; Rauschenberg, C.D.; Williams, C.R.; Bottenheim, J.W.; Chavez, F.P.; Halfacre, J.W.; Perovich, D.K.; Shepson, P.B.; et al. Variability of Atmospheric CO_2 over the Arctic Ocean: Insights from the O-Buoy Chemical Observing Network. *J. Geophys. Res. Atmos.* **2023**, *128*, e2022JD036437. [CrossRef]
65. Søgaard, D.H.; Thomas, D.N.; Rysgaard, S.; Glud, R.N.; Norman, L.; Kaartokallio, H.; Juul-Pedersen, T.; Geilfus, N.X. The relative contributions of biological and abiotic processes to carbon dynamics in subarctic sea ice. *Polar Biol.* **2013**, *36*, 1761–1777. [CrossRef]
66. Desservettaz, M.J.; Fisher, J.A.; Luhar, A.K.; Woodhouse, M.T.; Bukosa, B.; Buchholz, R.R.; Wiedinmyer, C.; Griffith, D.W.T.; Krummel, P.B.; Jones, N.B.; et al. Australian Fire Emissions of Carbon Monoxide Estimated by Global Biomass Burning Inventories: Variability and Observational Constraints. *J. Geophys. Res. Atmos.* **2022**, *127*, e2021JD035925. [CrossRef]
67. Su, M.; Shi, Y.; Yang, Y.; Guo, W. Impacts of different biomass burning emission inventories: Simulations of atmospheric CO_2 concentrations based on GEOS-Chem. *Sci. Total Environ.* **2023**, *876*, 162825. [CrossRef] [PubMed]
68. Gaubert, B.; Emmons, L.K.; Raeder, K.; Tilmes, S.; Miyazaki, K.; Arellano, A.F., Jr.; Elguindi, N.; Granier, C.; Tang, W.; Barré, J.; et al. Correcting model biases of CO in East Asia: Impact on oxidant distributions during KORUS-AQ. *Atmos. Chem. Phys.* **2020**, *20*, 14617–14647. [CrossRef] [PubMed]
69. Bastos, A.; O'Sullivan, M.; Ciais, P.; Makowski, D.; Sitch, S.; Friedlingstein, P.; Chevallier, F.; Rödenbeck, C.; Pongratz, J.; Luijkx, I.T.; et al. Sources of Uncertainty in Regional and Global Terrestrial CO_2 Exchange Estimates. *Glob. Biogeochem. Cycles* **2020**, *34*, e2019GB006393. [CrossRef]
70. Baker, D.F.; Law, R.M.; Gurney, K.R.; Rayner, P.; Peylin, P.; Denning, A.S.; Bousquet, P.; Bruhwiler, L.; Chen, Y.-H.; Ciais, P.; et al. TransCom 3 inversion intercomparison: Impact of transport model errors on the interannual variability of regional CO_2 fluxes, 1988–2003. *Glob. Biogeochem.* **2006**, *20*. [CrossRef]
71. Fung, I.; John, J.; Lerner, J.; Matthews, E.; Prather, M.; Steele, L.P.; Fraser, P.J. Three-dimensional model synthesis of the global methane cycle. *J. Geophys. Res. Atmos.* **1991**, *96*, 13033–13065., 1991. [CrossRef]
72. Kuhns, H.; Knipping, E.M.; Vukovich, J.M. Development of a United States–Mexico emissions inventory for the big bend regional aerosol and visibility observational (BRAVO) study. *J. Air Waste Manag. Assoc.* **2005**, *55*, 677–692. [CrossRef]
73. Lee, C.; Martin, R.V.; van Donkelaar, A.; Lee, H.; Dickerson, R.R.; Hains, J.C.; Krotkov, N.; Richter, A.; Vinnikov, K.; Schwab, J.J. SO_2 emissions and lifetimes: Estimates from inverse modeling using in situ and global, space-based (SCIAMACHY and OMI) 60 observations. *J. Geophys. Res. Atmos.* **2011**, *116*. [CrossRef]
74. Li, M.; Zhang, Q.; Kurokawa, J.-I.; Woo, J.-H.; He, K.; Lu, Z.; Ohara, T.; Song, Y.; Streets, D.G.; Carmichael, G.R.; et al. MIX: A mosaic Asian anthropogenic emission inventory under the international collaboration framework of the MICS-Asia and HTAP. *Atmos. Chem. Phys.* **2017**, *17*, 935–963. 935-2017. [CrossRef]
75. Maasakkers, J.D.; Jacob, D.J.; Sulprizio, M.P.; Turner, A.J.; Weitz, M.; Wirth, T.; Hight, C.; DeFigueiredo, M.; Desai, M.; Schmeltz, R.; et al. Gridded national inventory of US methane emissions. *Environ. Sci. Technol.* **2016**, *50*, 13123–13133. [CrossRef]
76. Oda, T.; Maksyutov, S. A very high-resolution (1 km × 1 km) global fossil fuel CO_2 emission inventory derived using a point source database and satellite observations of nighttime lights. *Atmos. Chem. Phys.* **2011**, *11*, 543–556. 543-2011. [CrossRef]
77. Sheng, J.-X.; Jacob, D.J.; Maasakkers, J.D.; Sulprizio, M.P.; Zavala-Araiza, D.; Hamburg, S.P. A high-resolution (0.1 × 0.1) inventory of methane emissions from Canadian and Mexican oil and gas systems. *Atmos. Environ.* **2017**, *158*, 211–215. [CrossRef]
78. Stettler, M.; Eastham, S.; Barrett, S. Air quality and public health impacts of UK airports. Part I: Emissions. *Atmos. Environ.* **2011**, *45*, 5415–5424. [CrossRef]
79. Takahashi, T.; Sutherl, S.C.; Wanninkhof, R.; Sweeney, C.; Feely, R.A.; Chipman, D.W.; Hales, B.; Friederich, G.; Chavez, F.; Sabine, C.; et al. Climatological mean and decadal change in surface ocean pCO_2, and net sea–air CO_2 flux over the global oceans. *Deep. Sea 85 Res. Part II Top. Stud. Oceanogr.* **2009**, *56*, 554–577. [CrossRef]

80. Van Donkelaar, A.; Martin, R.V.; Pasch, A.N.; Szykman, J.J.; Zhang, L.; Wang, Y.X.; Chen, D. Improving the accuracy of daily satellite-derived ground-level fine aerosol concentration estimates for North America. *Environ. Sci. Technol.* **2012**, *46*, 11971–11978. [CrossRef] [PubMed]
81. Vestreng, V.; Mareckova, K.; Kakareka, S.; Malchykhina, A.; Kukharchyk, T. Inventory Review 2007. In *Emission Data Reported to LRTAP Convention and NEC Directive, MSC-W Technical Report 1/07*; The Norwegian Meteorological Institute: Oslo, Norway, 2007.
82. Yevich, R.; Logan, J.A. An assessment of biofuel use and burning of agricultural waste in the developing world. *Glob. Biogeochem.* **2003**, *17*. [CrossRef]
83. Strong, K.; Roche, S.; Franklin, J.; Mendonca, J.; Lutsch, E.; Weaver, D.; Fogal, P.; Drummond, J.; Batchelor, R.; Lindenmaier, R. TCCON Data from Eureka (CA), Release GGG2014R3. TCCON Data Archive, Hosted by CaltechDATA. 2019. Available online: https://data.caltech.edu/records/m5vq1-3ga50 (accessed on 10 March 2023).
84. Notholt, J.; Schrems, O.; Warneke, T.; Deutscher, N.M.; Weinzierl, C.; Palm, M.; Buschmann, M. TCCON Data from Ny Ålesund, Spitsbergen (NO), Release GGG2014R1. TCCON Data Archive, Hosted by CaltechDATA. 2019. Available online: https://data.caltech.edu/records/vztb0-vsv44 (accessed on 10 March 2023).
85. Kivi, R.; Heikkinen, P.; Kyrö, E. TCCON Data from Sodankyla (FI), Release GGG2014R0. TCCON Data Archive, Hosted by CaltechDATA. 2014. Available online: https://data.caltech.edu/records/n2823-2yt07 (accessed on 10 March 2023).
86. Deutscher, N.M.; Notholt, J.; Messerschmidt, J.; Weinzierl, C.; Warneke, T.; Petri, C.; Grupe, P.; Katrynski, K. TCCON Data from Bialystok (PL), Release GGG2014R2. TCCON Data Archive, Hosted by CaltechDATA. 2015. Available online: https://data.caltech.edu/records/0cjh6-71m74 (accessed on 10 March 2023).
87. Notholt, J.; Petri, C.; Warneke, T.; Deutscher, N.M.; Buschmann, M.; Weinzierl, C.; Macatangay, R.; Grupe, P. TCCON Data from Bremen (DE), Release GGG2014R1. TCCON Data Archive, Hosted by CaltechDATA. 2019. Available online: https://data.caltech.edu/records/4hszb-99q28 (accessed on 10 March 2023).
88. Hase, F.; Blumenstock, T.; Dohe, S.; Gross, J.; Kiel, M. TCCON Data from Karlsruhe (DE), Release GGG2014R1. TCCON Data Archive, Hosted by CaltechDATA. 2015. Available online: https://data.caltech.edu/records/nhdv7-yfv69 (accessed on 10 March 2023).
89. Warneke, T.; Messerschmidt, J.; Notholt, J.; Weinzierl, C.; Deutscher, N.M.; Petri, C.; Grupe, P.; Vuillemin, C.; Truong, F.; Schmidt, M.; et al. TCCON Data from Orléans (FR), Release GGG2014R0. TCCON Data Archive, Hosted by CaltechDATA. 2014. Available online: https://data.caltech.edu/records/0n6jg-56q50 (accessed on 10 March 2023).
90. Sussmann, R.; Rettinger, M. TCCON Data from Garmisch (DE), Release GGG2014R2. TCCON Data Archive, Hosted by CaltechDATA. 2018. Available online: https://data.caltech.edu/records/7jdn6-vtg92 (accessed on 10 March 2023).
91. Morino, I.; Yokozeki, N.; Matzuzaki, T.; Horikawa, M. TCCON Data from Rikubetsu (JP), Release GGG2014R2. TCCON Data Archive, Hosted by CaltechDATA. 2018. Available online: https://data.caltech.edu/records/8db2k-rcp69 (accessed on 10 March 2023).
92. Wennberg, P.O.; Wunch, D.; Roehl, C.; Blavier, J.F.; Toon, G.C.; Allen, N.; Dowell, P.; Teske, K.; Martin, C.; Martin, J. TCCON Data from Lamont (US), Release GGG2014R1. TCCON Data Archive, Hosted by CaltechDATA. 2016. Available online: https://data.caltech.edu/records/cghqs-qp027 (accessed on 10 March 2023).
93. Goo, T.Y.; Oh, Y.S.; Velazco, V.A. TCCON Data from Anmeyondo (KR), Release GGG2014R0. TCCON Data Archive, Hosted by CaltechDATA. 2014. Available online: https://data.caltech.edu/records/dq9b4-19g61 (accessed on 10 March 2023).
94. Morino, I.; Matsuzaki, T.; Shishime, A. TCCON Data from Tsukuba (JP), 125HR, Release GGG2014R2. TCCON Data Archive, Hosted by CaltechDATA. 2018. Available online: https://data.caltech.edu/records/jgz9c-bwz67 (accessed on 10 March 2023).
95. Iraci, L.T.; Podolske, J.; Hillyard, P.W.; Roehl, C.; Wennberg, P.O.; Blavier, J.F.; Allen, N.; Wunch, D.; Osterman, G.B.; Albertson, R. TCCON Data from Edwards (US), Release GGG2014R1. TCCON Data Archive, Hosted by CaltechDATA. 2016. Available online: https://data.caltech.edu/records/2ssse-s7s41 (accessed on 10 March 2023).
96. Kawakami, S.; Ohyama, H.; Arai, K.; Okumura, H.; Taura, C.; Fukamachi, T.; Sakashita, M. TCCON Data from Saga (JP), Release GGG2014R0. TCCON Data Archive, Hosted by CaltechDATA. 2014. Available online: https://data.caltech.edu/records/q5ch6-1yp60 (accessed on 10 March 2023).
97. Morino, I.; Velazco, V.A.; Akihiro, H.; Osamu, U.; Griffith, D.W.T. TCCON Data from Burgos, Ilocos Norte (PH), Release GGG2014.R0. TCCON Data Archive, Hosted by CaltechDATA. 2018. Available online: https://data.caltech.edu/records/s3x4h-5pv98 (accessed on 10 March 2023).
98. Dubey, M.; Henderson, B.; Green, D.; Butterfield, Z.; Keppel-Aleks, G.; Allen, N.; Blavier, J.F.; Roehl, C.; Wunch, D.; Lindenmaier, R. TCCON Data from Manaus (BR), Release GGG2014R0. TCCON Data Archive, Hosted by CaltechDATA. 2014. Available online: https://data.caltech.edu/records/8pa5g-57r83 (accessed on 10 March 2023).
99. Griffith, D.W.; Deutscher, N.M.; Velazco, V.A.; Wennberg, P.O.; Yavin, Y.; Aleks, G.K.; Washenfelder, R.a.; Toon, G.C.; Blavier, J.F.; Murphy, C.; et al. TCCON Data from Darwin (AU), Release GGG2014R0. TCCON Data Archive, Hosted by CaltechDATA. 2014. Available online: https://data.caltech.edu/records/zmpsc-0ym23 (accessed on 10 March 2023).
100. De Mazière, M.; Sha, M.K.; Desmet, F.; Hermans, C.; Scolas, F.; Kumps, N.; Metzger, J.M.; Duflot, V.; Cammas, J.P. TCCON Data from Reunion Island (RE), Release GGG2014R0. TCCON Data Archive, Hosted by CaltechDATA. 2017. Available online: https://data.caltech.edu/records/k7c8c-6e426 (accessed on 10 March 2023).

101. Griffith, D.W.; Velazco, V.A.; Deutscher, N.M.; Murphy, C.; Jones, N.; Wilson, S.; Macatangay, R.; Kettlewell, G.; Buchholz, R.R.; Riggenbach, M. TCCON Data from Wollongong (AU), Release GGG2014R0. TCCON Data Archive, hosted by CaltechDATA. 2014. Available online: https://data.caltech.edu/records/aqkee-99p67 (accessed on 10 March 2023).
102. Sherlock, V.; Connor, B.J.; Robinson, J.; Shiona, H.; Smale, D.; Pollard, D. TCCON Data from Lauder (NZ), 125HR, Release GGG2014R0. TCCON Data Archive, Hosted by CaltechDATA. 2014. Available online: https://data.caltech.edu/records/2djsk-mpm66 (accessed on 10 March 2023).
103. Wennberg, P.O.; Roehl, C.; Wunch, D.; Toon, G.C.; Blavier, J.F.; Washenfelder, R.a.; Keppel-Aleks, G.; Allen, N.; Ayers, J. TCCON Data from Park Falls (US), Release GGG2014R1. TCCON Data Archive, Hosted by CaltechDATA. 2017. Available online: https://data.caltech.edu/records/ye7gn-2sc61 (accessed on 10 March 2023).
104. Blumenstock, T.; Hase, f.; Schneider, M.; Garcia, O.E.; Sepulveda, E. TCCON Data from Izana (ES), Release GGG2014R1. TCCON Data Archive, Hosted by CaltechDATA. 2017. Available online: https://data.caltech.edu/records/qf1ya-jjy76 (accessed on 10 March 2023).
105. Feist, D.G.; Arnold, S.G.; John, N.; Geibel, M.C. TCCON Data from Ascension Island (SH), Release GGG2014R0. TCCON Data Archive, Hosted by CaltechDATA. 2014. Available online: https://data.caltech.edu/records/rh1kp-b6c90 (accessed on 10 March 2023).

atmosphere

MDPI

Article

A Study on Radiological Hazard Assessment for Jordan Research and Training Reactor

Mohammad Talafha [1,2], Sora Kim [2] and Kyung-Suk Suh [1,2,*]

1 Department of Nuclear and Radiation Safety, University of Science and Technology, Daejeon 34113, Republic of Korea; mtalafha@kaeri.re.kr
2 Korea Atomic Energy Research Institute, Daejeon 34057, Republic of Korea; sorakim03@kaeri.re.kr
* Correspondence: kssuh@kaeri.re.kr

Abstract: Numerical simulations of atmospheric dispersion and dose assessment were performed for the Jordan Research and Training Reactor (JRTR) to evaluate its radiological effects on surrounding population and the environment. A three-dimensional atmospheric dispersion model was applied to investigate the behavior of the radionuclides released into the air, and a dose assessment model was used to estimate the radiological impact on the population residing in nearby cities around the JRTR. Considering full core meltdown an accidental scenario, most of the source term was assumed to be released from the JRTR. Simulations were performed to calculate the air and deposition concentrations of radioactive materials for July 2013 and January 2014. The monthly averaged values of concentrations, depositions, and dose rates were analyzed to identify the most harmful effects in each month. The results showed that relatively harmful effects occurred in January 2014, and the total annual dose rate was estimated to be approximately 1 mSv outside the 10 km radius from JRTR. However, the impact of a nuclear accident is not as severe as it might seem, as the affected area is not highly populated, and appropriate protective measures can significantly reduce the radiation exposure. This study provides useful information for emergency preparedness and response planning to mitigate the radiological consequences of a nuclear accident at the JRTR.

Keywords: atmospheric dispersion; nuclear accident; radionuclides; radiological impact

Citation: Talafha, M.; Kim, S.; Suh, K.-S. A Study on Radiological Hazard Assessment for Jordan Research and Training Reactor. *Atmosphere* **2023**, *14*, 859. https://doi.org/10.3390/atmos14050859

Academic Editor: Daniel Viúdez-Moreiras

Received: 6 April 2023
Revised: 8 May 2023
Accepted: 9 May 2023
Published: 11 May 2023

1. Introduction

The prompt and accurate evaluation of the environmental and health effects resulting from the release of radioactive materials into the atmosphere during nuclear accidents is crucial for implementing appropriate countermeasures to guarantee the safety of the surrounding population and environment [1]. The severity of these effects can vary widely depending on the release rate and duration of the radioactive materials, which can be transported over long distances and can contaminate wide areas around the accident sites, leading to long-term health problems and environmental damage. Therefore, an extensive environmental hazard assessment was performed to evaluate the potential radiological impact on the population residing in cities surrounding nuclear facilities.

Research reactors are constructed and operated for research and training purposes. They are typically used for the testing and development of nuclear fuel, reactor components, reactor instrumentation, and control systems, as well as for the training and education of personnel. They are also used for radioisotope production in medical and industrial applications. Research reactors operate at relatively low power levels, ranging from kilowatts to a few hundred megawatts [2].

The Jordan Research and Training Reactor (JRTR) was constructed and operated in Jordan in 2016. The JRTR is a 5 MW_{th} (upgradable to 10 MW_{th}) open-pool-type light-water reactor located at the Jordan University of Science and Technology (JUST), 70 km from the capital Amman in Jordan [3]. The JRTR is the first nuclear research reactor in Jordan and the

Middle East. It serves as a cornerstone for the development of Jordan's nuclear programs and is designed for use in three main areas: education and training, advanced nuclear research, and commercial and industrial services centered on radioisotope production [4]. The JRTR specifications are listed in Table 1 [5].

Table 1. JRTR specifications.

Reactor Type	Open-Tank-in-Pool
Thermal Power	5 MW_{th} (upgradable to 10 MW_{th})
Coolant and Cooling Method	Light water and downward, forced convection
Coolant Temperature	37 °C (inlet) and 44 °C (outlet)
Fuel	19.75 wt.% enriched U_3Si_2-Al
Moderator	Light water
Reflectors	Beryllium and heavy water
Absorber Materials	Hafnium and B_4C
Shielding	Water and heavy concrete

Although the JRTR operates at a lower power level than nuclear power plants, a risk assessment must be conducted to ensure public and environmental safety. In the event of an accident or incident involving a reactor, a large amount of radioactive material is released to the environment. Therefore, radiological hazard assessment is a major concern and requirement for research reactors. This assessment is used to evaluate the potential health risks associated with radiation exposure, and it involves various steps, such as characterization of the source term, evaluation of potential pathways of exposure, determination of potential doses and health effects, and evaluation of mitigation strategies [6,7]. The primary purpose of an environmental hazard assessment is to provide decision makers and stakeholders with information about potential consequences, and to ensure appropriate countermeasures for protecting the public and environment [8].

The JRTR Final Safety Analysis Report (FSAR) [3] and Radiation Environmental Report (RER) [9] identify the flow blockage of a fuel assembly as a limited accident that releases the highest source term to the environment. Another study by Xoubi [10] derived the source term of full core damage, and evaluated the atmospheric dispersion of released radionuclides for a downwind distance of 100 km using HotSpot code, which is a Gaussian plume model. The HotSpot code is a first-order approximation of the radiation effects associated with the atmospheric release of radioactive materials, designed for short-range (less than 10 km), and short-term (less than a few hours) predictions [11].

However, the Gaussian plume model has certain limitations under specific conditions, such as low wind speed, complex terrain, spatial and temporal changes in wind velocity, and deposition and transformation. In contrast, the particle-in-cell (PIC) model can handle the effects of time-varying wind and deposition processes, and incorporate terrain effects directly through adjustments in a three-dimensional grid system [12]. The characteristics of Gaussian plume and Lagrangian models are well described in some application research by the IAEA program [13,14].

In this study, the Hybrid Single-Particle Lagrangian Integrated Trajectory (HYSPLIT) model was applied to evaluate the dispersion patterns of radioactive materials released into the atmosphere and subsequence dose assessment. The HYSPLIT [15] can simulate the three-dimensional concentrations of radionuclides in air and their deposition on the ground due to dry and wet processes. After simulating the dispersion of the radionuclides, the calculated concentrations were utilized for radiological effect assessment in the dose module.

2. Materials and Methods

2.1. Site Location

The JRTR is located within JUST in Irbid City, Jordan. It is located approximately 70 km north of Amman and 13 km southeast of Irbid. The geographical coordinates of the

JRTR are 32.4627° N and 35.9730° E. Ground elevations of the site range from 560 m to 635 m [9]. Figure 1 illustrates the JRTR site and prominent cities within a radius of 80 km.

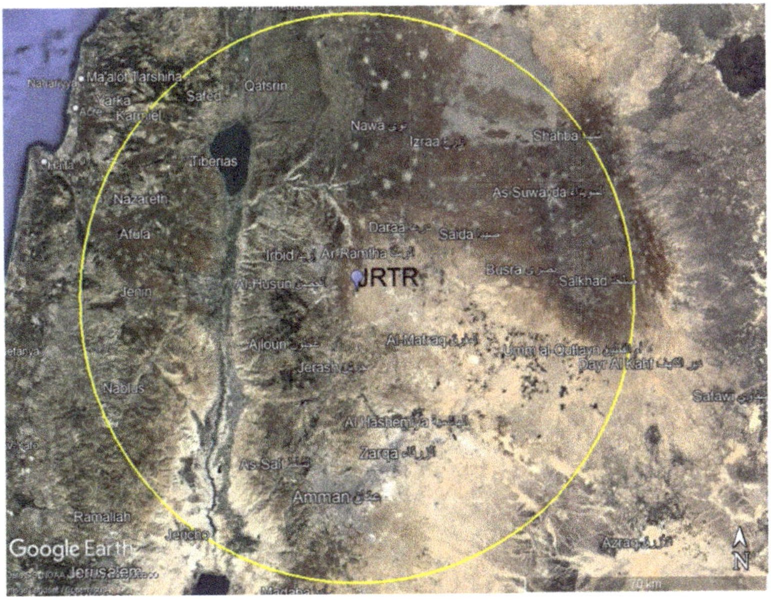

Figure 1. Location of JRTR and surrounding cities in the radius of 80 km.

The governorates of Irbid, Jarash, Ajlun, and Balqa are entirely situated within the 80 km radius of the JRTR. On the other hand, the governorates of Amman, Zarqa, and Mafraq are partially situated within the 80 km radius, with the majority of densely populated regions falling within this range. Table 2 shows the estimated population of Jordan by governorate at end-year 2013 and 2022 [16,17]. According to the below table, approximately 90% of Jordan's overall population resides within an 80 km proximity to the JRTR.

Table 2. Estimated population of Jordan by governorate at end-year 2013 and 2022.

Governorate	2013 [1]		2022	
	Population	Percentage (%)	Population	Percentage (%)
Amman	2,528,500	38.7	4,744,700	42.0
Balqa	437,500	6.7	582,100	5.2
Zarqa	972,900	14.9	1,616,000	14.3
Madaba	163,300	2.5	224,000	2.0
Irbid	1,162,300	17.8	2,095,700	18.5
Mafraq	306,900	4.7	651,100	5.8
Jarash	195,900	3.0	280,700	2.5
Ajlun	150,200	2.3	208,500	1.8
Karak	254,700	3.9	374,800	3.3
Tafiela	91,400	1.4	114,000	1.0
Ma'an	124,100	1.9	187,600	1.7
Aqaba	142,300	2.2	222,800	2.0
Total	6,530,000	100.0	11,302,000	100.0

[1] Excluding Syrian refugees.

2.2. *Atmosperic Dispersion Assessment*

The HYSPLIT model was used to calculate air concentrations and ground depositions for a nuclear accident in the JRTR. It is a complete system developed by the U.S. National Oceanic and Atmospheric Administration (NOAA) and is designed for atmospheric dispersion analysis by computing simple trajectories for complex dispersion and deposition simulations using either puff or particle approaches [15]. The input data were interpolated to an internal sub-grid centered to reduce memory requirements and enhance computational speed. The model requires the emission rate and physical characteristics of the pollutants to be defined. The model calculation method is a hybrid between Eulerian and Lagrangian approaches. Advection and diffusion calculations are performed in a Lagrangian framework, while concentrations are calculated in a Eulerian approach. Air concentration and depositions are associated with the mass of pollutant species, such as the release of puffs, particles, or a combination of both. The dispersion of a pollutant is calculated by assuming either a Gaussian or top-hat horizontal distribution within a puff, or from the dispersal of a fixed number of particles.

A Lagrangian model can compute air concentrations through either of two assumptions: puff or particle. In a puff model, the source is simulated by releasing pollutant puffs at regular intervals over the duration of the release. Each puff contains the appropriate fraction of the pollutant mass. The puff is advected according to the trajectory of its center position, while the size of the puff (both horizontally and vertically) expands in time to account for the dispersive nature of a turbulent atmosphere. Air concentrations are then calculated at specific points (or nodes on a grid) by assuming that the concentrations within the puff have a defined spatial distribution. In a Lagrangian particle model, the source can be simulated by releasing many particles over the duration of the release. In addition to the advective motion of each particle, a random component to the motion is added at each step according to the atmospheric turbulence at that time. In this way a cluster of particles released at the same point will expand in space and time simulating the dispersive nature of the atmosphere. Air concentrations are calculated by summing the mass of all the particles in a grid cell.

Dispersion in a Lagrangian model is computed following the particle or puff. The advection of a particle is computed independently of the dispersion calculation. The time integrated advection of each particle can be viewed as a simple trajectory which only requires the three-dimensional velocity field. The velocity fields from meteorological data are processed and interpolated to the internal model grid, and trajectories can be computed to test the advection components of the model. Detailed mathematical formulations of HYSPLIT are available in the literature [15,18–20].

2.3. Source Term

The FSAR [3] and RER [9] for JRTR have identified the complete blockage of a fuel assembly as a limited design basis accident that could result in the most significant release of fission products into the environment. Xoubi [10] additionally considered the scenario of one or more projectiles targeting the reactor building, leading to the destruction of the reactor confinement and pool, ultimately causing damage to all fuel assemblies. Given the absence of a containment building and the potential security threats resulting from the current situation in the Middle East, including the JRTR's proximity to the Syrian borders, a radiological impact assessment was conducted utilizing the source term provided by Xoubi [10].

For the computation of source term at the JRTR, the ORGEN-ARP code was used, and the cross-section library was generated for a JRTR fuel assembly using TRITON in SCALE [3,9,10]. The release fractions of radionuclides during the postulated severe accident were estimated in accordance with the International Atomic Energy Agency (IAEA) [6]. Table 3 lists the source terms released into the environment derived by Xoubi [10].

The total period of the accident applied to the dose calculation is assumed to be 31 days after the accident. Fission products in the core are immediately released to the environment following the accident. The release mode for accident is assumed to be a ground release. Noble gases are released in gas form, iodine isotopes released in various forms, including molecular iodine, aerosols, and organic compounds, and other radioactive nuclides are released in aerosol form.

Table 3. JRTR source term released to the environment due to a full core destruction by projectiles.

Nuclide	Inventory (Bq)	Nuclide	Inventory (Bq)	Nuclide	Inventory (Bq)
Kr-83m	7.97×10^{14}	I-134	4.84×10^{15}	Te-133m	2.61×10^{14}
Kr-85	3.49×10^{13}	I-135	3.93×10^{15}	Te-134	5.26×10^{14}
Kr-85m	1.84×10^{15}	Cs-134	7.87×10^{13}	Ba-139	1.97×10^{14}
Kr-87	3.71×10^{15}	Cs-134m	3.03×10^{13}	Ba-140	1.88×10^{14}
Kr-88	5.06×10^{15}	Cs-136	3.68×10^{13}	Ba-141	1.78×10^{14}
Kr-89	6.44×10^{15}	Cs-137	8.64×10^{13}	Sr-89	1.23×10^{14}
Xe-131m	5.39×10^{13}	Cs-138	3.09×10^{15}	Sr-90	5.39×10^{12}
Xe-133	9.91×10^{15}	Cs-139	2.91×10^{15}	Sr-91	1.69×10^{14}
Xe-135	2.86×10^{15}	Rb-88	1.53×10^{15}	Sr-92	1.74×10^{14}
Xe-135m	1.91×10^{15}	Rb-89	2.03×10^{15}	Sb-131	2.00×10^{14}
Xe-137	9.53×10^{15}	Te-127	1.42×10^{13}	Mo-99	2.40×10^{13}
Xe-138	9.61×10^{15}	Te-127m	2.00×10^{12}	Rh-105	5.39×10^{12}
Br-84	6.13×10^{14}	Te-129	4.57×10^{13}	Ru-103	1.24×10^{13}
I-130	1.43×10^{13}	Te-129m	8.16×10^{12}	Ru-105	5.98×10^{12}
I-131	1.86×10^{15}	Te-131	2.04×10^{14}	Tc-99m	2.13×10^{13}
I-132	2.77×10^{15}	Te-131m	3.60×10^{13}	Np-239	1.48×10^{13}
I-133	4.18×10^{15}	Te-132	3.43×10^{14}		

2.4. Meteorological Data

Meteorological normals and extremes for precipitation, humidity, and wind speed at Ramtha Station and wind direction at Irbid Station are shown in Table 4 [9]. The data were based on measurements from 1976 to 2000.

Table 4. Histological meteorological normals and extremes at Ramatha and Irbid stations near JRTR site between 1976 and 2000.

Months	Precipitation (mm)			Relative Humidity (%)	Wind		
					Speed (m/s)		Prevailing Direction (Deg.)
	Mean	Max. Monthly	Max. in 24 h	Mean Daily	Mean	Max. Daily	
1	49.4	63.9	38.5	73.3	2.8	20.6	207
2	44.6	125.5	36.5	65.2	3.1	15.4	244
3	41.9	116.6	60	64.4	3.3	15.4	252
4	8.5	29.9	14.0	49.5	2.9	20.6	269
5	3.2	23.4	25.3	40.4	2.9	36.0	280
6	0.8	0.2	14.8	46.8	3.4	12.3	282
7	0.0	0.0	0.0	50.2	3.7	14.4	285
8	0.0	0.0	0.0	52.5	2.8	12.3	282
9	0.2	2.3	2.9	52.4	2.7	16.5	279
10	7.8	20	19.8	55	2.8	13.4	274
11	25.9	1.6.4	40.2	59.1	2.9	11.3	227
12	38.7	91.3	32.0	68.5	2.8	11.8	234
Period	1976~2000	1976~2000	1976~2000	1976~2000	1981~2000	1985~2000	NA

Radionuclide dispersion proportionally decreased with high humidity and precipitation and increased with the high wind speed, while deposition proportionally increased with high humidity and precipitation and decreased with the low wind speed. Table 4 shows that January had the highest mean precipitation and relative humidity, with a relatively low mean wind speed. Therefore, the release of radioactive materials in January is anticipated to result in the highest radionuclide concentrations in areas near the JRTR site. July had zero precipitation and the highest mean wind speed, which resulted in the dispersion of radionuclides far from the JRTR site.

Radiological hazard assessments in the JRTR FSAR and RER [3,9] were conducted using meteorological data from May 2013 to April 2014. Therefore, in this study, July 2013 and January 2014 were selected to evaluate the atmospheric dispersion and radiological effects of relatively large and small depositions, respectively. The meteorological data required to run the HYSPLIT model were obtained from a data archive consisting of the Global Data Analysis System (GDAS) of NOAA.

2.5. Dose Assessment

Radiation exposure is a major factor in the management of nuclear emergencies. In terms of accuracy and reliability, actual measurements are the optimal strategy for obtaining information concerning the dose received by individuals. However, in practice, obtaining measurements from all individuals affected by nuclear accidents over the entire period of exposure is a major challenge [21]. Therefore, a dose assessment model has been applied to estimate the radiological exposure of the public to a nuclear accident [22]. Radionuclides released to the environment is a potential risk to humans and the environment; therefore, major efforts have been concentrated on developing an internationally acceptable system [23–25].

Radiation assessment in humans can be evaluated by considering the exposure pathways. Human exposure to radiation can be classified as external and internal doses. External radiation from radioactive materials directly affects the human body (i.e., high-energy radiation, such as gamma radiation, which penetrates the human body). This radiation can be emitted from facilities, air-suspended radionuclides, or radionuclides deposited on the ground. External exposure is typically the principal route of radioactive effluents released from nuclear plants, whereas internal exposure is caused by inhalation, ingestion, or absorption through intact or broken skin. Ingestion is usually the principal route of radioactive effluent intake. The external and internal exposures can be calculated using the equations described below [22].

Inhalation of radionuclides in the plume:

$$H_{E(inh),i} = C_{a,i}(x) \times D_{inh,i} \times I_{inh} \tag{1}$$

$H_{E(inh),i}$: Effective dose from inhalation (Sv);
$C_{a,i}(x)$: Average activity concentration of radionuclide i at location x (Bq/m^3);
$D_{inh,i}$: Inhalation dose coefficient (Sv/Bq);
I_{inh}: Inhalation rate (m^3/a).

External exposure from the cloud:

$$H_{E(ex,cloud),i} = C_{a,i}(x) \times D_{(ex,cloud),i} \times (O_{out} + (1 - O_{out})L_{cloud}) \tag{2}$$

$H_{E(ex,cloud),i}$: Effective dose from external exposure to radionuclide i (Sv);
$C_{a,i}(x)$: Time integrated activity concentration of radionuclide i in air at location x (Bq·s/m^3);
$D_{(ex,cloud),i}$: External dose coefficient for immersion in the cloud (Sv per (Bq·s)/m^3);
O_{out}: Fraction of time spent outdoors;
L_{cloud}: Dimensionless factor considering the shielding effect of buildings.

External exposure from deposited radionuclides:

$$H_{E(ex,deposit),i} = C_{s,i}(x) \times D_{ex,\ deposit} \times (O_{out} + (1 - O_{out})L_{deposit}) \tag{3}$$

$H_{E(ex,deposit),i}$: Effective dose from deposited radionuclide i (Sv);
$C_{s,i}(x)$: Time integrated activity concentration of radionuclide i on soil at location x (Bq·s/m^2);
$D_{ex,deposit}$: External dose coefficient from the deposit (Sv per (Bq·s)/m^2);
O_{out}: Fraction of time spent outdoors;
$L_{deposit}$: Dimensionless factor that considers the shielding effect of buildings.

The calculation of the total effective dose resulting from the release of radionuclides during an accident necessitates the use of dose coefficients. The U.S. Environmental Protection Agency's Federal Guidance Reports (FGR) [26,27] provided all the necessary dose coefficients for every exposure pathway associated with the released radionuclides.

2.6. Other Input Parameters

HYSPLIT [15] offers three distinct mechanisms for removing pollutants, namely dry deposition, wet depletion, and radioactive decay. Dry deposition is characterized by a deposition velocity (v_d) that varies depending on the surface type and isotope. For this study, two deposition velocities were employed: 0.3 cm/s for respirable particles and 8 cm/s for non-respirable particles [10].

Wet depletion encompasses two processes: the first involves the continuous ingestion of polluted air into a cloud from a polluted boundary layer, followed by transportation within the cloud (in-cloud); the second process involves rain falling through a polluted layer (below-cloud). For soluble gases, wet removal is defined by specifying its Henry's Law constant. Gaseous wet removal occurs only for the fraction of the pollutant that lies below the cloud top. Particle wet removal is defined by a scavenging ratio within the cloud and an explicit scavenging coefficient for pollutants below the cloud base. The relevant values used in this study were obtained from the suggested values in the HYSPLIT user's guide [28].

3. Results and Discussion

In January, a release of radioactive materials was anticipated to result in the highest radionuclide concentrations in areas surrounding the JRTR site. July had the highest mean wind speed, which resulted in the dispersion of radionuclides far from the JRTR site (Table 4). In particular, simulations were performed in July 2013 and January 2014, which was the period used for the radiological hazard assessment in the JRTR FSAR and RER [3,9].

Figure 2 shows the monthly integrated air concentration map for July 2013, where all the radionuclides listed in Table 3 are assumed to be released into the atmosphere. As demonstrated in Table 4, most radioactive materials in July moved eastward owing to the westerly wind. The GDAS of NOAA in July exhibited similar wind patterns, and the measured wind fields in the JRTR report maintained a constant westerly wind during summer.

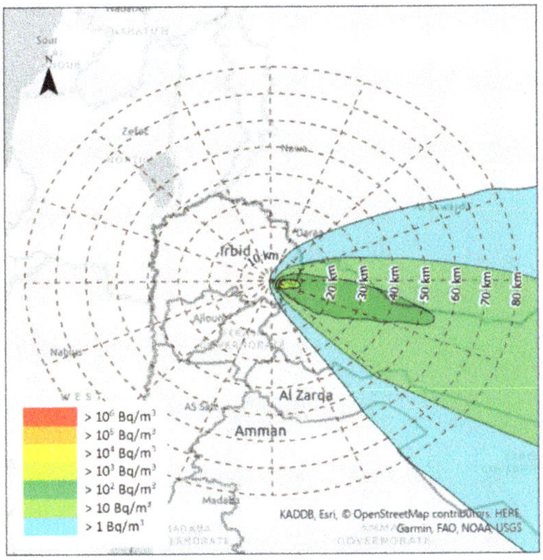

Figure 2. Integrated air concentration in July 2013.

Figure 3 illustrates the monthly integrated air concentration map in January 2014 and the radioactive materials transported in various directions owing to the complicated wind patterns in winter. The concentrations of radionuclides in the air were relatively higher near the JRTR site in January than in July because they did not disperse widely in various areas in January owing to the relatively large precipitation and low wind speed in winter.

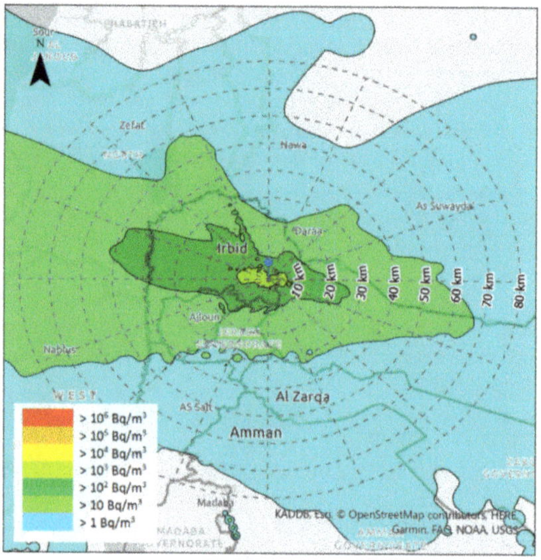

Figure 3. Integrated air concentration in January 2014.

Figures 4 and 5 depict the total ground deposition in July 2013 and January 2014, respectively. These depositions exhibited the same patterns as the integrated air concentrations illustrated in Figures 2 and 3. More deposits were recorded near the JRTR in January than in July.

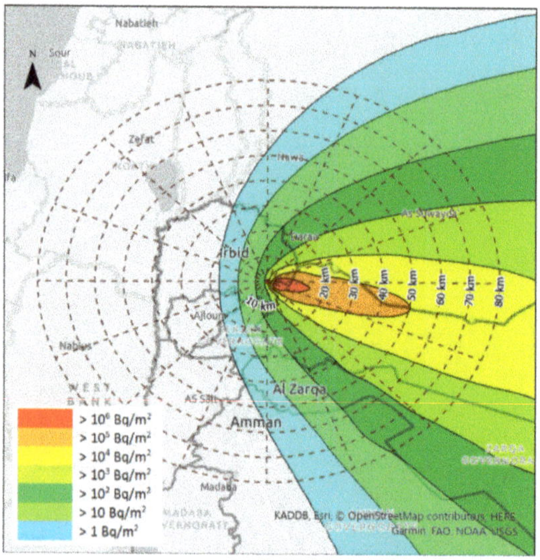

Figure 4. Total ground deposition for the whole month of July 2013.

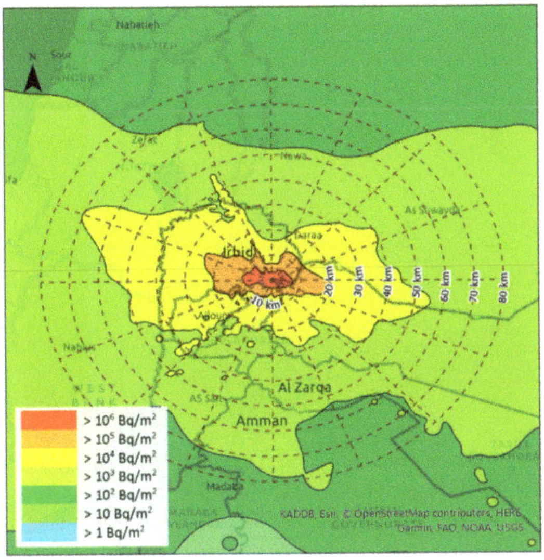

Figure 5. Total ground deposition for the whole month of January 2014.

The calculated monthly dose, which accounts for the internal and external doses, as presented in Equations (1)–(3), is shown in Figures 6 and 7. The maximum dose for the whole month of July 2013 was estimated to be approximately 3.6×10^{1} mSv at a radius of 5 km from the JRTR, whereas the minimum dose was approximately 2.6×10^{-7} mSv at a radius of 80 km from the JRTR. The maximum and minimum doses in January 2014 were estimated to be approximately 5.0×10^{1} mSv and 4.9×10^{-5} mSv. As mentioned previously, January has relatively high humidity and precipitation; therefore, a large amount of radioactive material is deposited and high dose rates occur near the JRTR.

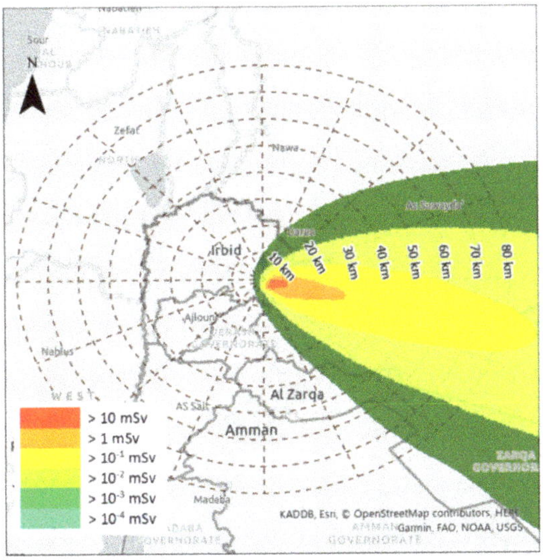

Figure 6. Total dose received in July 2013.

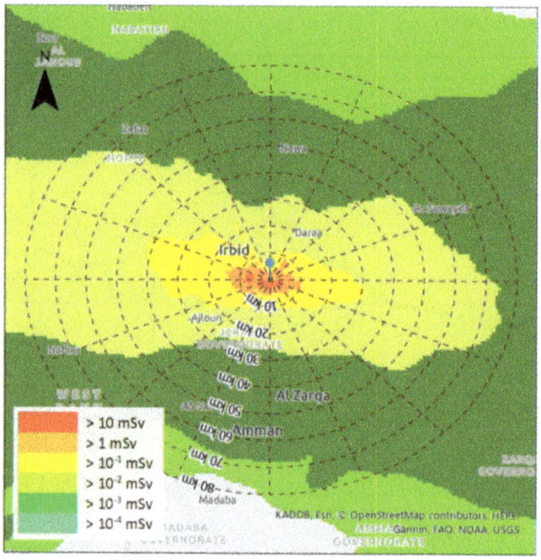

Figure 7. Total dose received in January 2014.

The thyroid dose by inhalation and deposition was estimated for the three age groups by using the calculated air concentrations and ground depositions in July 2013 and January 2014. The maximum monthly thyroid doses were 80 mSv, 62 mSv and 40 mSv at a radius of 5 km from the JRTR for 1-year old infants, 10-year old children, and adults in July 2013. In addition, they showed 110 mSv, 86 mSv and 55 mSv for 1-year old infants, 10-year old children, and adults in January 2014.

According to Figure 6, although a dose of 1 mSv was calculated more than 20 km to the east of the JRTR site, the affected areas consist of low populated lands, with only a few small communities located within this region. With the exception of the Mafraq governorate, all other governorates received a minimal dose of less than 0.1 mSv. Figure 7 shows that the dose in the area within a 10 km radius of the JRTR site exceeds 1 mSv, impacting the Bani Obeid district, but posing no significant danger to the densely populated Irbid Qasabah district or other districts in the Irbid governorate. Some areas in the Irbid and Mafraq governorates are estimated to receive a dose greater than 1 mSv, while all other governorates receive a very low dose of less than 0.1 mSv. Both Figures 6 and 7 show that even if a severe accident was to damage the JRTR confinement and core, only a limited area is estimated to be at significant risk of high exposure to radioactive materials. Densely populated cities and regions are estimated to receive a minimal dose.

The aforementioned estimated doses were calculated without considering any protective measures. Consequently, if appropriate protective actions are implemented, the doses are expected to decrease substantially. The radiological effects on people living near the JRTR differed depending on the air and deposition concentrations in their local areas, which varied according to meteorological conditions. Individuals residing in the eastern area of the JRTR presented high radiological risks in July, while some individuals living in the western and eastern areas of the JRTR faced high risks in January. In this regard, developing a protective plan for people in each direction for a nuclear accident could minimize the radiological risks.

In this study, the maximum dose was estimated to be 50 mSv for the month of January, approximately 5 km away from the JRTR site. This value, when compared to the JRTR Final Safety Analysis Report (FSAR) [3], was found to be reached at approximately 120 m from the JRTR site after 30 days of an accidental release. It should be noted that the JRTR source term in the FSAR was calculated based on only one fuel assembly meltdown. Conversely, Xoubi [10] estimated that the total effective dose of 50 mSv could be reached for a distance of more than 10 km, depending on the Pasquil stability class, with the maximum dose of approximately 200 Sv for most stability classes. This significant difference in total effective dose values is attributed to several parameters that have a significant impact on radiological hazard assessment, such as the source term, meteorological conditions, and dispersion models used for risk hazard assessment.

4. Conclusions

To mitigate the environmental and health risks of nuclear accidents, effective emergency response plans must be established. These plans should encompass the establishment of evacuation routes and zones, communication systems, and emergency shelters. Furthermore, educational programs should be conducted to raise public awareness about the potential health and safety risks associated with nuclear accidents.

In this study, numerical simulations were performed for the summer and winter seasons to estimate the occurrence of relatively high and low radiological effects near the JRTR. A higher dose was estimated in winter than in summer owing to meteorological conditions, such as humidity, precipitation, and wind patterns. However, even if we assumed that the maximum calculated dose in winter or summer was sustained for one year, the expected annual dose for densely populated areas would be below the annual limit for the public. Consequently, even in the event of a severe nuclear accident in the JRTR, the radiological risks for people residing in nearby cities are considerably low.

To ensure safe operation and maintenance of nuclear facilities, safety regulations and standards should be established and enforced. Furthermore, effective emergency response plans must be organized to provide technical assistance for the safe operation of nuclear facilities. By implementing these measures, the potential risks of nuclear accidents can be minimized, thereby promoting the overall safety and well-being of the general public.

Author Contributions: Conceptualization, M.T. and K.-S.S.; methodology, M.T. and K.-S.S.; software, M.T.; validation, M.T. and S.K.; formal analysis; M.T., S.K. and K.-S.S.; resources, M.T. and K.-S.S.; data curation, M.T., S.K. and K.-S.S.; writing—original draft preparations, M.T., S.K. and K.-S.S.; writing—review and editing, M.T. and K.-S.S.; visualization, M.T.; supervision, K.-S.S.; project administration, S.K. and K.-S.S.; funding acquisition, S.K. and K.-S.S.; investigation, M.T., S.K. and K.-S.S. All authors have read and agreed to the published version of the manuscript.

Funding: This research was supported by a National Research Foundation of Korea (NRF) grant funded by the Korean government (Ministry of Science and ICT (RS-2023-00144350)), the research program of the Korea Atomic Energy Research Institute, and the Korea Research Environment Open NET-work (KREONET).

Institutional Review Board Statement: Not applicable.

Informed Consent Statement: Not applicable.

Data Availability Statement: Not applicable.

Conflicts of Interest: The authors declare no conflict of interest.

References

1. Suh, K.S.; Han, M.H.; Jung, S.H.; Lee, C.W. Three-dimensional numerical modeling of pollutant transport at local-scale complex terrain. *Ann. Nucl.* **2008**, *35*, 1016–1023. [CrossRef]
2. Jawerth, N.; Mattar, E. Exploring research reactors and their use. *IAEA Bull. Res. React.* **2019**, *60–64*, 4–5.
3. Korea Atomic Energy Research Institute. *Jordan Research and Training Reactor Final Safety Analysis Report*; Korea Atomic Energy Research Institute: Daejeon, Republic of Korea, 2016.
4. Xoubi, N. Jordan's First Research Reactor Project: Driving Forces, Present Status and the Way Ahead. In Proceedings of the European Research Reactor Conference, Rome, Italy, 20–24 March 2011; pp. 26–30.
5. Kim, K.-O.; Jun, B.J.; Lee, B.; Park, S.-J.; Roh, G. Comparison of first criticality prediction and experiment of the Jordan research and training reactor (JRTR). *Nucl. Eng. Technol.* **2020**, *52*, 14–18. [CrossRef]
6. IAEA. *Derivation of the Source Term and Analysis of the Radiological Consequences of Research Reactor Accidents, Safety Reports Series No. 53*; IAEA: Vienna, Austria, 2008.

7. IAEA. *Preparedness and Response for a Nuclear or Radiological Emergency, Safety Standards Series No. GSR Part 7*; IAEA: Vienna, Austria, 2015.
8. IAEA. *Criteria for Use in Preparedness and Response for a Nuclear or Radiological Emergency, Safety Standards Series No. GSG-2*; IAEA: Vienna, Austria, 2011.
9. Korea Atomic Energy Research Institute. *Jordan Research and Training Reactor Radiation Environmental Report*; Korea Atomic Energy Research Institute: Daejeon, Republic of Korea, 2015.
10. Xoubi, N. Source term derivation and radioactive release evaluation for JRTR research reactor under severe accident. *Sci. Technol. Nucl. Install.* **2020**, *2020*, 1868965. [CrossRef]
11. Homann, S.G.; Aluzzi, F. *HotSpot Health Physics Codes Version 3.1.2 User's Guide*; Lawrence Livermore National Laboratory: Livermore, CA, USA, 2020.
12. Till, J.E.; Grogan, H. *Radiological Risk Assessment and Environmental Analysis*; Oxford University Press: Oxford, UK, 2008.
13. Raúl, P.; Kathleen, M.T.; Sohan, L.C.; Francesco, M.; Emilie, N.; Gert, S.; Dejan, T. Mid-range atmospheric dispersion modelling. Intercomparison of simple models in EMRAS-2 project. *J. Environ. Radioact.* **2016**, *162–163*, 225–234. [CrossRef]
14. Periáñez, R.; Bezhenar, R.; Brovchenko, I.; Duffa, C.; Iosjpe, M.; Jung, K.T.; Kim, K.O.; Kobayashi, T.; Liptak, L.; Little, A.; et al. Marine radionuclide transport modelling: Recent developments, problems and challenges. *Environ. Model. Softw.* **2019**, *122*, 104523. [CrossRef]
15. Stein, A.F.; Draxler, R.R.; Rolph, G.D.; Stunder, B.J.B.; Cohen, M.D.; Ngan, F. NOAA's HYSPLIT atmospheric transport and dispersion modeling system. *Bull. Amer. Meteor. Soc.* **2015**, *96*, 2059–2077. [CrossRef]
16. Jordan Department of Statistics. *Jordan Estimated Population of 2022*; Jordan Department of Statistics: Amman, Jordan, 2023.
17. Jordan Department of Statistics. *Jordan Statistical Yearbook 2013*; Jordan Department of Statistics: Amman, Jordan, 2013.
18. Draxler, R.R. *HYSPLIT4 User's Guide*; NOAA Technical Memorandum ERL ARL-230; NOAA Air Resources Laboratory: Silver Spring, MD, USA, 1999.
19. Draxler, R.; Hess, G. An overview of the HYSPLIT_4 modelling system for trajectories. *Aust. Meteorol. Mag.* **1998**, *47*, 295–308.
20. Draxler, R.R.; Hess, G. *Description of the HYSPLIT4 Modeling System*; NOAA Tech. Memo. ERL ARL-224; NOAA Air Resources Laboratory: Silver Spring, MD, USA, 1997; p. 24.
21. Takahara, S.; Iijima, M.; Watanabe, M. Assessment Model of Radiation Doses from External Exposure to the Public after the Fukushima Daiichi Nuclear Power Plant Accident. *Health Phys.* **2020**, *118*, 664–677. [CrossRef] [PubMed]
22. UNSCEAR. *Methodology for Estimating Human Exposures Due to Radioactive Discharges*; UNSCEAR: Vienna, Austria, 2013.
23. ICRP. Age-dependent dose to members of the public from intake of radionuclides: Part 5 compilation of ingestion and inhalation dose coefficients. *Ann. ICRP* **1996**, *26*, 1–91. [CrossRef] [PubMed]
24. ICRP. Human respiratory tract model for radiological protection. *Ann. ICRP* **1994**, *24*, 1–482. [CrossRef]
25. ICRP. Age-dependent doses to members of the public from intake of radionuclides: Part 4 Inhalation dose coefficients. *Ann. ICRP* **1995**, *25*, 1–402. [CrossRef]
26. Eckerman, K.F.; Ryman, J.C. *Federal Guidance Report No. 12: External Exposure to Radionuclides in Air, Water, and Soil*; U.S. Environmental Protection Agency: Washington, DC, USA, 1993.
27. Eckerman, K.F.; Wolbarst, A.B.; Rischardson, A.C.B. *Federal Guidance Report No. 11: Limiting Values of Radionuclide Intake and Air Concentration and Dose Conversion Factors for Inhalation, Submersion, and Ingestion*; U.S. Environmental Protection Agency: Washington, DC, USA, 1988.
28. Draxler, R.; Stunder, B.; Rolph, G.; Stein, A.; Taylor, A.; Zinn, S.; Loughner, C.; Crawford, A. *HYSPLIT User's Guide*; NOAA Air Resources Laboratory: Silver Spring, MD, USA, 2022.

atmosphere

Article

Dispersion and Radiation Modelling in ESTE System Using Urban LPM

Ľudovít Lipták *, Peter Čarný, Michal Marčišovský, Mária Marčišovská, Miroslav Chylý and Eva Fojciková

ABmerit, s.r.o., Hornopotočná 1, P.O. Box 200, 917 01 Trnava, Slovakia; carny@abmerit.sk (P.Č.); marcisovsky@abmerit.sk (M.M.); marcisovska@abmerit.sk (M.M.); chyly@abmerit.sk (M.C.); fojcikova@abmerit.sk (E.F.)

* Correspondence: liptak@abmerit.sk; Tel.: +421-33-5513-345

Abstract: In cases of accidental or deliberate incidents involving a harmful agent in urban areas, a detailed modelling approach is required to include the building shapes and spatial locations. Simultaneously, when applied to crisis management, a simulation tool must meet strict time constraints. This work presents a Lagrangian particle model (LPM) for computing atmospheric dispersion. The model is implemented in the nuclear decision support system ESTE CBRN, a software tool developed to calculate the atmospheric dispersion of airborne hazardous materials and radiological impacts in the built-up area. The implemented LPM is based on Thomson's solution for the nonstationary, three-dimensional Langevin equation model for turbulent diffusion. The simulation results are successfully analyzed by testing compatibility with Briggs sigma functions in the case of continuous release. The implemented LPM is compared with the Joint Urban 2003 Street Canyon Experiment for instantaneous puff releases. We compare the maximum concentrations and peak times measured during two intensive operational periods. The modeled peak times are mostly 10–20% smaller than the measured. Except for a few detector locations, the maximum concentrations are reproduced consistently. In the end, we demonstrate via calculation on single computers utilizing general-purpose computing on graphics processing units (GPGPU) that the implementation is well suited for an actual emergency response since the computational times (including dispersion and dose calculation) for an acceptable level of result accuracy are similar to the modeled event duration itself.

Keywords: Lagrangian particle model; urban modelling; CBRN events; emergency response; Joint Urban 2003 Experiment

Citation: Lipták, Ľ.; Čarný, P.; Marčišovský, M.; Marčišovská, M.; Chylý, M.; Fojciková, E. Dispersion and Radiation Modelling in ESTE System Using Urban LPM. *Atmosphere* **2023**, *14*, 1077. https://doi.org/10.3390/atmos14071077

Academic Editor: Daniel Viúdez-Moreiras

Received: 3 May 2023
Revised: 18 June 2023
Accepted: 20 June 2023
Published: 26 June 2023

1. Introduction

A crisis management mission is a quick, confident and complex response to harmful events. One such mission is the application of adequate mitigating measures. This goal is achievable if the decision making is based on a detailed understanding of the current situation. Therefore, modelling tools are very helpful in these situations.

The ESTE CBRN software tool is a nuclear decision support system designed to model the radiological impacts of a radiological or nuclear event in urban areas. The software aims to identify and calculate the corresponding wind field; specifically, this software is used to perform atmospheric dispersion and radiological impact calculations. In the original approach, the atmospheric modelling in ESTE CBRN was based on Gaussian modelling in combination with a puff trajectory model. This approach is also applied in several other fast-response systems, e.g., in SCIPUFF (the Second Order Closure Integrated Puff model) [1]. The Gaussian approach has short simulation times and low requirements for implementation in urban area because it models the urban environment only partially and does not include 3-dimensional models of buildings. In ESTE CBRN, the Gaussian formulation of urban atmospheric dispersion is expressed by urban sigma functions [2,3].

Dispersion modelling characterized as 3-dimensional is achievable using the Eulerian or Lagrangian approach. Using one of these approaches, with the building locations and their shapes, the calculation can be performed in detail when required. An existing limitation is the amount of computational resources available and the requirement on the calculation duration.

In addition to the Gaussian model of the ESTE system, a 3D approach for the dispersion modelling in ESTE was implemented using a Eulerian model [4]. This 3D model of the ESTE system was validated in the UDINEE project (Urban Dispersion International Evaluation Exercise, [5,6]) against the Joint Urban Experiment JU2003 performed in Oklahoma City (USA). In that project, we compared the urban wind field computational model and the Eulerian approach implemented in ESTE with measured meteorological data and concentrations of SF_6 [4]. The compliance of the modeled and measured data was moderate. Many variables compared, such as puff arrival time, are generally well reproduced and fulfil the urban modelling criteria. However, the calculation duration using the Eulerian approach did not comply with emergency requirements. This disadvantage motivated us to replace the Eulerian model with the Lagrangian dispersion model in the ESTE CBRN to satisfy the requirements to perform a 3D calculation approach for emergency purposes.

The Lagrangian particle model implemented is based on Thomson's simplest solution for the nonstationary, three-dimensional Langevin equation model [7]. This approach (and similar models) is applied in other software systems, such as QUIC-PLUME [8], UrbanLS [9,10] and P-SPRAY [11].

In the present work, we focus on validating the LPM as a proper atmospheric model tool in an urban environment. The validation is performed mainly on the JU2003 experimental data, where we show that the LPM implemented in ESTE CBRN is appropriate for urban dispersion modelling.

The ESTE CBRN system is not only an atmospheric transport modelling system; it is also a system for radiological consequence calculations, with the impact calculation assuming all exposure pathways. The calculation of external effective doses due to ground shine and cloud shine are performed on the same 3D computational domain of the modeled urban area on which the corresponding atmospheric transport calculation is performed. This method of implementation enables us to take into account 3D modelling of buildings in three phases (calculation of wind field, calculation of dispersion and calculation of external exposure). Three-dimensional modelling of buildings is the key for including their shielding effects in the calculation of external exposure.

The ESTE CBRN system models several types of CBRN events, but it focuses on the modelling of the radiological dispersal device (RDD). Events with the RDD represent a special type of radiological event. The database of RDD events implemented in ESTE CBRN consists of (i) a model for a classical radiological dispersal device (CRDD), (ii) a model for an improvised burned-up nuclear fuel dispersion device (INFDD), (iii) a model for a mortar dispersing pulverized chemical compounds of radioactive nuclides, and (iv) a model for an improvised nuclear device (IND).

In this study, we focus on the creation of complex system for emergency response. Such systems must fulfill strict computational time constraints (results could be available in, for example, 15–20 min), and acceptable accuracy of results. One option to achieve these goals is the application of large computer clusters, which provide the capability for fast computing and the capability to cover large geographical areas [12]. Another option is to utilize the technology of GPGPU, which offers a massive parallel computing approach. We show that a Lagrangian particle dispersion model coupled with a radiological impact calculation, both implemented using GPGPU technology, is a time-efficient method to perform accurate and short-duration calculations.

In the present study, we do not analyze the calculation of the urban wind field. Some original approaches are designed to run fast [13], a compromise between computational accuracy and a real-time approach. Another approach represents the direct application of CFD (Computational Fluid Dynamics) techniques. In the case of ESTE CBRN, the urban wind field is calculated through a CFD model using Reynolds-averaged Navier–Stokes (RANS). The CFD approaches are more accurate but also more time-consuming. Here, one option is an application of supercomputing resources. Another option is to have a well-designed set of pre-calculated wind fields for a particular urban area prepared in advance. The application of pre-calculated wind fields was realized for the Operational Canadian Urban Dispersion Modelling (CUDM) System, in the case of the principal Canadian towns [14]. A pre-calculated database of wind fields is also applied in the ESTE CBRN.

The structure of the present work is as follows: In Section 2 (Modelling Approach), we describe the applied Lagrangian particle models together with a short description of the wind field calculation applied. Furthermore, we describe the radiological model for the calculation of the basic radiological parameters caused by the relevant exposure pathways. In Section 3, we present the validation tests of the LPM implemented in ESTE CBRN. The first test is a simple comparison against empirical formulas of Briggs sigma functions; the second test is a comparison against the JU2003. In this section, we also analyze the application of the whole calculation model, including the Lagrangian particle model and radiological impact model, as a system running on GPU as a real-time running system for emergency response. Finally, in Section 4 we provide a summary.

2. Modelling Approach

2.1. Modelling of Dispersion in Urban Atmosphere

Lagrangian particle models describe transport and dispersion by modelling particle motion inside a wind field composed of two components, a mean wind and a turbulent wind. The new positions of the particles (in general noted as $\vec{x}$) are given by the following equations:

$$x(t+\Delta t) = x(t) + u\,\Delta t = x(t) + U\,\Delta t + u_f \Delta t, \tag{1}$$

$$y(t+\Delta t) = y(t) + v\,\Delta t = y(t) + V\,\Delta t + v_f \Delta t, \tag{2}$$

$$z(t+\Delta t) = z(t) + w\,\Delta t = z(t) + W\,\Delta t + w_f \Delta t. \tag{3}$$

where Δt is the time step; u, v and w (in general noted as $\vec{u}$) are components of Lagrangian velocity; U, V and W (in general noted as $\vec{U}$) are components of the mean (Eulerian) wind field; and u_f, v_f and w_f (in general noted as $\vec{u_f}$) are the fluctuating components of Lagrangian velocity. The fluctuating velocity component follows Thomson's equations [15]:

$$du_f = a_1\left(\vec{x}, \vec{u}, t\right)dt + b_{1j}\left(\vec{x}, \vec{u}, t\right)dW_j, \tag{4}$$

$$dv_f = a_2\left(\vec{x}, \vec{u}, t\right)dt + b_{2j}\left(\vec{x}, \vec{u}, t\right)dW_j, \tag{5}$$

$$dw_f = a_3\left(\vec{x}, \vec{u}, t\right)dt + b_{3j}\left(\vec{x}, \vec{u}, t\right)dW_j. \tag{6}$$

where dWj(t) is an incremental of the Wiener process; a_i is the i-component of the deterministic term; and b_{ij} are the stochastic coefficients. In the equations we assume summation over j, from 1 to 3.

The implemented LPM is based on Thomson's simplest solution, which represents a simplification of Thomson's solution for the nonstationary, three-dimensional Langevin equation model for turbulent diffusion. In the case of rotating the coordinate system into the direction of the mean wind field, the fluctuating velocity components have the forms (here, assume rotation into the x direction; thus, $V_r = W_r = 0$, where the subscript r means variable in the rotated system) [7]:

$$\begin{aligned} du_{rf} &= \left[-\tfrac{C_0\varepsilon}{2}\left(\lambda_{11}(u_r - U_r) + \lambda_{13}w_r\right) + \tfrac{\partial U_r}{\partial z}w_r + \tfrac{1}{2}\tfrac{\partial \tau_{13}}{\partial z}\right]dt, \\ &+\left[\tfrac{\partial \tau_{11}}{\partial z}\left[\lambda_{11}(u_r - U_r) + \lambda_{13}w_r\right] + \tfrac{\partial \tau_{13}}{\partial z}\left[\lambda_{13}(u_r - U_r) + \lambda_{33}w_r\right]\right]\tfrac{w_r}{2}dt + (C_0\varepsilon)^{\frac{1}{2}}dW_1(t) \end{aligned} \tag{7}$$

$$dv_{rf} = \left[-\frac{C_0 \varepsilon}{2} \left(\lambda_{22} v_r \right) + \frac{\partial \tau_{22}}{\partial z} \left(\lambda_{22} v_r \right) \frac{w_r}{2} \right] dt + \left(C_0 \varepsilon \right)^{\frac{1}{2}} dW_2(t), \tag{8}$$

$$\begin{aligned} dw_{rf} &= \left[-\tfrac{C_0 \varepsilon}{2} \left(\lambda_{13} (u_r - U_r) + \lambda_{33} w_r \right) + \tfrac{1}{2} \tfrac{\partial \tau_{33}}{\partial z} \right] dt \\ &+ \left[\tfrac{\partial \tau_{13}}{\partial z} [\lambda_{11} (u_r - U_r) + \lambda_{13} w_r] + \tfrac{\partial \tau_{33}}{\partial z} [\lambda_{13} (u_r - U_r) + \lambda_{33} w_r] \right] \tfrac{w_r}{2} dt + \left(C_0 \varepsilon \right)^{\frac{1}{2}} dW_3(t) \,. \end{aligned} \tag{9}$$

where C_0 is the universal constant for the Lagrangian structure function. While sources refer to various ranges for this constant, we applied the value 5.6 [7]. ε is the mean rate of turbulence kinetic energy dissipation. dWj are uncorrelated, normally distributed variables with mean values of zero and standard deviations equal to 1. The components of the inverse tensor λ_{ij} are given by the following formulas:

$$\lambda_{11} = \left(\tau_{11} - \frac{\tau_{13}^2}{\tau_{33}} \right)^{-1}, \lambda_{22} = \tau_{22}^{-1}, \lambda_{33} = \left(\tau_{33} - \frac{\tau_{13}^2}{\tau_{11}} \right)^{-1}, \lambda_{13} = \left(\tau_{13} - \frac{\tau_{11} \tau_{33}}{\tau_{13}} \right)^{-1}. \tag{10}$$

where τ_{ij} are the components of the Reynolds stress tensor, which are defined using the variances of the velocity $\tau_{11} = \sigma_u{}^2$, $\tau_{22} = \sigma_v{}^2$, $\tau_{33} = \sigma_w{}^2$ and the Reynolds shear stress $\tau_{13} = \tau_{uw}$.

In our approach, the mean field, $\vec{U}$, the Reynolds stress tensor, τ_{ij}, and kinetic energy dissipation, ε, are calculated numerically as a solution of the Reynolds-averaged Navier–Stokes (RANS) equations of buoyant and incompressible fluid with the Boussinesq hypothesis. For not assuming the temperature variances, the complete fluid equations were reduced to the following form:

$$\frac{\partial U_i}{\partial t} + U_j \frac{\partial}{\partial x_j} (U_i) = \frac{\partial}{\partial x_j} \left(\nu \frac{\partial U_i}{\partial x_j} \right) - \frac{1}{\rho} \frac{\partial p}{\partial x_i} + g_i + \frac{\partial}{\partial x_j} \tau_{ij}, \tag{11}$$

$$\frac{\partial U_i}{\partial x_i} = 0, \tag{12}$$

$$\tau_{ij} = \nu_t \left(\frac{\partial U_i}{\partial x_j} + \frac{\partial U_j}{\partial x_i} \right) - \frac{2}{3} \left(k + \nu_t \frac{\partial U}{\partial x_i} \right) \delta_{ij}. \tag{13}$$

where ν is the kinematic viscosity; ν_t is the turbulent kinematic viscosity; ρ is the density; δ_{ij} is the Kronecker delta symbol; and k is the turbulent kinetic energy. Our approach focuses on the steady-state solution. Thus the time derivate term in Equation (11) is omitted. Equations (11)–(13) are solved using the approach of the SIMPLE algorithm (Semi-Implicit Method for Pressure-Linked Equations, [16]). The turbulence closure considered is the standard k–ε turbulence model. The calculation is performed using the OpenFOAM framework (available at URL https://www.openfoam.com/ (accessed on 23 June 2023)). The description of the boundary conditions depends on the known wind speed, friction velocity and Obukhov length, which are determined from the measured meteorological conditions [4].

The locally determined Reynolds tensor from (11)–(13) is regularized by transforming the Reynolds tensor into the coordinate system in which the main direction is defined by the mean wind field, and then by fitting it to the following form:

$$\tau_{11} = \sigma_u^2 = k_1 u_*^2, \ \ \tau_{22} = \sigma_v^2 = k_1 u_*^2, \ \tau_{33} = \sigma_w^2 = k_3 u_*^2. \tag{14}$$

where we assume $k_1 = 4$, $k_3 = 1.3^2 = 1.7$. This setting is well defined for a neutral condition [3] and applicable for stable and unstable conditions as an acceptable approximation. A similar parametrization is applied in [8].

In a real situation, we are faced with various types of events leading to various properties of airborne particles. A basic characteristic of a dispersed particle is that its size has a significant impact on the particle's motion in the turbulent wind field. Three categories of particles are assumed [3]:

(i) Particles with radii smaller than 5 μm and gases: Gravitational settling can be neglected due to their small particle size; thus, the flow in the mean wind field and the dispersion are dominant and considered in the calculation, while gravitational settling is not.

(ii) Particles with radii between 5 μm and 80 μm: Gravitational settling is non-negligible; thus, it is considered in the flow in the mean wind field and in the dispersion. Gravitational settling is represented by the terminal velocity v_t, given as (Stoke's law):

$$v_t = 2\,r^2 g \frac{\rho_{part}}{9\nu}\,. \tag{15}$$

(iii) Particles with radii greater than 80 μm: The main effect determining the vertical motion is gravitational settling. The gravitational fall begins at zero velocity and is described by the equation (Stoke's law for a small sphere):

$$\Delta v = \left(\frac{F_d}{m} - g\right)\Delta t = \left(9\nu\frac{v_d}{2\rho_{part}r^2} - g\right)\Delta t, \tag{16}$$

where Δv is the change in the vertical part of the velocity in time step, Δt; r is the radius; ρ_{part} is the density of the particles; ν is the dynamic viscosity (equal to 0.000018); and v_d is the air flow relative to the particle.

Deposition on surfaces is also treated with respect to the particle size:

(i) Particles with radii smaller than 80 μm: Each particle has its settling velocity. For particles larger than 5 μm, the settling velocity is approximated using their terminal velocity. For particles smaller than 5 μm, the settling velocity is approximated with a curve expressing the dependence between the particle radius and settling velocity (according to [3]). When the bottom face of the particle's cell is a ground surface, the particle leaves one part of its activity on the surface. This part of the activity is expressed using the settling velocity as:

$$\Delta A(t) = A(t)\left(1 - \exp\left(-\frac{\mathrm{v_d}\,\Delta t}{h}\right)\right). \tag{17}$$

where h is the height of the ground cell in which the particle is present. This approach is also called the mass reduction approach (e.g., please see the description of the FLEXPART model at URL https://www.flexpart.eu/ (accessed on 23 June 2023)). The vertical surfaces are treated in the same way, although a certain level of probability that the particle does not produce deposition on the walls is also assumed. This value is set to 85%, based on the observational data reported in [17].

(ii) Particles with radii greater than 80 μm: When crossing a horizontal surface, the particle is deposited totally because its contact with the surface is due to gravitational fall. When crossing a vertical surface, it is reflected by a specified probability. Otherwise is deposited totally. The probability of reflection is set to 85% (based on [17]).

2.2. Applied Computational Domain

In our study, with a focus on validation and analysis of applicability for emergency responses, we performed calculations for several sites. For each site, we applied the corresponding computational domain. All the calculation domains were prepared to perform a realistic calculation of atmospheric transport for the given modeled locations.

The simplest location was the area of the multi-unit nuclear power plant (NPP) in Mochovce, Slovakia. The location consists of several large buildings, including eight cooling towers, four reactor buildings, and many smaller buildings. The calculation domain, already applied in our study [18], has 377,474 cells and covers an area of 2.8 km × 1.4 km × 1 km. The horizontal dimensions of the cell are 20 m × 20 m. The parameters of the computational models are listed in Table 1.

Table 1. Basic parameters of the applied computational domains.

Parameter	NPP Mochovce	Košice	Oklahoma City
Discretization	70 × 140 × 39	185 × 242 × 39	275 × 330 × 46
Total number of cells	377,474	1,684,925	4,029,620
Horizontal resolution	20 m × 20 m	7 m × 7 m	5 m × 5 m
Maximum height above ground [m]	1000 m	189 m	194 m
Height resolution [m]	4.0 to 50 m	1.5 to 12 m	1.0 to 16 m

The largest model is represented by the calculation model of downtown Oklahoma City (OK, USA). It has 4,029,620 cells and covers an area of 1.4 km × 1.65 km × 194 m. The horizontal dimensions of the cell are 5 m × 5 m. The height of the cells varied between 1 m (for the layer near the terrain) and 16 m for the uppermost layer. This computational model, applied already in our previous study [4] within the UDINEE project, reflects the urban situation in 2003, since its main purpose is to provide a comparison with JU2003.

The third location studied is the historical center of Kosice, Slovakia. It is an urban area with a large number of middle-sized buildings and many streets. The computational domain was created using the OSM Buildings data. The domain contains 1,684,925 cells (with a cell size of 7 m) and covers an area of 1.3 km × 1.7 km × 189 m. Its parameters are summarized in Table 1. The actual location of the modeled area is shown in Figure 1, and the prepared computational domain is shown in Figure 2. In the study, the modeled historical center of Kosice was used to test the applied dosimetric algorithms.

(a) (b)

Figure 1. Views of the city center of Kosice. (**a**) A closer view directly on the modeled area; (**b**) A distant view with a visualized domain area (in blue).

A special calculation domain was prepared for the comparison to Briggs sigma functions. Its resolution is 20 m in the horizontal direction and 5 m in vertical resolution. The domain size was 5000 m × 1000 m × 500 m (=length × width × height). The domain does not include any buildings.

2.3. Modelling of Urban Conditions Using Empirical Briggs Formulas

Modelling of atmospheric transport in an urban environment can also be performed in a simplified approach when the real positions of the buildings are considered. In such a case, the wind field is defined uniformly in the horizontal direction and with adequate analytic formulas for vertical profiles (or with a measured vertical profile).

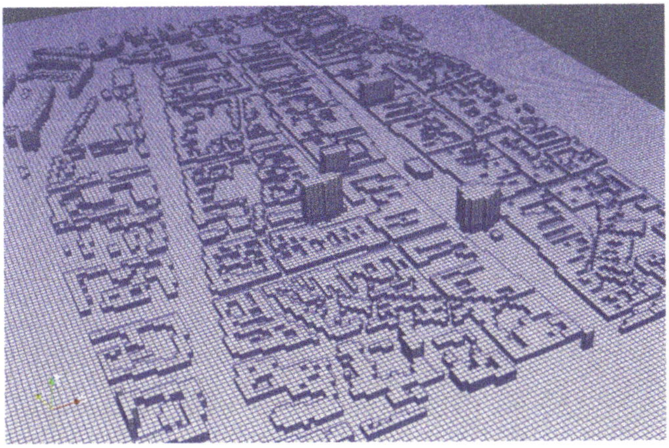

Figure 2. The city center of Kosice as visualized in the orthogonal computational mesh, using [19].

In this simplified approach, we analyze our implemented LPM. We perform calculations for neutral and stable conditions. The wind fields and turbulent fields in the computational domain at neutral and stable conditions are defined by equations [3]:

$$\frac{\kappa z}{u_*}\frac{\partial u}{\partial z} = h_m \tag{18}$$

$$\varepsilon = \frac{u_*^3}{\kappa z} e_m, \tag{19}$$

$$\tau_{11} = 4u_*^2 R_m^2, \tau_{12} = \tau_{23} = 0, \tau_{13} = u_*^2 R_m^2, \tau_{22} = 4u_*^2 R_m^2, \tau_{33} = 1.69 u_*^2 R_m^2. \tag{20}$$

where the constant κ = 0.4 is von Karman's constant; u_* is the friction velocity; z is the height above the terrain; z_0 is the surface roughness; and u is the wind at the height of z. The component of the Reynolds stress tensor, R, under neutral conditions is expressed as parametrized in [3]. R_m is equal to 1 in neutral conditions (neglecting the Coriolis corrections for assuming small heights), and it is equal to $(1 - z/z_i)$ in stable conditions. The parameter z_i represents the inversion height. h_m is equal to 1 in neutral conditions, and it is equal to $(1 + 5z/L)$ in stable conditions. e_m is equal to 1 in neutral conditions, and it is equal to $(1 + 4z/L)$ in stable conditions. L is the Monin-Obukhov length.

The surface roughness represents a parameter with which we can model the presence of buildings very generally, although it is not really in the model. We vary the surface roughness between 0.07 (corresponding to a locality with an almost flat terrain) and 1.0 (corresponding to an area with buildings).

In such a setting, we model a continuous release in a specific height above the ground. The atmospheric transport modeled through LPM under the described simplified urban and rural conditions has to be consistent with the Briggs plume sigma functions for urban and rural conditions [2]. These functions represent widely accepted plume behavior. They express the functional dependencies of standard deviations of concentration on the distance from the release point in vertical and lateral directions, as summarized in Table 2.

2.4. JU2003 Experiment

The primary comparison of the implemented LPM was performed using the experimental data from Joint Urban 2003, conducted July 2003 in Oklahoma City, USA e.g., [20,21]. The experiment consisted of ten Intensive Observation Periods (IOPs) during which SF_6 was released as a tracer gas and its concentration was measured at various locations in downtown Oklahoma City.

Table 2. The dependence on the downwind distance of the Briggs sigma functions for various stability categories and environmental conditions (rural, urban).

Pasquill Stability	Urban σ_y (m)	Urban σ_z (m)	Rural σ_y (m)	Rural σ_z (m)
A (unstable)	$0.32 \times (1 + 0.0004x)^{-0.5}$	$0.24 \times (1 + 0.001x)^{-0.5}$	$0.22 \times (1 + 0.0001x)^{-0.5}$	0.20x
D (neutral)	$0.16 \times (1 + 0.0004x)^{-0.5}$	$0.14 \times (1 + 0.0003x)^{-0.5}$	$0.08 \times (1 + 0.0001x)^{-0.5}$	$0.06 \times (1 + 0.0015x)^{-0.5}$
F (stable)	$0.11 \times (1 + 0.0004x)^{-0.5}$	$0.08 \times (1 + 0.00015x)^{-0.5}$	$0.04 \times (1 + 0.0001x)^{-0.5}$	$0.016 \times (1 + 0.0003x)^{-1}$

Our validation study analyzed the instantaneous puff releases of IOP 3 and IOP 5. The basic input data is taken from [20], including the release setup, meteorological data and tracer measurements. The puff releases of IOP 3 were performed through balloon bursting, with the corresponding SF_6 mass of 1000 g at the height of 2 m above the ground. The focus of our study was the first three puff releases, which started at 9:00, 9:20 and 9:40, respectively (all times are in CDT, central time zone). The puff releases of IOP 5 were conducted the same way, but the applied release masses were 500 g. The corresponding start times were 15:00, 15:20 and 15:40, respectively. Both IOPs were situated at the same location, near the Botanical Gardens (see Figure 3).

Figure 3. Downtown Oklahoma City. Displayed are the locations of the TGA tracers for IOP 3 and the release points (yellow circles).

The tracer measurement data included locations of samplers and their maximum reported concentrations (in pptv, parts per trillion volume). The instantaneous concentrations were measured using Trace Gas Analyzers (TGAs) with a frequency of 2 Hz. The upper measurement range varied between 11,000 pptv and 23,000 pptv. The lower limit was about 150 pptv (MLOD).

The meteorological data used in our study were also taken from [20], where the 15-min averages of sodar winds at 100-m AGL are provided. The meteorological measurements were used to define the boundary conditions based on [3,22]. The wind and turbulence fields for the computational domain of Oklahoma City were calculated by applying computational fluid dynamics techniques. Specifically, the wind fields resulted as solutions of the Reynolds-averaged Navier–Stokes equations for buoyant, incompressible fluid using the Boussinesq approximation [4]. The turbulence is governed by the standard k–ε model.

This approach to the urban wind field calculation was applied in our earlier analysis of JU2003 [4]. There, we showed that our method of wind field calculation was applicable. The ratio of the modeled to the measured value of the wind speed measurements inside the downtown area was close to unity, and the count of this ratio, being between 0.5 and 2.0 (expressed as the variable factor of 2, FAC2), was equal to 80%. The turbulent kinetic energy was generally underestimated by 30–40%.

2.5. Modelling of Radiological Effects

A set of radiological parameters is modeled and calculated. The most important for radiological impacts are effective doses from external exposure and committed effective doses from inhalation. The radiation exposure occurs: (i) during the initial phase and passage of the cloud and (ii) after the passage of the cloud. The first group of radiological parameters include external irradiation from the cloud, external irradiation from the deposited material and internal irradiation via inhalation of airborne radioactive particles and gases. After the passage of the cloud, external exposure from deposited radionuclides on the ground and walls and internal exposure from inhalation of resuspended radioactive material are considered.

The dose rate DR_{inhal} of the committed effective dose by inhalation is calculated for a considered radionuclide n as:

$$DR_{inhal}\ (n, age, x, t) = C_t(n,x)\ CF_{inhal}\ (age,n)\ BR(age). \tag{21}$$

where $C_t(n,x)$ is the short-time average concentration at the point of interest x in time t. BR is the breathing rate, depending on age. CF_{inhal} is the conversion factor for the committed effective dose due to inhalation, depending on age and nuclide. The integral dose is calculated by using the corresponding time-integrated concentration. In a numerical approach, the concentration in a given location is equal to the concentration value in the corresponding computational cell.

The dose rate DR_{depo} of the external dose by fallout is calculated for a considered radionuclide n as:

$$DR_{depo}\ (n,x,t) = \Sigma_{domain,y}\ D_t\ (n,y)\ CF_{depo}(n,x,y)\ SF(x,y). \tag{22}$$

where $D_t(n,x)$ is the deposition value (in Bq/m^2) at time t and location x. The calculation represents the summation over all the terrain and building surfaces of the calculation domain. The $CF_{depo}(n,x,y)$ is the external dose conversion factor for deposition (external dose at point x from the surface at point y). The $SF(x,y)$ is the shielding factor, representing the shielding effect of the buildings. In our approach, it is equal either to 1, if there is no obstacle (e.g., a building), or to 0, if there is at least one obstacle along a straight line between point x and the surface of point y. The corresponding integral dose is calculated by the summation $\Sigma_t\ \Delta t\ DR_{depo}(n,x,t)$, where we sum the corresponding time interval.

To calculate deposition doses on the calculation domain, the surface deposition $D_t(n,x)$ in the given location equals the surface activity value on the corresponding computational surface tile (=ground or building). The shielding factor $SF(x,y)$ is calculated by applying the Muller-Trumbore intersection algorithm [23]. This algorithm is applied in order to reduce computational time. The calculation considers each pair of tile centers where one center is the source and the other is the receiver, with each building surface tile used as a tested shielding object. The $CF_{depo}(n,x,y)$ is represented by a library of pre-calculated factors. This factor is equal to the impact (=effective dose) caused by a tile of unit surface activity located between 1 m and 2000 m from the point of interest. The library was prepared using numerical simulations in MCNP 5 [24].

The dose rate DR_{cloud} of the external dose by cloud shine is calculated for the considered radionuclide n as:

$$DR_{cloud}\ (n,x,t) = \Sigma_{domain,y}\ C_t(n,y)\ CF_{cloud}(n,x,y)\ SF(x,y). \tag{23}$$

The $CF_{cloud}(n,x,y)$ is the cloud analogy to $CF_{depo}(n,x,y)$. It is the conversion factor for external exposure at point x caused by airborne activity at point y. Similar to $CF_{depo}(n,x,y)$, $CF_{cloud}(n,x,y)$ is represented by a library of pre-calculated factors. The calculation is performed the same way for ground shine. The volume concentration $C_t(n,y)$ in the given location is equal to the cell concentration value. The calculation considers each cell center the source, each bottom surface tile the impacted point and each building surface tile a tested shielding object.

3. Results

3.1. Model Comparison to Briggs Formulas

The first test of our implementation was performed in a quasi-urban approach. The surface was represented by a plain without a building. We applied the calculation domain with a flat terrain, described in Section 2.2. The turbulence and wind fields were determined using analytic formulas for vertical profiles, defined by the Equations (18)–(20). The modeled dispersion results were compared to the Briggs plume sigma functions for urban and rural conditions [2]. To validate the implemented LPM, we focused on neutral conditions and stable conditions. In the case of neutral conditions, we generated several states of meteorological conditions. We modeled a continuous release at the height of 50 m above ground. We generated a couple of meteorological conditions to show that the LPM aligns with the Briggs sigma functions. In the case of neutral conditions, we analyzed two cases for the wind speed: one moderate value of wind speed (2.4 m/s) and one higher value (4.8 m/s). Three distinct values of surface roughness were assumed, 0.07, 0.3 and 1.0, covering areas from flat terrain through small buildings and large buildings.

For the modeled plume, we evaluated the first and second moment (i.e., the mean and variance) of the spatial distribution of the concentration in air as a function of the distance from the release point separately in the horizontal lateral direction and in the vertical direction. The resulting variance σ was compared to the Briggs formulas. They are expressed in a general form of $Ax(1 + Bx)^c$, where c is either -0.5 or -1, depending on the weather conditions (see Table 2). The analysis is focused on parameter A, which primarily determines the behavior at small and medium distances.

For surface roughness 1.0 at neutral conditions, we obtained A of about 0.15 in the lateral direction, which is close to the value of 0.16 for the urban σ_y. In the vertical direction, the interpolated value of A is about 0.11, which is again close to the urban Briggs σ_z value of 0.14. For the surface roughness of 0.07 at a neutral condition, we obtained A of about 0.063 in the lateral direction and 0.042 in the vertical direction. These values are consistent with the coefficients for rural conditions, which are 0.08 and 0.06 (in lateral and vertical directions, respectively). In all cases, the results show low dependence on wind speed. The order of magnitude of interpolated values of parameter B is comparable with the Briggs formula. Interpolated B is satisfactorily small. A more accurate determination of B is beyond the range of our analysis. An overview of the meteorological conditions considered and the resulting interpolated coefficients are listed in Table 3.

The dependencies on the distances from the release point modeled by LPM and predicted by the Briggs formulas are presented in Figure 4. The results of LPM implemented in ESTE for high surface roughness reproduce the analytical dependencies for urban conditions sufficiently. Similarly, the modeled results for low surface roughness reproduce the analytical behavior in rural conditions. The case with an intermediate surface roughness value represents the middle ground between the urban and rural conditions. As for neutral conditions, we obtained a similar consistency between the Briggs sigma functions and the modeled variance of stable conditions (see Table 3).

In Figure 4, we identify one interfering effect influencing dispersion in the vertical direction. The release is situated at a height of 50 m. The performed analyses can be relevant only to the distances where the plume reaches the ground or the upper boundary. The plume reflects effectively by reaching such a surface. The calculated variance beyond this distance does not represent the situation expressed by the analytical formulas. In the case of urban conditions, Briggs z-sigma is 25 m at the approximate distance of 150 m; at this distance, the plume reaches the boundaries ($2 \times \sigma_z = 50$ m). Therefore, the curve of the modeled variance is more deviated from the analytical lines in the case of dispersion in the vertical direction, and the modeled curve bends beyond the distance of about 200 m considerably. This is not the case with the horizontal dispersion because the test computational domain is sufficiently wide to avoid reflecting in the y-direction. This allows us to test the behavior of the plume up to a downwind distance of about 600–700 m.

Table 3. Input parameters for model comparison with Briggs sigma functions and the resulting interpolated values of coefficients of Briggs sigma functions.

Case	Pasquill Stability	Wind Speed [m/s]	Surface Roughness	Monin–Obukhov [m]	Friction Velocity	Interpolated Parameters for *y*	Interpolated Parameters for *z*
1	D	2.4	1.0	-	0.4169	A = 0.147 B = 0.0005	A = 0.113 B = −0.0018
2	D	4.8	1.0	-	0.8338	A = 0.147 B = 0.0005	A = 0.116 B = −0.0019
3	D	2.4	0.3	-	0.2736	A = 0.097 B = 0.0001	A = 0.066 B = −0.0003
4	D	4.8	0.3	-	0.5472	A = 0.095 B = 0.0000	A = 0.064 B = −0.0001
5	D	2.4	0.07	-	0.1935	A = 0.063 B = −0.0006	A = 0.042 B = 0.0011
6	D	4.8	0.07	-	0.3870	A = 0.064 B = −0.0005	A = 0.041 B = 0.0011
7	F	1.9	1.0	610	0.32	A = 0.953 B = 0.0009	A = 0.067 B = −0.0013
8	F	1.9	0.07	610	0.15	A = 0.059 B = 0.0007	A = 0.032 B = −0.0001

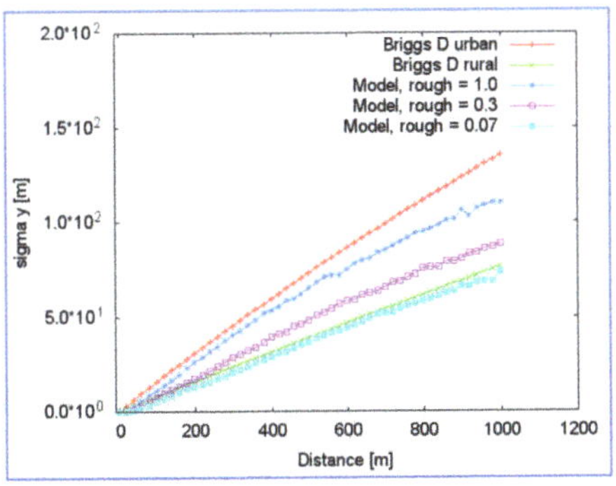

(a)

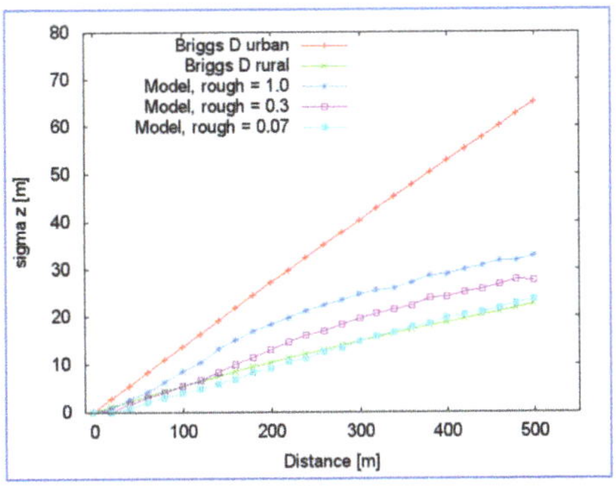

(b)

Figure 4. Comparison of Briggs sigma functions for stability category of D, with modeled dispersion by LPM at various levels of surface roughness. (**a**) Comparison of dispersion in lateral (y) direction; and (**b**) Comparison of dispersion in vertical (z) direction.

3.2. Model Comparison to JU2003 Experiment

Our comparison to the JU2003 experiment consists of two steps. First, we determined the urban wind and turbulence fields for the given IOP. Second, we performed the atmospheric transport calculations and finished with comparisons of modeled detector responses with the measured responses. We focused on the calculation of the maximum values of the detectors and the times when the maximum concentrations were reached (i.e., the peak times).

We calculated the steady-state flow during the specific hours when the puff releases occurred; specifically, we calculated the mean hour flow field from 9:00 to 10:00 h for IOP 3 and the mean hour flow field from 15:00 to 16:00 h for IOP 5. The application of the hourly mean field has the effect that our modeled tracer measurements reflect: the mean maximum values over all three puff releases occurred in a specific hour. Similarly, the modeled peak times are compared to the mean peak times. We omit the approach to model each separate puff release because modelling each puff requires working with more detailed meteorological data that is unavailable in our analyses. The puff movements lasted about 5 min in the neighboring vicinity where the detectors were distributed. The applied meteorological data [22] were 15-min averages. Instead of assuming negligible fluctuations (such an assumption is potentially false) of the wind field within 15 min when performing a direct comparison with separate puff releases, we assumed that three puff releases could represent an ensemble to evaluate the mean behavior corresponding to the hourly mean meteorological conditions. To support this approach, we emphasize that the releases occurred close to the edges of buildings, dividing the puff transport into different streets. Therefore, unmeasured small time-scale fluctuations could negatively affect comparison on the level of separate puff releases.

We applied 30,000,000 particles in the validation. Such a high number was required to gain the statistical resolution comparable with the lower measurement. The measurement range was from 150 pptv to about 12,000 pptv. The achieved resolution of the concentration calculation was on the level of 100–200 pptv.

The modeled values of maximum tracer responses are summarized in Table 4. To help our analysis, we follow detectors' positions compared to the modeled puff transport. Figure 5 visualizes the time integral of concentration in air 2 m above the ground. Here, the puff trajectory at this vertical level is called the approximate continuous region of the highest time integrated concentration, starting at the release point, continuing along several streets, and ending when leaving the domain. Naturally, the puff movement follows the street canyons first and the mean wind field later when moving in the region without a dominant built-up area. In all cases, we evaluated the uncertainty of the modeled values in addition to the estimation of the statistical error. The uncertainty is based on the variation of the calculated maximum concentration in the cell defined by the location of a detector and its neighboring cells. This estimated variance (=uncertainty) is given in Table 4.

In the case of IOP 3, the closest tracers to the puff trajectory are measurements 0, 4, 6 and 7. The mean measured maximum values for 4, 6 and 7 are bounded only from the bottom (outside the range of the TGA sampler), and the corresponding values modeled by the LPM are also above the reported range of the TGA sampler. Measurement 0 is sufficiently far from the release point, and here the mean maximum value is measured and modeled consistently. Measurements 1 and 8 are slightly farther from the puff transport, and we obtained underestimated values for them. Measurements 3 and 9 are far from the modeled puff trajectory. The actual measurements provided oscillated maxima, which suggest that they were on the edge of the real puff transport. They were once affected and once not. The modeled values are comparable with those for some puffs, but the mean values were again underestimated. Measurements 2 lies farther away, and the measured and modeled values are at the level of 0 pptv.

In the case of IOP 5, the closest tracer to the release point is measurement 9, and both modeled and measured maxima are correspondingly above the detection capabilities. Measurements 2 and 8 are the farthest; they are located at the end of the highly built-up area and in the direction of the mean wind from the release point. The modeled maximum concentrations are consistent with the mean measured value. In the cases of measurements 6 and 7, the modeled maxima are close to the medium maximum value (the middle value when ordered from the smallest to the largest), but far when compared to the mean value.

Measurements 1 and 4 lie visibly far from the expected puff transport. Both are modeled as having a negligible maximum concentration, which is the case for measurement 1 but not for measurement 4. Measurement 2 is located farther away, and the measured and modeled values are at the level of 0 pptv. Finally, measurement 0 has three almost stable maxima, and the modeled value is underestimated by more than a factor of 2.

Table 4. Maximum measured TGA-sampled concentrations (pptv), taken from [21], compared to modeled values.

IOP 5—TGA No.	Puff 1-Concentration [pptv]	Puff 2-Concentration [pptv]	Puff 3-Concentration [pptv]	Mean-Concentration [pptv]	Modeled-Concentration [pptv]
0	9890	5810	5800	7166	1060 +/− 100
1	-	-	663	221	132 +/− 26
2	13,500	4570	1870	6646	5310 +/− 510
4	>25,300	4380	2250	≈10,643	0 +/− 0
6	48	6710	>23,100	≈9952	7440 +/− 26
7	0	8870	>24,500	≈11,123	4320 +/− 1500
8	12,100	4290	2020	6137	10,300 +/− 800
9	9210	>12,100	>12,200	≈>11,170	110,700 +/− 6400
IOP 3—TGA No.	**Puff 1-Concentration [pptv]**	**Puff 2-Concentration [pptv]**	**Puff 3-Concentration [pptv]**	**Mean-Concentration [pptv]**	**Modeled-Concentration [pptv]**
0	12,600	12,300	19,600	14,833	12,700 +/− 2100
1	3130	0	4180	2437	0 +/− 20
2	0	0	202	67	0 +/− 0
3	10,100	1710	12,500	8100	2920 +/− 100
4	>13,000	>12,800	>25,800	>12,800	45,700 +/− 5300
6	>11,900	21,600	>23,700	≈21,600	35,900 +/− 200
7	>12,200	>12,100	>11,900	>11,900	390,000 +/− 98,000
8	4808	163	10,700	5224	530 +/− 0
9	0	329	11,700	4010	530 +/− 200

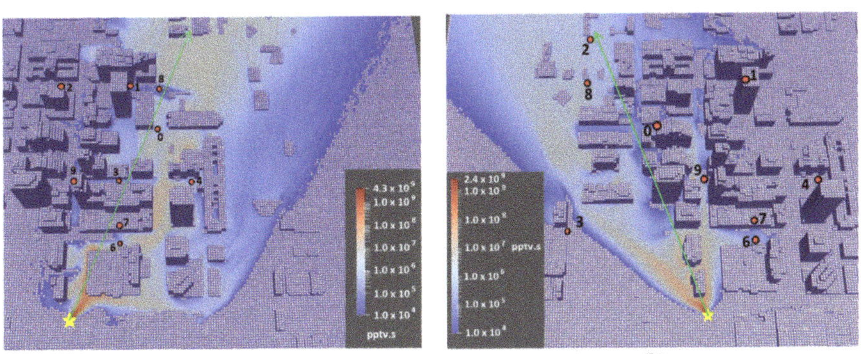

(a) (b)

Figure 5. The modeled time integral of concentration [pptv.s] for IOP 3 (**a**) and for IOP 5 (**b**). The release points (yellow star), the locations of the TGA tracers (red circles) and hourly mean wind direction (green arrow) for both IOPs are displayed too.

To summarize, for the close measurements, which lie along the puff movements, the measured maxima are clearly above the detection range, and the modeled results were also above said range. For the distant measurements in the direction of the mean wind, the modeled maxima are mostly within a factor of 2. The measurements lying aside are reproduced partially (i.e., in some cases, the measurements fluctuate, and these fluctuations have no footprint in the modeled results). They could not be present because the wind field calculation as a one-hour mean already removes them. An additional finding is that the model has a small agreement in the case of cross streets, such as measurement 0 for IOP 5 (underestimated 7-fold) or measurement 3 for IOP 3 (underestimated 2.5-fold).

The second quantity we compared was the times when the maximum values of the tracer concentration were reached, the so-called peak time. Here, we made some objective data selections because not all measurements were fully applicable for the peak time analyses because either the modeled or/and the measured peaks were present (such as measurement 2 for IOP 3).

For IOP 3, we analyzed measurements 0, 1, 3, 4 and 6. In measurement 1, the modeled cloud missed the measurement location, as mentioned above; thus, a comparison was not done. For measurements 0 and 3, the ratio between the modeled and measured times is 0.8. For measurements 4 and 6, the time of the maximum value is estimated due to the maximum value overreaching the limit of the detector. These estimated ratios between the modeled and measured peak times range from 0.4 to 0.7.

The same comparison was performed for IOP 5. For measurements 0, 2 and 8, the ratios between the modeled and measured times are 0.8, 1.1 and 1.0, respectively. All these measurements lie close to the line of the wind direction. For measurements 6, 7 and 9, the ratios of the modeled and measured peak times are 0.4, 0.7 and 0.6, respectively. As specified above, these three measurements lie more or less aside, and the modeled peak times are underestimated. These ratios were approximated because the time series had missing values when the concentration was above the upper limit. Measurement 4 was not analyzed because the modeled response was zero. In measurement 1, there were neither measured nor modeled actual peak times.

In general, we see that the peak times were reproduced acceptably. When the peak time is measured and determined, we obtain the ratio model vs. observation approximately equal to 0.9; in other words, we underestimated the peak time by 10%. When the exact position of the peak time is not determined through measurement (missing values in time series), these ratios lie in the range of 0.4 to 0.7. If we included the shape of the time series, the actual values would shift toward higher ratios (to 0.7). The model still reproduces the observation within the factor of 2.

A similar comparison was performed in our earlier analysis [4] within the UDINEE project. The meteorological fields were computed in said project using the same approach. However, we used the Eulerian model for the dispersion calculation in [4]. Within the UDINEE project, only the cases in which the peak time was measured and determined were considered. The model vs. observation ratio of the peak time was equal to about 1.1 on average. It corresponds to the value of 0.9 modelled by LPM. Our presented results of LPM are comparable to our earlier Eulerian dispersion analyses.

3.3. Computational Time Analysis

We also analyzed the feasibility of applying the developed model within a real-time decision support system. This means that we focused on the conditions under which the model is applicable to respond to a real situation. The analyzed model in this section includes two parts: the above-validated Lagrangian particle model and the dosimetry model for calculating the corresponding radiological impacts.

The dosimetry model for radiological impacts includes the calculation of three pathways, as defined in Section 2.5 (Modelling of Radiological Effects): (a) impacts due to inhalation of airborne radioactive particles and gases, (b) impacts due to external exposure from deposited material (so-called ground shine), and (c) impacts due to external exposure from airborne activity directly (so-called cloud shine).

In the case of an immediate release of a radioactive contaminant, the duration of the event, including the atmospheric transport through the impacted urban area, is at the level of minutes. Therefore, modelling the corresponding event as a part of an emergency response must be on the same time scale. Consequently, we can require that the total calculation time of an event, including the evaluation of radiological impacts, shall be between several minutes and 30 min.

In order to test our complex calculation model, we prepared three locations of various sizes, as described in Section 2.3 (Applied Computational Domains). Their parameters are summarized in Table 1.

Both calculation models, LPM and radiological model, were implemented in C# using parallel computing to run on a single multi-core processor (CPU) and graphic card (GPGPU). The CPU calculations were performed on a computer with an Intel Core i7-1255U 12th Gen processor, 1700MHz, and 32 GB of RAM. The GPGPU calculations were performed on NVIDIA RTX A2000 12 GB.

In our analysis, we tested separately the time performance of the LPM and the radiological model. In the case of the LPM, we determined the time required to evaluate 1 min of atmospheric transport. We modeled an immediate release such that all particles were released simultaneously at the start of modelling of the atmospheric dispersion. Only a negligible part of the particles left the computational domain in the first few minutes. In the comparisons, we present the average duration of a 1-min atmospheric transport. All test calculations were done for the same defined time step, whose value was 0.05 s. This value was below the Courant limit. This time step was also used to compare the computational performance in the same conditions for all three domains. Thus, all results in 1 min after the start of atmospheric transport correspond to 1200 steps. The resulting computational times for the applied 3D models and various numbers of particles are summarized in Table 5.

Table 5. Time performance of implemented LPM for the applied urban models for 1-min atmospheric transport. Results for CPU as well as GPGPU.

	Oklahoma City	City of Košice	NPP Mochovce
Total cell number	4,029,620	1,684,925	377,474
Number of particles	Time [min] CPU/GPGPU	Time [min] CPU/GPGPU	Time [min] CPU/GPGPU
10,000	0.9/0.1	0.4/0.1	0.1/0.1
100,000	1.4/0.3	0.8/0.2	0.5/0.2
250,000	2.2/0.4	1.6/0.4	1.3/0.4
500,000	3.9/0.7	2.9/0.7	2.4/0.7
1,000,000	6.8/1.3	5.8/1.2	5.0/1.2

The performance of the implemented model was much better when it ran on GPU. This outcome is evident because the computational performance of GPGPUs is much higher than that of CPUs. The times for the GPGPU showed a very low dependence on the size of the computational domain compared to the CPU. The almost linear dependence of the computational times on the number of particles for GPGPU was observed for up to 30,000,000 particles. Modelling 1 min of atmospheric transport using 30,000,000 particles lasted 34 min.

The results summarized in Table 5 must be compared with some requirements about the accuracy of results. A compact urban area, such as a historical center, is typically between 2 and 4 km^2 (examples include the old town of Vienna or Prague or downtown Oklahoma City), equal to a domain side length of 1.5 km to 2 km. Therefore, we can assume the mean travel distance of the radioactive cloud is about 1000 m.

We can utilize the urban Briggs sigma functions to estimate the size of the cloud in such distances from the event location. The mean size (over category stability) of a cloud could be approximated by the value of $2 \times \sigma$, which is about 340 m in the distance of 1000 m. If we further assume a mean cell volume of 200 m^3 (=10 m $\times$ 10 m $\times$ 2 m), the cloud covers approximately 400,000 cells. This means that there are about 50,000 cells in the 1-sigma inner region and 350,000 cells in the outer region. In the inner region, we have 2/3 of all particles; thus, if we assume about 500,000 particles, then there are 6–7 particles in one cell, on average. Similarly, we would obtain a half particle per cell for the outer region. Globally, the minimal requirement to see the puff (its volume) on the level of 90–95% in the distance of 1000 m is to have at least one million particles. In such a case, the number of particles per cell is about one particle. We have to remark that the air concentration (similarly the deposition on terrain and other quantities) will decrease about 5 to 6 orders of magnitude from their original values at the event point.

To visualize the estimated numbers of particles, we performed a calculation of the deposited materials in our computational model of Oklahoma City. In the test case, we assumed uniform small particles—all particles had the same activity and deposition velocity. We studied the ground deposit after the passage of a radioactive cloud to observe the effect of the number of particles applied. We applied three different numbers of particles, 10^5, 10^6 and 5×10^6. Computation of the 10-min dispersion of a radioactive cloud lasted 3 min, 12 min and 60 min, respectively. The example of differences in deposited patterns between 10^5 and 10^6 particles is visualized in Figure 6.

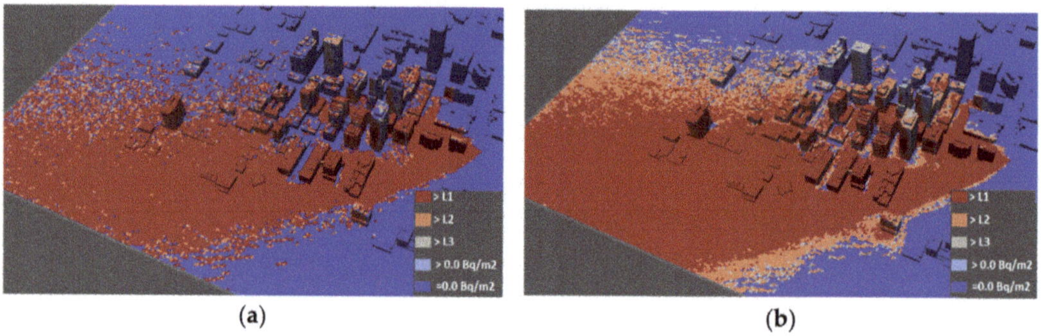

Figure 6. Example of deposit on terrain for the model of Oklahoma City. (**a**) The figure corresponds to the run with 100,000 particles, and (**b**) the figure for 1,000,000 particles. The surfaces impacted by deposition are displayed using the categorization of L1, L2, L3 and non-zero.

The maximum value (=D_{max}) is in all three cases the same, fluctuating by 2–3%. The area of the region with deposits above the level L1 (=$10^{-3} \times D_{max}$) changes by less than 1% from 10^6 to 5×10^6, and by 10% from 10^5 to 10^6. The area of the region with deposits above the level L2 (=$10^{-4} \times \mathrm{D}_{\mathrm{max}}$) changes by about 2% from 10^6 to 5×10^6, and by 16% from 10^5 to 10^6. Finally, the area of the region with deposits above the level L3 (=$10^{-5} \times D_{max}$) changes by about 7% from 10^6 to 5×10^6, and by 24% from 10^5 to 10^6.

Applying about 1,000,000 particles reduces the statistical error of the main parameters that were modeled by 1–2%. The results were obtained within the approximate time of 10 min. The conclusion is that the implemented LPM is generally applicable in an emergency response in which results are required in short time periods.

The second part of the simulation is the radiological impact calculation. In the case of the computational domain of Kosice, we evaluated the three primary effects of the dose on the affected population. We tested two possible computing configurations (CPU and GPGPU), the same as for the dispersion analyses described above.

The radiological impact calculation has a different scale dependence than the dispersion calculation. In the radiological impact calculation, the computational time depends only on the number of cells (mainly in horizontal directions) or the number of cell surfaces; in other words, the computational time depends on the domain characteristics and not on the number particles. In our implemented computational approach for cloud shine (the ground shine and inhalation doses do not require such a technique), we work with an approximation in which the cells are grouped into larger supercells forming from one central cell and neighbors or even next-to-neighbors. These various levels of grouping are performed when the cells are beyond specified distances from the impacted point to minimize the approximation error. Therefore, the larger computational domain of Oklahoma City has approximately two times longer performance times for radiological impacts. We always consider buildings' shielding effect, as described in Section 2.2 (Modelling of Radiological Effects).

The resulting performance is summarized in Table 6. For both configurations, we obtained a similar result. The whole calculation lasted less than 30 s. The radiological impacts are relevant only in a few time intervals after the event (e.g., 1 min, 3 min and 5 min and 10 min after the event). Therefore, the computational time for radiological impacts is low: 2 or 3 min in total. We can conclude analogously that the radiological impact calculation for dispersion is performed in real-time.

Table 6. Time performance in seconds of radiological impact calculations for the model of Košice.

	GPGPU Time [s]	CPU Time [s]
Inhalation	<1	<1
Groundshine	3	4
Cloudshine	11	13

4. Conclusions and Summary

In this study, we implemented a Lagrangian particle model in the nuclear decision support system ESTE CBRN. The LPM is based on Thomson's solution to the nonstationary, three-dimensional Langevin equation model for turbulent diffusion [7].

The validation was performed in two experimental environments. First, we modeled the cloud dispersion resulting from a continuous release, which is well described by Briggs sigma functions. In this comparison, a quasi-urban test field was modeled because the effect of buildings was included only implicitly in the vertical wind profiles. These wind profiles define the wind and turbulence fields in the entire computational domain without the actual buildings. Thus, this approach is more of a consistency test than a real validation. In this test case, we showed that the Briggs sigma functions are reproduced through LPM calculation satisfactorily. That means that the vertical wind profiles with high surface roughness corresponding to the presence of large buildings lead to a plume behavior similar to the urban Briggs sigma functions. Analogously, the profiles with low surface roughness lead to downwind dependence similar to that shown by Briggs sigma functions for rural conditions.

The implemented LPM was compared to the Joint Urban 2003 Street Canyon Experiment for instantaneous puff releases. This comparison setup represents an objective validation approach. We calculated the hourly mean wind fields for two IOPs, and we performed a comparison of the modeled atmospheric dispersion to mean behavior over three realized releases during the specific hour. We compared the maximum concentrations and peak times measured during two intensive operational periods of the JU2003 Experiment, IOP 3 and IOP 5. The modeled peak times are mostly 10 to 20% smaller than the measured ones. The maximum concentrations are reproduced moderately. The best modeled results were obtained when the measurements were laid in the approximate path of the puffs. Less accurate results were at locations farther from the main puff transport area, which are generally difficult places for modelling. These values were usually underestimated by more than a factor of 2. Thus, under the conditions we applied in the presented comparison to JU2003, the results are in moderate and adequate agreement.

The main concern of the work was to build a system potentially applicable for crisis situations to perform urban atmospheric transport calculations and radiological impact calculations within strict time requirements. These requirements assume that the whole calculation will be performed with a speed similar to the time scale of real events. Therefore, the urban model created is based on a parallel computation technique whereby particles are modeled simultaneously. The whole system is created as an application running on GPU technology.

In the final comparison, we prepared several computational 3D models of urban areas. We studied the computational times of the atmospheric dispersion calculation and the radiological parameter calculations. The focus is on the calculation on a GPU, which enables the calculation to be performed using a single computer. We show that the computational time is at the level of minutes, which copes with the need for crisis management. At the same time, a sufficient number of particles is applied to obtain acceptably accurate results. In our implementation, with about 1,000,000 or more modeled particles applied, the errors of resulting parameters are at several per cent.

Similarly, we show that the radiological parameters can be calculated quickly, which again fully complies with the needs of crisis management. The most time-demanding part is the calculation of effective dose from airborne activity, which is defined computationally as the sum of contributions from domain cells, including the shielding effect of the building walls. We show that contributions of all exposure pathways can be calculated in tens of seconds. Overall, we demonstrated that the whole computation of atmospheric dispersion and radiological impacts on the urban environment can be performed using a powerful computer in a time scale equal to the time scale of the real event.

Author Contributions: Conceptualization, Ľ.L., P.Č., E.F., M.M. (Michal Marčišovský), M.M. (Mária Marčišovská), M.C.; methodology, Ľ.L., P.Č., M.M. (Michal Marčišovský); software, Ľ.L., M.C.; validation, Ľ.L., P.Č., E.F.; formal analysis, Ľ.L., P.Č., E.F., M.M. (Michal Marčišovský), M.M. (Mária Marčišovská); investigation, Ľ.L., P.Č., E.F., M.M. (Michal Marčišovský), M.M. (Mária Marčišovská); resources, Ľ.L., P.Č., E.F., M.M. (Michal Marčišovský), M.M. (Mária Marčišovská); writing—original draft preparation, Ľ.L., P.Č., E.F., M.M. (Michal Marčišovský), M.M. (Mária Marčišovská); visualization, Ľ.L., E.F., M.C.; supervision, Ľ.L., P.Č., E.F., M.M. (Michal Marčišovský), M.M. (Mária Marčišovská); All authors have read and agreed to the published version of the manuscript.

Funding: This research was funded by Slovak Research and Development Agency (The Ministry of Education, Science, Research and Sport of the Slovak Republic), grant number APVV-20-0628. The research was co-funded also from internal resources of the company ABmerit, Ltd., where the authors are employed.

Institutional Review Board Statement: Not applicable.

Informed Consent Statement: Not applicable.

Data Availability Statement: The data of this analysis are publicly not available.

Conflicts of Interest: The authors declare no conflict of interest.

References

1. Sykes, R.I.; Parker, S.F.; Henn, D.S.; Cerasoli, C.P.; Santos, L.P. *PC-SCIPUFF Version 1.3—Technical Documentation*; ARAP Report No. 725; Titan Corporation: San Diego, CA, USA, 2000.
2. Briggs, G.A. *Diffuion Estimation for Small Emissions, ATDL Contribution File No. 79*; Atmospheric Turbulence and Diffusion Laboratory: Oak Ridge, TN, USA, 1973.
3. Hanna, S.R.; Briggs, G.A.; Hosker, R.P. *Handbook on Atmospheric Diffusion*; Technical Report DOE-TIC-11223; Office of Energy Research, US Department of Energy: Washington, DC, USA, 1982.
4. Liptak, L.; Fojcíková, E.; Carny, P. Comparison of the ESTE CBRN Model with the Joint Urban 2003 Experiment. *Bound. Layer Meteorol.* **2019**, *171*, 439–464. [CrossRef]

5. Hernández-Ceballos, M.A.; Hanna, S.; Bianconi, R.; Bellasio, R.; Mazzola, T.; Chang, J.; Andronopoulos, S.; Armand, P.; Benbouta, N.; Čarný, P.; et al. UDINEE: Evaluation of Multiple Models with Data from the JU2003 Puff Releases in Oklahoma City. Part I: Comparison of Observed and Predicted Concentrations. *Bound. Layer Meteorol.* **2019**, *171*, 323–349. [CrossRef]
6. Hernández-Ceballos, M.A.; Hanna, S.; Bianconi, R.; Bellasio, R.; Chang, J.; Mazzola, T.; Andronopoulos, S.; Armand, P.; Benbouta, N.; Čarný, P.; et al. UDINEE: Evaluation of Multiple Models with Data from the JU2003 Puff Releases in Oklahoma City. Part II: Simulation of Puff Parameters. *Bound. Layer Meteorol.* **2019**, *171*, 351–376. [CrossRef]
7. Rodean, H.C. *Stochastic Lagrangian Models of Turbulent Diffusion*; The American Meteorological Society: Boston, MA, USA, 1996; 82p.
8. Williams, M.D.; Brown, M.J.; Boswell, D.; Singh, B.; Pardyjak, E.M. Testing of the QUIC-Plume Model with Wind-Tunnel Measurements for a High-Rise Building. In Proceedings of the 26th Agricultural and Forest Meteorology/13th Air Pollution/5th Urban Environment/16th Biometeorology and Aerobiology, January 2004. Available online: https://ams.confex.com/ams/AFAPURBBIO/webprogram/Paper78636.html (accessed on 2 May 2023).
9. Benbouta, N. *Canadian Urban Flow and Dispersion Model (CUDM): Developing Canadian Capabilities for CBRN Hazard Dispersion in Urban Areas*; CMOE Dorval TR 2007-001 Technical Report; Environment Canada: Toronto, ON, Canada, 2007.
10. Hogue, R.; Benbouta, N.; Mailhot, J.; Belair, S.; Lemonsu, A.; Yee, E.; Lien, F.-S.; Wilson, J.D. The development of a prototype of Canadian urban flow and dispersion modeling system. In Proceedings of the AMS 89 Annual Conference, Joint Session 19 Urban Transport and Dispersion Modeling, Phoenix, AZ, USA, 14 January 2009.
11. Oldrini, O.; Armand, P.; Duchenne, C.; Olry, C.; Moussafir, J.; Tinarelli, G. Description and preliminary validation of the PMSS fast response parallel atmospheric flow and dispersion solver in complex built-up areas. *Environ. Fluid Mech.* **2017**, *17*, 997–1014. [CrossRef]
12. Oldrini, O.; Armand, P.; Duchenne, C.; Perdriel, S. Parallelization Performances of PMSS Flow and Dispersion Modeling System over a Huge Urban Area. *Atmosphere* **2019**, *10*, 404. [CrossRef]
13. Röckle, R. Bestimmung der Strömungsverhältnisse im Bereich komplexer Bebauungsstrukturen. Ph.D. Thesis, Techn. Hochschule Darmstadt, Darmstadt, Germany, 1990.
14. Ek, N.; Benbouta, N.; Bourgouin, P.; Gauthier, J.-P.; Gauthier, N.; Hogue, R.; Holly, M.; Trudel, S.; Zaganescu, C. The Canadian urban dispersion modelling (CUDM) system: An operational prototype. In Proceedings of the 34th AMOP Technical Seminar on Environmental Contamination and Response, Banff, AB, Canada, 4–6 October 2011; pp. 603–623.
15. Thomson, D.J. Criteria for the selection of stochastic models of particle trajectories in turbulent flows. *J. Fluid Mech.* **1987**, *180*, 529–556. [CrossRef]
16. Patankar, S.V.; Spalding, D.B. A calculation procedure for heat, mass and momentum transfer in three-dimensional parabolic flows. *Int. J. Heat Mass Transf.* **1972**, *15*, 1787–1806. [CrossRef]
17. Roed, J.; Andersson, K.G. *Early Measurements in Scandinavia following the Chernobyl Accident (IAEA-TECDOC—1131)*; International Atomic Energy Agency (IAEA): Vienna, Austria, 2000.
18. Lipták, Ľ.; Fojcíková, E.; Krpelanová, M.; Fabová, V.; Čarný, P. The ESTE Decision Support System for Nuclear and Radiological Emergencies: Atmospheric Dispersion Models. *Atmosphere* **2021**, *12*, 204. [CrossRef]
19. ParaView—Open Source Post-Processing Visualization Engine. Available online: https://www.paraview.org/ (accessed on 25 June 2023).
20. Clawson, K.L.; Carter, R.G.; Lacroix, D.J.; Biltoft, C.A.; Hukari, N.F.; Johnson, R.C.; Rich, J.D. *Joint Urban 2003 (JU03) SF6 Atmospheric Tracer Field Tests*; NOAA Technical Memorandum OARARL-254; Air Resources Lab.: College Park, MD, USA, 2005.
21. Allwine, K.J.; Leach, M.J.; Stockham, L.W.; Shinn, J.S.; Hosker, R.P.; Bowers, J.F.; Pace, J.C. Overview of Joint Urban 2003—An atmospheric Dispersion Study in Oklahoma City. Preprints, Symposium on Planning, Nowcasting and Forecasting in the Urban Zone, American Meteorological Society. 2004. Available online: www.researchgate.net (accessed on 23 June 2023).
22. Wesson, G.; Katul, G.; Lai, C.-T. Sensible heat flux estimation by flux variance and half-order time derivative methods. *Water Resour. Res.* **2001**, *37*, 2333–2343. [CrossRef]
23. Möller, T.; Trumbore, B. Fast, Minimum Storage Ray-Triangle Intersection. *J. Graph. Tools* **1997**, *2*, 21–28. [CrossRef]
24. Los Alamos National Laboratory. MCNP—A General Monte Carlo N-Particle Transport Code, Version 5, Software Product Produced by Los Alamos National Laboratory. Available online: https://mcnp.lanl.gov/ (accessed on 2 February 2021).

MDPI
St. Alban-Anlage 66
4052 Basel
Switzerland
www.mdpi.com

Atmosphere Editorial Office
E-mail: atmosphere@mdpi.com
www.mdpi.com/journal/atmosphere

www.ingramcontent.com/pod-product-compliance
Lightning Source LLC
Chambersburg PA
CBHW061436210926
51850CB00008B/1531
* 9 7 8 3 0 3 6 5 8 7 5 5 4 *